DEDICATION

To my wife, Benay,
whose patience and understanding
made this book possible.

PREFACE

This book is designed to help students and professionals who use mathematics in their daily routine learn *Mathematica*, a computer system designed to perform complex mathematical calculations. My approach is simple: learn by example. Along with easy to read descriptions of the most widely used commands, I have included a collection of over 750 examples and solved problems, each specifically designed to illustrate an important feature of the *Mathematica* software.

I have included those commands and options which are most commonly used in algebra, trigonometry, calculus, differential equations, and linear algebra. Most examples and solved problems are short and to the point. Comments have been included, where appropriate, to clarify what might be confusing to the reader.

The reader is encouraged not only to replicate the output shown in the text, but to make modifications and investigate the resulting effect upon the output. I have found this to be the most effective way to learn the syntax and capabilities of this truly unique program.

The first three chapters serve as an introduction to the syntax and style of *Mathematica*. The structure of the remainder of the book is such that the reader need only be concerned with those chapters of interest to him or her. If, on occasion, a command is encountered which has been discussed in a previous chapter, the index may be used to conveniently locate the command's description.

Without a doubt you will be impressed with *Mathematica*'s capabilities. It is my sincere hope that you will use the power built into this software to investigate the wonders of mathematics in a way that would have been impossible just a few years ago.

I would like to take this opportunity to thank the staff at McGraw-Hill for their help in the preparation of this book and a special note of thanks to Mr. Joel Lerner for his encouragement and support of this project.

EUGENE DON

CONTENTS

CHAPTER 1

Getting Acquainted

1.1 NOTATION AND CONVENTIONS

Mathematica is a language that is best learned by experimentation. Therefore, the reader is urged to try as many examples and problems as possible and experiment by changing options and parameters. In fact, this chapter may be considered a tutorial for those readers who want to get their hands on *Mathematica* right away.

New commands are introduced with a ■ bullet and options associated with them are bulleted with a • symbol for easy reference.

In keeping with *Mathematica*'s conventions, all commands and instructions will be written in Courier boldface type and *Mathematica* output in Courier regular face type.

This line is written in Courier boldface type.

This line is written in Courier regular face type.

Menu commands in this text are described using double arrows (⇒). For example, Format⇒Style⇒Input, written in Arial font, means go to the Format menu, then to the Style submenu, and then click on Input. (See page 2.)

Mathematica occasionally uses a special symbol, `, which we call a backquote. Do not confuse this with an apostrophe.

Finally, most *Mathematica* commands use an arrow, →, to specify options within the command. You may use −> (− followed by >) as an alternate, if you wish.

The examples used in this book were executed using *Mathematica* version 4. You may notice some differences on your computer if you are using version 3 or 5. Most noticeably, the base of the natural logarithm may be represented, on output, as E and the imaginary unit will be written I, as opposed to e and i, respectively. Versions 4 and 5 will also convert the sequence − > automatically into → and ! = into ≠. There are additional enhancements in version 5 which, for the most part, will be transparent to the novice user.

```
※ Mathematica 4 - [Untitled-1 *]
 File  Edit  Cell  Format  Input  Kernel  Find  Window  Help
 Untitled-1 *          Style                          ▶    Title              Alt+1
                       Screen Style Environment       ▶    Subtitle           Alt+2
                       Printing Style Environment     ▶    Subsubtitle        Alt+3
                                                           Section            Alt+4
                       Show Expression    Shift+Ctrl+E      Subsection         Alt+5
                       Option Inspector... Shift+Ctrl+O     Subsubsection      Alt+6
                       Remove Options:...                   Text               Alt+7
                                                           SmallText          Alt+8
                       Style Sheet                    ▶  ✓ Input              Alt+9
                       Edit Style Sheet...                 Output
                                                           InlineFormula
                       Font                           ▶    DisplayFormula
                       Face                           ▶    Program
                       Size                           ▶    Other...           Alt+0
                       Text Color                     ▶
                       Background Color               ▶
                       Choose Font:..

                       Text Alignment                 ▶
                       Text Justification             ▶
                       Word Wrapping                  ▶
                       Cell Dingbat                   ▶
                       Horizontal Lines               ▶

                       Show Ruler
                       Show ToolBar
                       Show Page Breaks
                       Magnification                  ▶
```

1.2 THE KERNEL AND THE FRONT END

The *kernel* is the computational engine of *Mathematica*. You input instructions and the kernel responds with answers in the form of numbers, graphs, matrices, or other appropriate displays. The kernel works silently in the background and, for the most part, is invisible.

The interface between the user and the kernel is called the *front end* and the medium of the front end is the *Mathematica notebook*. The notebook not only enables you to communicate with the kernel, but is a convenient tool for documenting your work.

To execute an instruction, you type the instruction and then press [ENTER]. Most PCs have two [ENTER] keys, but only the [ENTER] key to the *far right* of the keyboard will execute instructions. The other [ENTER] key must be used with the [SHIFT] key; otherwise, you will merely get a new line. If you are using a Macintosh computer, do not confuse the [ENTER] key with the [RETURN] key.

The picture in Example 1 below shows the standard *Mathematica* display. The symbols on the right-hand side form the BasicInput palette and allow access by mouse-click to the most common mathematical symbols. (If you don't see the palette on your screen, click on File⇒Palettes⇒BasicInput and it should appear.) Other palettes are available for specialized purposes and can be accessed via File⇒Palettes.

Each symbol is accessed by clicking on the palette. If you use the palette, your notebooks will look like pages from a math textbook. Most examples in this book take full advantage of the BasicInput palette. However, each *Mathematica* symbol has an alternative descriptive format which can be typed. For example, π can be represented as Pi and $\sqrt{5}$ can be written Sqrt[5]. These representations are useful for experienced *Mathematica* users who prefer not to use the mouse.

The notebook in Example 1, labeled "Untitled-1," is where you input your commands and where *Mathematica* places the result of its calculations. The picture below shows the input and output of Example 1. (The display on a Macintosh computer will look slightly different.)

EXAMPLE 1

Add 2 and 3.

Notice that the kernel has assigned In[1] to the input expression and Out[1] to the output. This enables you to keep track of the order in which the kernel evaluates instructions. These labels are important because the order of evaluation does not always correspond to the physical position of the instruction within the notebook. In this book, however, we shall *not* include In and Out labels in our examples.

In working out the examples and problems in this book, you may find that your answers do not agree with the answers given in the text. This may occur if you have defined a symbol to have a specific value. For example, if x has been defined as 3, all occurrences of x will be replaced by 3. You should clear the symbol (see Section 1.5) and try the problem again. All examples and problems assume that symbols have been cleared prior to execution.

You can work on several different notebooks in a single *Mathematica* session. However, if you are using only one kernel, changes to symbols in one notebook will affect identical symbols in all notebooks.

There are times when you may wish to evaluate only part of an expression. To do this, select the portion of the expression that you wish to evaluate. Then press [CTRL] + [SHIFT] + [ENTER] on a PC or [COMMAND] + [RETURN] on a Mac.

EXAMPLE 2

Suppose we wish only to perform the multiplication in the expression 2 * 3 + 5.

First select 2 * 3:

2 * 3 + 5

Then press [CTRL] + [SHIFT] + [ENTER] (PC) or [COMMAND] + [RETURN] (Mac)

6 + 5

Occasionally you may introduce an instruction which takes an excessively long time to execute, or you may inadvertently create an infinite loop. To abort a calculation, go to Kernel⇒Abort Evaluation. Alternatively, you may press [ALT] + . to abort. ([COMMAND] + . on the Macintosh). If this does not work, you will have to terminate the kernel by going to Kernel⇒Quit Kernel⇒Local. However, by doing so, you will lose all your defined symbols and values. Your *Mathematica* notebook will not be lost, however, so they can easily be restored.

As with all computer software, there are times when *Mathematica* will completely crash. The only remedy is to close *Mathematica* and reload it. On rare occasions, you may have to reboot your computer. In either event your *Mathematica* notebook will be lost. It is therefore extremely important to **BACK UP YOUR NOTEBOOK OFTEN!**

Finally, there may be times when you wish to include comments within your *Mathematica* commands. Anything written within (* and *) is ignored by the *Mathematica* kernel.

EXAMPLE 3

```
12 + (* these words will be ignored by the kernel *) 3
15
```

SOLVED PROBLEMS

1.1 Multiply 12 by 17 and then add 9.

SOLUTION

```
12 * 17 + 9
213
```

1.2 Multiply only the 12 by 17 in Problem 1.1, but do not add the 9.

SOLUTION

```
12 * 17 + 9          ←Select 12*17 with the mouse.
```

[CTRL] + [SHIFT] + [ENTER] or [COMMAND] + [RETURN] on a Mac

```
204 + 9
```

1.3 The following program is an infinite loop. Execute it and then abort the evaluation.

```
x = 1;
While [x > 0, x = x + 1]
```

SOLUTION

```
x = 1;
While[ x > 0,x = x + 1]
[ALT] + .

$Aborted
```

1.4 Multiply 17.2 by 16.3 and then add 4.7.

SOLUTION

```
17.2 * 16.3 + 4.7
285.06
```

1.5   Multiply 17.2 by the sum of 16.3 and 4.7.

SOLUTION

```
17.2 * (16.3 + 4.7)
361.2
```

1.6   Compute the sum of $2x + 3$, $5x + 9$, and $4x + 2$.

SOLUTION

```
(2 x + 3) + (5 x + 9) + (4 x + 2)
14 + 11 x
```

1.3 *MATHEMATICA* QUIRKS

Mathematica is extremely case sensitive.

For example, `Integrate` and `integrate` are different.   All *Mathematica* commands begin with a capital letter. Some commands, such as `FindRoot`, use more than one capital letter. To avoid conflicts, it is a good idea for all user-defined symbols to begin with a lowercase letter.

Different brackets for different purposes.

- Square brackets are used for function arguments: `Sin[x]` *not* `Sin(x)`.
- Round brackets (parentheses) are used for grouping: `(2 + 3) * 4` means add $2 + 3$ first, then multiply by 4.   Never type `[2 + 3] * 4`.
- Curly brackets (braces) are used for lists: `{1, 2, 3, 4}`.   More about lists in Chapter 3.

Use `E`, not `e`, for the base of the natural logarithm.

Since every *Mathematica* symbol begins with a capital letter, the base of the natural logarithm is `E`. This causes a bit of confusion, so be careful.   Similarly, `I` (not `i`) is the imaginary unit.   ℯ and ⅈ from the BasicInput palette may be freely used if desired.

Polynomials are not written in "standard" form.

Mathematica writes polynomials with the constant term first and increasing powers from left to right. Thus, the polynomial $x^2 + 2x - 3$ would be written $-3 + 2x + x^2$.

SOLVED PROBLEMS

1.7   Compute $\sqrt{81}$ using the `Sqrt` function.   What happens if you do not use a capital S?

SOLUTION

```
Sqrt[81]
9
sqrt[81]

General::spell1 : Possible spelling error : new symbol
  name "sqrt" is similar to existing symbol "Sqrt".

sqrt[81]
```

1.8   Use parentheses to multiply the sum of 2 and 3 by the sum of 5 and 7.   What happens if you use square brackets?

SOLUTION

```
(2 + 3)(5 + 7)
60
[2 + 3][5 + 7]

Syntax::txntxi:"[2+3]" is incomplete; more input is needed.
[2 + 3][5 + 7]
```

1.9 Use the `Sin` function to compute $\sin(\pi/2)$. What happens if you use round parentheses?

SOLUTION

```
Sin[Pi/2] or Sin[π/2]
1
Sin(Pi/2)
```

$$\frac{\pi \text{Sin}}{2}$$

> *Mathematica* thinks you want to multiply the symbol `Sin` by π and divide by 2.

1.10 Lindsay typed `[4 + 1] * [6 + 2]` during a *Mathematica* session. Why didn't she get an answer of 40?

SOLUTION

Square brackets cannot be used for grouping. Round parentheses must be used.

1.11 Why didn't Alexis get an answer of 3 when she typed `sqrt[9]`?

SOLUTION

Mathematica functions must begin with a capital letter.

1.12 Why didn't Ariel get an answer of 1 when she typed `Cos(0)`?

SOLUTION

Square brackets, not round parentheses, must be used to contain arguments of functions.

1.4 *MATHEMATICA* GIVES EXACT ANSWERS

Mathematica is designed to work as a mathematician works: exactly. So you do not get the 10- or 12-digit numerical approximation a calculator would give, but a mathematical expression which is 100% accurate.

EXAMPLE 4

```
√12
2√3
```

EXAMPLE 5

```
1/3 + 3/5 - 5/7 + 2/11
```

$$\frac{463}{1115}$$

EXAMPLE 6

$\pi + \pi$
2π

EXAMPLE 7

$\sqrt{-1}$
i

SOLVED PROBLEMS

1.13   Simplify $\sqrt{2} + \sqrt{8} + \sqrt{18}$.

SOLUTION

$\sqrt{2} + \sqrt{8} + \sqrt{18}$ or Sqrt[2] + Sqrt[8] + Sqrt[18]
$6\sqrt{2}$

1.14   Compute the sum of the reciprocals of 3, 5, 7, 11, and 13.

SOLUTION

$\frac{1}{3} + \frac{1}{5} + \frac{1}{7} + \frac{1}{11} + \frac{1}{13}$ or 1/3 + 1/5 + 1/7 + 1/11 + 1/13

$\frac{12673}{15015}$

1.15   Compute the square root of π *exactly* using the `Sqrt` function.

SOLUTION

Sqrt[Pi]
$\sqrt{\pi}$ ← This is the only way to represent the square root of π <u>exactly</u>.

1.16   Multiply $\sqrt{8}$ by $\sqrt{2}$.

SOLUTION

$\sqrt{8}\sqrt{2}$ or Sqrt[8] * Sqrt[2]
4

1.17   Simplify $\sqrt{3} + \sqrt{12} + \sqrt{27} + \sqrt{48}$, leaving your answer in radical form.

SOLUTION

$\sqrt{3} + \sqrt{12} + \sqrt{27} + \sqrt{48}$
$10\sqrt{3}$

1.5 *MATHEMATICA* BASICS

In this section we will discuss some of the simpler concepts within *Mathematica*. Each will be explained in greater detail in a subsequent chapter.

Symbols are defined using any sequence of alphanumeric characters (letters, digits, and certain special characters) not beginning with a digit. Once defined, a symbol retains its value until it is changed, cleared or removed.

Arithmetic operations are performed in the obvious manner using the symbols $+$, $-$, $*$, and /. Exponentiation is represented by a caret, $^\wedge$, so $\mathbf{x^\wedge y}$ means x^y. Just as in algebra, a missing symbol implies multiplication, so $\mathbf{2a}$ is the same as $\mathbf{2*a}$. Be careful, however, when multiplying two symbols, since $\mathbf{ab}$ represents the *single* symbol beginning with $\mathbf{a}$ and ending with $\mathbf{b}$. To multiply $\mathbf{a}$ by $\mathbf{b}$ you must separate the two letters with $*$ or $\times$ (on the BasicInput palette) or a space: $\mathbf{a*b}$ or $\mathbf{a \times b}$ or $\mathbf{a\,b}$.

EXAMPLE 8

```
a = 2
2
b = 3
3
c = a + b
5
```

Notice that the result of *each* calculation is displayed. This is sometimes annoying, and can be suppressed by using a semicolon (;) to the right of the instruction.

EXAMPLE 9

```
a = 2;
b = 3;
c = a + b

5
```

Operations are performed in the following order: (*a*) exponentiation, (*b*) multiplication and division, (*c*) addition and subtraction. If the order of operations is to be modified, parentheses, ( ), must be used. Be careful not to use [ ] or { } for this purpose.

EXAMPLE 10

```
2 + 3 * 5
17
(2 + 3) * 5
25
```

Each symbol in *Mathematica* represents something. Perhaps it is the result of a simple numerical calculation or it may be a complicated algebraic expression.

EXAMPLE 11

$$a = 3;$$

$$b = \sqrt{\frac{x^2 + 1}{2x + 3}};$$

Here, $\mathbf{a}$ is a symbol representing the numerical value 3 and $\mathbf{b}$ is a symbol representing an irrational algebraic expression.

If you ever forget what a symbol represents, simply type $\mathbf{?}$ followed by the symbol name to recall its definition.

EXAMPLE 12 (continuation of example 11)

```
?a
Global `a
a = 3

?b
Global `b
```

$$b = \sqrt{\frac{1 + x^2}{3 + 2x}}$$

To delete a symbol, so that it can be used for a different purpose, the **Clear** or the **Remove** command can be used.

- **Clear[*symbol*]** clears the *symbol's* definition and values, but does not clear its attributes, messages, or defaults. *symbol* remains in *Mathematica*'s symbol list.
- **Remove[*symbol*]** removes *symbol* completely. *symbol* will no longer be recognized unless it is redefined.

symbol=. will also clear the definition of *symbol*.

EXAMPLE 13 (continuation of example 12)

```
Clear[a]
?a                    ←?a recalls information about the symbol a.
Global `a

Remove[b]
?b

Information::notfound : Symbol b not found.
```

The **N** command allows you to compute a numerical approximation.

- **N[*expression*]** gives the numerical approximation of *expression* to six significant digits (*Mathematica*'s default).
- **N[*expression*, n]** attempts[1] to give an approximation accurate to n significant digits.

Another convenient technique is to use **//N** to the right of the expression being approximated. *expression* **//N** is equivalent to **N[*expression*]**. **//** can be used for other *Mathematica* commands as well.

- *expression* **//Command** is equivalent to **Command[*expression*]**.

EXAMPLE 14

```
N[π] or π//N
3.14159
N[π, 50]
3.1415926535897932384626433832795028841971693993751
```

The *Mathematica* kernel keeps track of the results of previous calculations. The symbol **%** returns the result of the previous calculation, **%%** gives the result of the calculation before that, **%%%**, and so forth. Using **%** wisely can save a lot of typing time.

---

[1]Unless the numbers in ***expression*** are exact or of sufficiently high precision, *Mathematica* may be unable to return n significant digits of approximation.

EXAMPLE 15

To construct, $\sqrt{\pi + \sqrt{\pi + \sqrt{\pi}}}$, we could type: `Sqrt[Pi+Sqrt[Pi+Sqrt[Pi]]]`. A less confusing way of accomplishing this is to type

```
Sqrt[Pi];          ← The semicolon suppresses the output of the intermediate calculations.
Sqrt[Pi + %];
Sqrt[Pi + %]
```

$$\sqrt{\pi + \sqrt{\sqrt{\pi} + \pi}}$$

Using the BasicInput palette, we can type

```
√π;
√π + %;
√π + %
```

$$\sqrt{\pi + \sqrt{\sqrt{\pi} + \pi}}$$

SOLVED PROBLEMS

1.18 Define $a = 3$, $b = 4$, and $c = 5$. Then multiply the sum of a and b by the sum of b and c. Print only the final answer.

SOLUTION

```
a = 3;
b = 4;
c = 5;
(a + b) * (b + c)
63
```

1.19 Let $a = 1$, $b = 2$, $c = 3$ and add a, b, and c. Then clear a, b, and c from the kernel's memory and add again.

SOLUTION

```
a = 1;
b = 2;
c = 3;
a + b + c
6
Clear[a,b,c]
a + b + c
a + b + c
```

1.20 Obtain a 25-decimal approximation of e, the base of the natural logarithm.

SOLUTION

```
N[E,26] or N[e,26]      ← 26 significant digits gives 25 decimal places.
```

```
2.7182818284590452353602875
```

1.21 (a) Express $\dfrac{1}{7} + \dfrac{2}{13} - \dfrac{3}{19} + \dfrac{1}{23}$ as a single fraction.

(b) Obtain an approximation accurate to 20 decimal places.

SOLUTION

`1/7 + 2/13 - 3/19 + 1/23`

$$\frac{7249}{39767}$$

`N[%,20]`
`0.18228682073075665753`

1.22 Compute $\sqrt{968}$ (*a*) exactly (*b*) approximately to 25 significant digits.

SOLUTION

$\sqrt{968}$ or `Sqrt[968]`
$22\sqrt{2}$

`N[%,25]`
`31.11269837220809107363715`

1.23 Multiply 12 by 6.   Then multiply 15 by 7.   Then use % and %% to add the two products.

SOLUTION

`12 * 6`
`72`
`15 * 7`
`105`
`% + %%`
`177`

1.24 Compute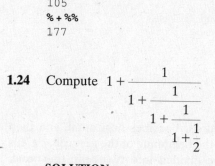

SOLUTION

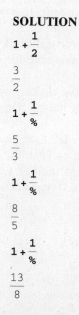

1.25   Compute the value of $1 + (1 + (1 + (1 + (1 + 1)^2)^2)^2)^2$.

SOLUTION

```
1 + 1
2
1 + %^2
5
1 + %^2
26
1 + %^2
677
1 + %^2
458330
```

*Percent Mean the answer
that you have*

1.6   CELLS

Cells are the building blocks of a *Mathematica* notebook.   Cells are indicated by blue brackets at the right-hand side of the notebook.   (Most likely you have already noticed these brackets and were wondering what they meant.)   Cells can contain subcells, which may in turn contain sub-subcells, and so forth.

The kernel evaluates a notebook on a cell-by-cell basis, so if you have several instructions within a single cell, they will all be executed with a single press of the [ENTER] key.

EXAMPLE 16

```
a = 1 + 2
b = 2 + 7
c = a + b

3
9
12
```

← All three lines are contained within a single cell. [ENTER] is pressed only once.

A new cell can be formed by moving the mouse until the cursor becomes horizontal, and then clicking.   A horizontal line will appear across the screen to mark the beginning of the new cell.   Cells can be divided by going to Cell⇒Divide Cell.   The cell will be divided into two cells, the break occurring at the point where the cursor is positioned.

Cells can be combined (merged) by selecting the appropriate cell brackets (a vertical black line should appear) and then clicking on Cell⇒Merge Cells.   Keyboard shortcuts are available for each of these operations and are listed in the menu.

To avoid extremely long notebooks, cells can be closed (or compressed) by double-clicking on the cell bracket.   The bracket will change appearance, looking something like a fish hook.   Double-clicking a second time will open the cell.

There are different types of cells for different purposes.   Only input cells can be fed to the kernel for evaluation.   Text cells are used for descriptive purposes.   Other cell types such as Title, Subtitle, Section, Subsection, etc., can by found in the menu Format⇒Style.   A convenient toolbar can be displayed, if you wish, by clicking on Format⇒Show Toolbar.   This will display the cell type at the top of your notebook window.

SOLVED PROBLEMS

1.26   Let $a = 2x + 3$ and $b = 5x + 6$.   Then compute $a + b$.

(*a*)   Place each instruction in a separate cell and execute them individually.

(*b*)   Place all three instructions in a single cell and execute them simultaneously.

SOLUTION

This is what the output looks like *after* execution.

(*a*)

```
a = 2 x + 3
3 + 2 x
b = 5 x + 6
6 + 5 x
a + b
9 + 7 x
```

(*b*)

```
a = 2 x + 3
b = 5 x + 6
a + b
3 + 2 x
6 + 5 x
9 + 7 x
```

1.27   Let $a = 2x + 3y + 4z$, $b = x + 3y + 5z$, and $c = 3x + y + z$ and compute the sum of a, b, and c. Place four lines within a single cell and execute, printing only the final result.

SOLUTION

```
a = 2 x + 3 y + 4 z;
b = x + 3 y + 5 z;
c = 3 x + y + z;
a + b + c
6 x + 7 y + 10 z
```

1.7   GETTING HELP

There are many sources of help in *Mathematica*. First and foremost is the *Mathematica* manual. Second, and probably more convenient, is the help browser built into the software (see p. 14). There you will find all available commands grouped by topic; simply click to get the information you need. The entire *Mathematica* manual is built into the software as well.

The help files contain numerous examples which you may want to play with. Feel free to make any changes in the help files without fear of modifying their content. These files are protected and your changes will not be permanent.

If you know the name of the command you want, you can use a question mark, ?, followed by the name of the command to determine its syntax. More extensive information about the command, including attributes and options, can be obtained using ??.

Occasionally, when you make an error, *Mathematica* will beep. If you are not sure what you did to cause this, you can sometimes get a clue by going to Help⇒Why The Beep?

EXAMPLE 17

Suppose you know that the command **Plot** graphs a function, but you cannot remember its syntax.

```
?Plot
Plot [f, {x, xmin, xmax}] generates a plot of
  f as a function of x from xmin to xmax. Plot [{f1,
  f2, . . . }, {x, xmin, xmax}] plots several functions fi.
```

If information is needed about attributes or optional settings (and their defaults), ?? can be used.

```
??Plot
Plot [f, {x, xmin, xmax}] generates a plot of
  f as a function of x from xmin to xmax. Plot [{f1,
  f2, . . . }, {x, xmin, xmax}] plots several functions fi.
```

```
Attributes [Plot] = {HoldAll , Protected }
```
$$\text{Options [Plot]} = \left\{ \text{AspectRatio} \to \frac{1}{\text{GoldenRatio}} , \text{ Axes} \to \text{Automatic} , \right.$$
```
AxesLabel →None, AxesOrigin →Automatic , AxesStyle →Automatic ,
Background →Automatic , ColorOutput →Automatic ,
Compiled →True , DefaultColor →Automatic , Epilog → {},
Frame →False, FrameLabel →None , FrameStyle →Automatic ,
FrameTicks →Automatic , GridLines →None ,
ImageSize →Automatic , MaxBend →10., PlotDivision →30.,
PlotLabel →None , PlotPoints →25 , PlotRange →Automatic ,
PlotRegion →Automatic , PlotStyle →Automatic ,
Prolog → {}, RotateLabel →True, Ticks →Automatic ,
DefaultFont :→$DefaultFont , DisplayFunction :→$DisplayFunction ,
FormatType :→$FormatType , TextStyle :→$TextStyle }
```

Options can also be obtained using the **Options** command.   This is useful if you want to specify an option but cannot remember its name.

EXAMPLE 18

```
Options[Solve]
```
```
{InverseFunctions→Automatic, MakeRules→False, Method→3,
 Mode →Generic, Sort →True, VerifySolutions→ Automatic,
 WorkingPrecision→∞}
```

Very often you may remember part of a symbol name, but not the whole name.   If you know the beginning is Arc, for example, type in the part you know and then press [CTRL] + K.   This will generate a menu of all commands and functions beginning with Arc. Then click on the one you want. If you are using a Macintosh computer, use [COMMAND] + K. (The [COMMAND] key is the key with the apple on it.)

EXAMPLE 19

Type in `Arc` and then press [CTRL] + K or [COMMAND] + K.

Another way of determining symbol names is to use ? together with wildcards. The character * acts as a wildcard and takes the place of any sequence of characters. Wildcards can be used anywhere, at the beginning, middle, or end of a symbol.

EXAMPLE 20

(a) Find all commands beginning with 'Inv'.

```
?Inv*
```

Inverse	InverseJacobiDC
InverseBetaRegularized	InverseJacobiDN
InverseEllipticNomeQ	InverseJacobiDS
InverseErf	InverseJacobiNC
InverseErfc	InverseJacobiND
InverseFourier	InverseJacobiNS
InverseFunction	InverseJacobiSC
InverseFunctions	InverseJacobiSD
InverseGammaRegularized	InverseJacobiSN
InverseJacobiCD	InverseSeries
InverseJacobiCN	InverseWeierstrassP
InverseJacobiCS	

(*b*)   Find all commands ending with 'in'.

```
? * in
ArcSin            GroupPageBreakWithin Sin
AxesOrigin        Join                 StackBegin
Begin             Min                  StringJoin
CoefficientDomain PageBreakWithin      $MachineDomain
ConstrainedMin    Plain
```

(*c*)   Find all commands with 'our' in the middle.

```
? * our *
ButtonSource     ContourShading     InverseFourier
ContourGraphics  ContourSmoothing   LightSources
ContourLines     ContourStyle       ListContourPlot
ContourPlot      Fourier            MovieContourPlot
Contours
```

Wildcards can also be used to determine which symbols have been used thus far by the kernel. Typing ?`* returns a list of all symbols that have been defined during your *Mathematica* session. The character ` (back quote) stands for global—you want a list of all *global* symbols. (See Appendix Sect. A.2.)

EXAMPLE 21

Note: The results of this example may be slightly different on your computer.

```
a = 3;
b2xy = 4;
xyz7 = 5;
?`*
a   b2xy   xyz7
```

`Clear["`*"]` will clear all global symbols.   `Remove["`*"]` removes all global symbols.

EXAMPLE 22

```
Remove["`*"]
?`*                    ←Check to see if any symbols remain.
Information::nomatch : No symbol matching `* found.
```

SOLVED PROBLEMS

1.28   Obtain basic information about the *Mathematica* command `Simplify`.

SOLUTION

```
?Simplify
Simplify[expr] performs a sequence of algebraic
   transformations on expr, and returns the simplest form it
   finds.
```

1.29   Obtain extended information about the *Mathematica* command `Simplify` including default settings for options.

SOLUTION

```
??Simplify
```

Simplify[expr] performs a sequence of algebraic
   transformations on expr, and returns the simplest form it
   finds.
Attributes[Simplify] = {Protected}

Options[Simplify] = {ComplexityFunction→Automatic,
TimeConstraint→300, Trig→True}

1.30 Obtain help on the *Mathematica* command **Factor** and then factor $x^3 - 6x^2 + 11x - 6$.

SOLUTION

```
?Factor
```

Factor[poly] factors a polynomial over the integers.
   Factor[poly, Modulus→p] factors a polynomial modulo a
   prime p.   Factor[poly, Extension→{a1, a2, ... }] factors a
   polynomial allowing coefficients that are rational
   combinations of the algebraic numbers ai.

```
Factor[x³ – 6x² + 11x – 6]
```
(– 3 + x) (– 2 + x) (– 1 + x)

1.31 Find all *Mathematica* commands beginning with Abs.

SOLUTION

```
?Abs *
```

Abs   AbsoluteThickness
AbsoluteDashing   AbsoluteTime
AbsolutePointSize

1.32 Find all *Mathematica* commands beginning with Sin and ending with al.

SOLUTION

```
?Sin * al
```

SinhIntegral SinIntegral

1.33 Find all *Mathematica* commands beginning with Fi.

SOLUTION

```
?Fi *
```

Fibonacci   Find
File   FindList
FileBrowse   FindMinimum
FileByteCount   FindRoot
FileDate   First
FileInformation   Fit
FileName   FixedPoint
FileNames   FixedPointList
FileType

1.34 Find all *Mathematica* commands beginning with Fi and ending with t.

SOLUTION

```
?Fi * t
FileByteCount        Fit
FindList             FixedPoint
FindRoot             FixedPointList
First
```

1.8 PACKAGES

There are many specialized functions and procedures which are not loaded when *Mathematica* is initially invoked.   Rather, they must be loaded separately from files in the *Mathematica* directory on the hard drive.   These files are of the form *filename*.m.

EXAMPLE 23

A circle of radius 1 has an implicit equation $x^2 + y^2 = 1$. To plot implicit functions, the command **ImplicitPlot** is needed. However, **ImplicitPlot** is only available as part of the package **Graphics**. (This information can be determined by going to the help index under add-ons.)   To load this graphics command simply type

```
<<Graphics`ImplicitPlot`
```

If your computer has sufficient memory, you can access the whole graphics package by typing **<<Graphics`** and *Mathematica* will load each part of the package as it is called.   The appropriate graphics command can then be accessed.

```
ImplicitPlot[x² + y² = = 1, {x, -1, 1}];
```

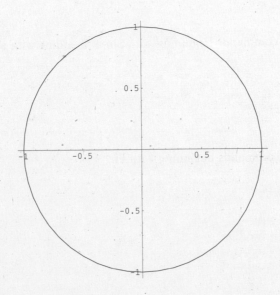

Note the double equal sign in the `ImplicitPlot` command.   We will discuss `ImplicitPlot` in more detail in Chapter 4, Two-Dimensional Graphics.

Once a package is loaded you can get a list of the functions it contains by using the **Names** command.

EXAMPLE 24

```
<<Miscellaneous`Calendar`
```
←Loads the package Miscellaneous`Calendar`.
```
Names["Miscellaneous`Calendar`*"]
```

```
{"Calendar","CalendarChange","DayOfWeek","DaysBetween",
"DaysPlus","EasterSunday","EasterSundayGreekOrthodox",
"Friday","Gregorian","Islamic","JewishNewYear","Julian",
"Monday","Saturday","Sunday","Thursday","Tuesday","Wednesday"}
```

EXAMPLE 25

A common error when using packages is to type a command *before* the appropriate package has been loaded. For example, if we attempt to execute **ImplicitPlot** before loading the package nothing happens:

```
ImplicitPlot[x²+y²==1,{x,-1,1}]
```
```
ImplicitPlot[x²+y²==1,{x,-1,1}]
```

If we now attempt to load the Graphics package, we get an error message.

```
<<Graphics`ImplicitPlot`
```
```
ImplicitPlot::shdw :
  Symbol ImplicitPlot appears in multiple contexts <<1>>;
      definitions in context Graphics`ImplicitPlot` may
      shadow or be shadowed by other definitions.
```

Typing the command again does not give us our graph because we now have two symbols ImplicitPlot, and the one introduced in error before loading the package takes precedence. The trick is to eliminate it from the symbol table and then plot our graph.

```
Remove[ImplicitPlot]
```
```
ImplicitPlot[x²+y²==1,{x,-1,1}];
```

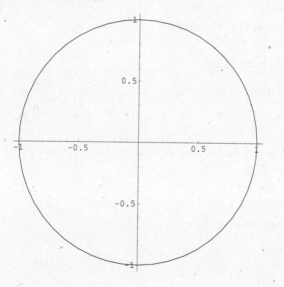

SOLVED PROBLEMS

1.35 The package **DiscreteMath`Combinatorica`** contains over 230 functions in combinatorics and graph theory. One of these is **KSubsets**, which lists all subsets of size *k* of a given set.   Load the package and execute `Ksubsets[{1,2,3,4,5},3]`.   This will return a list of all subsets of $\{1, 2, 3, 4, 5\}$ containing three elements.

SOLUTION

```
<<DiscreteMath`Combinatorica`
KSubsets[{1, 2, 3, 4, 5}, 3]
{{1, 2, 3}, {1, 2, 4}, {1, 2, 5}, {1, 3, 4}, {1, 3, 5},
  {1, 4, 5}, {2, 3, 4}, {2, 3, 5}, {2, 4, 5}, {3, 4, 5}}
```

1.36   The function `DayOfWeek` appears in the package `Miscellaneous`Calendar`` and gives the day of the week of any date in the calendar.   Load the package, obtain help to determine its syntax, and then determine which day of the week January 1, 2000 was.

SOLUTION

```
<<Miscellaneous`Calendar`

? DayOfWeek
DayOfWeek[{y, m, d}, cal] gives the day of the week for
   year y, month m, and day d in calendar cal. The default
   calendar is the usual American calendar.   The date can also
   be given in {y, m, d, h, m, s} form.

DayOfWeek[2000, 1, 1]
Saturday
```

1.37   The package `Miscellaneous`ChemicalElements`` contains functions which give important information about their chemical and physical properties.   Among the functions included in this package are `AtomicWeight` and `AtomicNumber`, whose definitions are self-explanatory. Compute the atomic weight and atomic number of titanium.

SOLUTION

```
<<Miscellaneous`ChemicalElements`

?Titanium
Titanium is a chemical element.

?AtomicWeight
AtomicWeight[element] gives the atomic weight of the
   specified element.

? AtomicNumber
AtomicNumber[element] gives the atomic number of the
   specified element.

AtomicWeight[Titanium]
47.867

AtomicNumber[Titanium]
22
```

1.9   A PREVIEW OF WHAT'S TO COME

If you have just purchased your copy of *Mathematica*, you probably cannot wait to give it a test run.   Included below is a collection of problems for you to try.   What follows are some basic commands.   To keep things simple, options have been omitted and *Mathematica*'s default values are used exclusively. We will discuss modifications to these commands in subsequent chapters, but for now, just have fun!

SOLVED PROBLEMS

1.38 Obtain a 50-significant-digit approximation to $\sqrt{\pi}$.

SOLUTION

`N[`$\sqrt{\pi}$`, 50]` or `N[Sqrt[Pi], 50]`

1.7724538509055160272981674833411451827975494561224

Observe that this approximation gives only 49 decimal places.

1.39 Solve the algebraic equation $x^3 - 2x + 1 = 0$.

SOLUTION

`Solve[x`^{3}` - 2x + 1 == 0]` or `Solve[x^3 - 2x + 1 == 0]`

$$\left\{\{x \to 1\}, \left\{x \to \frac{1}{2}\left(-1 - \sqrt{5}\right)\right\}, \left\{x \to \frac{1}{2}\left(-1 + \sqrt{5}\right)\right\}\right\}$$

1.40 Express $(1 + x)^{10}$ in polynomial form.

SOLUTION

`Expand[(1 + x )`^{10}`]` or `Expand[(1+x)^10]`

$1 + 10x + 45x^2 + 120x^3 + 210x^4 + 252x^5 + 210x^6 + 120x^7 + 45x^8 + 10x^9 + x^{10}$

1.41 What is the 1000th prime?

SOLUTION

`Prime[1000]`

7919

1.42 Plot the graph of $y = \sin x$ from 0 to 2π.

SOLUTION

`Plot[Sin[x], {x, 0, 2`π`}];` or `Plot[Sin[x], {x, 0, 2 * Pi}];`

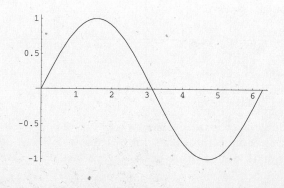

1.43   Sketch the graphs of $y = \sin x$, $y = \sin 2x$, and $y = \sin 3x$, $0 \le x \le 2\pi$, on one set of axes.

SOLUTION

```
Plot[{Sin[x],Sin[2x],Sin[3x]},{x,0,2π}];
```

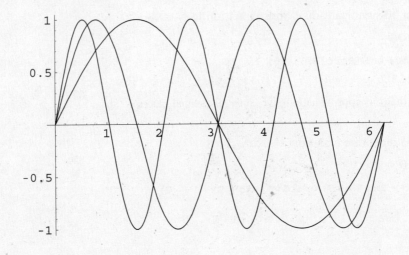

1.44   Sketch the graph of $(x^2 + y^2)^2 = xy$, $-2 \le x \le 2$.

SOLUTION

Since this is an implicit equation, we must first import **ImplicitPlot** from the **Graphics** package.

```
<<Graphics`ImplicitPlot`
ImplicitPlot[(x²+y²)²==x*y,{x,-2,2}]; or
ImplicitPlot[(x^2+y^2)^2==x*y,{x,-2,2}];
```

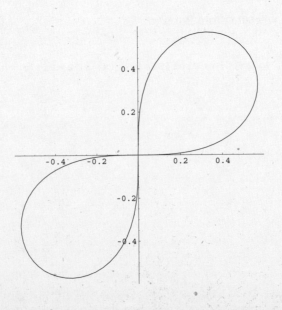

1.45 Sketch the three-dimensional surface defined by $z = (x^2 + 3y^2)e^{-(x^2 + y^2)}$.

SOLUTION

```
Plot3D[(x² + 3y²)E^-(x²+y²),{x, - 3,3},{y, - 3,3}]; or
Plot3D[(x^2 + 3y^2)*Exp[-(x^2 + y^2)],{x,-3,3},{y,-3,3}];
```

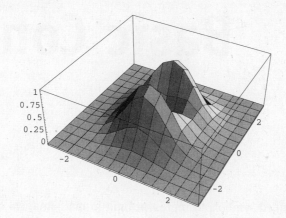

CHAPTER 2

Basic Concepts

2.1 CONSTANTS

Mathematica uses predefined symbols to represent built-in mathematical constants.

- `Pi` or π is the ratio of the circumference of a circle to its diameter.
- `E` or `e` is the base of the natural logarithm.

Both `Pi` and `E` are treated symbolically and do not have values, as such. However, they may be approximated to any degree of precision.

EXAMPLE 1

`N[π,50]` will compute a 50-significant-digit approximation to π (49 decimal places).

```
N[π,50]
3.1415926535897932384626433832795028841971693993751
```

- `Degree` is equal to `Pi/180` and is used to convert degrees to radians.
- `GoldenRatio` has the value $(1 + \sqrt{5})/2$ and has a special significance with respect to Fibonacci series. It is also used in *Mathematica* as the default width to height ratio of two-dimensional plots.
- `Infinity` or ∞ is a constant with special properties. For example, $\infty + 1 = \infty$.
- `EulerGamma` is Euler's constant and is approximately 0.577216. It has applications in integration and in asymptotic expansions.
- `Catalan` is Catalan's constant and is approximately 0.915966. It is used in the theory of combinatorial functions.

EXAMPLE 2

How much is $\infty + \infty$?

```
∞ + ∞
∞
```

SOLVED PROBLEMS

2.1 Approximately how many radians are in 90 degrees?

SOLUTION

```
90 Degree //N          ← expression//N is the same as N[expression].
1.5708
```

24

2.2   Show that `GoldenRatio` satisfies the algebraic equation $x^2 - x - 1 = 0$.

SOLUTION

```
x = GoldenRatio;
x^2 - x - 1//N
0
```

2.3   What happens if you try to subtract ∞ from ∞?

SOLUTION

```
∞ - ∞
∞::indet :
   Indeterminate expression -∞+∞ encountered.
Indeterminate
```

2.4   Compute a 20-decimal-place approximation to e, the base of the natural logarithm.

SOLUTION

```
N[E, 21]
2.71828182845904523536
```

2.2   "BUILT-IN" FUNCTIONS

In this section we discuss some of the more commonly used functions *Mathematica* offers. Because of the vast number of functions available, no attempt is made toward completeness. Additional functions are discussed in detail in later chapters.

Standard mathematical functions can be accessed by name or by clicking on its symbol in a *Mathematica* palette. For example, the square root of a number can be obtained using either the function **Sqrt** or, alternatively, by using the $\sqrt{\ }$ symbol from the BasicInput palette. Remember that the argument of a function must be contained within square brackets, **[ ]**.

- **Sqrt[x]** or $\sqrt{x}$ gives the nonnegative square root of x.

EXAMPLE 3

```
Sqrt[1521] or √1521
39
```

Higher-order roots can be computed by recalling that $\sqrt[n]{x} = x^{1/n}$. The symbol $\sqrt[\square]{\ }$ on the BasicInput palette may also be used. Notice that higher-order roots of negative numbers are given in a special format.

EXAMPLE 4

The cube root of 8 is given directly, but the cube root of -8 is given in terms of $\sqrt[3]{-1}$.

```
8^(1/3) or ∛8
2
```

```
(-8)^(1/3) or ∛-8
2 (-1)^(1/3)
```

EXAMPLE 5

```
N[√2]
1.41421
N[√2,50]
1.4142135623730950488016887242096980785696718753769
```

The function which returns the absolute value of x, $|x|$, is **Abs**.

■   **Abs[x]** returns x if $x \geq 0$ and $-x$ if $x < 0$.

The function Abs can also be applied to complex numbers. If z is the complex number $x + y\,I$, Abs[z] returns its modulus, $\sqrt{x^2 + y^2}$.

EXAMPLE 6

```
Abs[5]
5
Abs[-5]
5
Abs[5 + 12 I]
13
```

It is sometimes useful to have a function which determines the sign of a number.

■   **Sign[x]** returns the values -1, 0, or 1 depending upon whether x is negative, 0, or positive, respectively.

EXAMPLE 7

```
Sign[-27.5]
-1
Sign[0]
0
Sign[6.254]
1
```

The factorial of a positive integer, n, represented $n!$ in mathematical literature, is the product of the integers $1, 2, 3, \ldots, n$. By definition, $0! = 1$. For noninteger values of n, $n!$ is defined by $\Gamma(n + 1)$ where Γ is Euler's gamma function.

■   **Factorial[n]** or $n!$ gives the factorial of n.

EXAMPLE 8

```
5!
120
0!
1
Factorial[3.5]
11.6317
```

Mathematica has a built-in random number generator. This is a useful function in probability theory and statistical analysis, e.g., random walks and Monte Carlo methods.

■   **Random[ ]** gives a uniformly distributed real pseudorandom number in the interval $[0, 1]$.
■   **Random[*type*]** returns a uniformly distributed pseudorandom number of type *type*, which is either Integer, Real, or Complex. Its values are between 0 and 1, in the case of Integer or Real, and contained within the rectangle determined by 0 and $1 + i$, if *type* is Complex.

- **Random[*type*, *range*]** gives a uniformly distributed pseudorandom number in the interval or rectangle determined by *range*. *range* can be either a single number or a list of two numbers such as {a,b} or {a+bI,c+dI}. A single number m is equivalent to {0,m}.
- **Random[*type*, *range*, n]** gives a uniformly distributed pseudorandom number to n significant digits in the interval or rectangle determined by *range*.

EXAMPLE 9 (your answers will be different from those shown)

Random[Integer]
0
← returns 0 or 1 with equal probability

Random[Real]
0.386297
← returns a six-significant-digit real number between 0 and 1

Random[Complex]
0.420851 + 0.382187 i
← returns a complex number in the rectangle whose opposite vertices are 0 and 1 + i

Random[Real, 5]
1.83872
← returns a real number uniformly distributed in the interval [0,5]

Random[Real, {3,5}]
3.95386
← returns a real number uniformly distributed in the interval [3, 5]

Random[Real, {3,5}, 10]
4.014673296
← returns a real number uniformly distributed in the interval [3,5] to 10 significant digits

Random[Integer, {1,10}]
7
← returns an integer between 1 and 10 with equal probability 1/10

Random[Complex, {2 + I, 5 + 6 I}]
2.61319 + 4.30869 i
← returns a complex number in the rectangle whose opposite vertices are the complex numbers 2 + i and 5 + 6i

A positive integer is prime if it is divisible only by itself and 1. For technical reasons, 1 is not considered prime; the smallest prime is 2.

- **Prime[n]** returns the nth prime.

EXAMPLE 10

Find the 7th prime.

Prime[7]
17

The Fibonacci numbers are defined by

$$f_1 = 1,$$
$$f_2 = 1,$$
$$f_n = f_{n-2} + f_{n-1} \quad n \geq 3$$

Thus the first few Fibonacci numbers are 1, 1, 2, 3, 5, 8, 13, 21, ...

- The nth Fibonacci number is returned by **Fibonacci[n]**.

EXAMPLE 11

Fibonacci[7]
13

There are three *Mathematica* functions which convert real numbers to nearby integers.

- **Round[x]** returns the integer closest to x. If x lies exactly between two integers (e.g., 5.5), Round returns the nearest even integer.

- **Floor[x]** returns the greatest integer which does not exceed x. This is sometimes known as the "greatest integer function" and is represented in many textbooks by $\lfloor x \rfloor$.
- **Ceiling[x]** returns the smallest integer not less than x. Many textbooks represent this by $\lceil x \rceil$.

EXAMPLE 12

```
Round[5.75]
6
Floor[5.75]
5
Ceiling[5.75]
6
```

A decimal number can be broken up into two parts, the integer portion (number to the left of the decimal point) and the fractional portion.

- **IntegerPart[x]** gives the integer portion of x (decimal point excluded).
- **FractionalPart[x]** gives the fractional portion of x (decimal point included).

Observe that `IntegerPart[x] + FractionalPart[x] = x`.

EXAMPLE 13

```
IntegerPart[4.67]
4
FractionalPart[4.67]
0.67
IntegerPart[4.67] + FractionalPart[4.67]
4.67
```

If m and n are positive integers, there exist unique integers q and r such that

$$m = qn + r \qquad \text{with } 0 \le r < n$$

This result is known as the *Division Algorithm*. q is called the *quotient* and r the *remainder*. The *Mathematica* functions **Quotient** and **Mod** return the quotient and remainder, respectively.

- **Quotient[m,n]** returns the quotient when m is divided by n.
- **Mod[m,n]** returns the remainder when m is divided by n.

EXAMPLE 14

```
Quotient[17, 3]
5
Mod[17, 3]
2
```

Suppose a and b are two integers. If there exists an integer k such that $a = kb$, we say that b divides a. Alternatively, a is a multiple of b.

Let m and n be two integers. If b divides both m and n, we say that b is a common divisor of m and n. The largest common divisor of m and n is called their *greatest common divisor*.

If a is a multiple of *both* m and n, we say a is a common multiple of m and n. The smallest common multiple of m and n is called their *least common multiple*.

- **GCD[m,n]** returns the greatest common divisor of m and n.
- **LCM[m,n]** returns the least common multiple of m and n.

EXAMPLE 15

Find the greatest common divisor and least common multiple of 48 and 72.

```
GCD[48,72]
24
LCM[48,72]
144
```

The functions **GCD** and **LCM** extend to more than two arguments.

EXAMPLE 16

Find the greatest common divisor and least common multiple of 24, 40, and 48.

```
GCD[24, 40, 48]
8
LCM[24, 40, 48]
240
```

The *Fundamental Theorem of Arithmetic* guarantees that every positive integer can be factored into primes in a unique way.

- The function **FactorInteger[n]** gives the prime factors of n together with their appropriate exponents.

EXAMPLE 17

```
FactorInteger[2381400]
{{2, 3}, {3, 5}, {5, 2}, {7, 2}}
```

The prime factors of 2,381,400 are 2, 3, 5, and 7 with exponents, respectively, 3, 5, 2, and 2. In other words, $2{,}381{,}400 = 2^3 3^5 5^2 7^2$. The result of this operation produces a nested sequence of *lists*. (A list is a *Mathematica object*, enclosed within braces, { }, which will be discussed in detail in Chapter 3.)

In order to estimate computational efficiency, it is useful to be able to determine how long an operation or sequence of operations takes to execute.

- **Timing[*expression*]** evaluates *expression*, and returns a list of time used, together with the result obtained.

Timing counts only the CPU time spent in the *Mathematica* kernel. It does not include overhead time spent in the front end.

EXAMPLE 18

How long does it take the kernel to compute the 1,000,000th prime?

```
Prime[1000000]//Timing          ← This is equivalent to Timing[Prime[1000000]].
{0.51 Second,15485863}
```

Of course, the actual time taken will vary, depending upon the speed of the CPU.

Logarithms and exponential functions to any base can be easily computed.

- **Log[x]** represents the natural logarithm. If a base, *b*, other than *e* is required, the appropriate form is **Log[b, x]**.
- The function **Exp[x]** is the natural exponential function. Other equivalent forms are **E^x** and **Eˣ**. Lowercase e cannot be used, but the special symbol e from the BasicInput palette may be used instead. Exponential functions to the base b are computed by **b^x** or **bˣ**.

EXAMPLE 19

Compute ln 100, the natural logarithm of 100.

`Log[100]`
Log[100]

```
┌─────────────────────────────────────────┐
│ Observe that Mathematica always gives exact answers. │
│ Approximations are supplied only when requested.     │
└─────────────────────────────────────────┘
```

`Log[100]//N`
4.60517

EXAMPLE 20

Compute $\log_2 100$.

`Log[2, 100]`

$$\frac{\text{Log}[100]}{\text{Log}[2]}$$ ← This is the exact value of $\log_2 100$, expressed in terms of natural logarithms.

`Log[2, 100]//N`
6.64386

EXAMPLE 21

To compute a numerical approximation of e^2, we can write

`Exp[2]//N` or `E²//N` or `e²//N`
7.38906

- The six basic trigonometric functions, sine, cosine, tangent, secant, cosecant, and cotangent, are represented in *Mathematica* by `Sin`, `Cos`, `Tan`, `Sec`, `Csc`, and `Cot`, respectively.

Mathematica assumes the arguments of trigonometric functions to be in radians. Problems involving degrees must first be converted to radians if trigonometric functions are involved. For this purpose, one can use the built-in constant, `Degree`, whose value is $\pi/180$. The symbol °, located on the BasicInput palette, may be used as well.

EXAMPLE 22

60° is equivalent to $\pi/3$ radians. To compute its sin using radian measure, we write

`Sin [`$\frac{\pi}{3}$`]` or `Sin[Pi/3]`

$$\frac{\sqrt{3}}{2}$$

If we wish to compute its sin using degree measure, we can type

`Sin[60 Degree]` or `Sin[60°]`

$$\frac{\sqrt{3}}{2}$$

Care must be taken with trigonometric powers. The square of $\sin x$ in trigonometric form is traditionally written $\sin^2 x$, but *Mathematica* will accept only `Sin[x]²` or `Sin[x]^2`.

EXAMPLE 23

Compute the square of sin 60°.

`Sin[60°]²` or `Sin[60 Degree]^2`

$$\frac{3}{4}$$

- The inverse trigonometric functions are **ArcSin**, **ArcCos**, **ArcTan**, **ArcSec**, **ArcCsc**, and **ArcCot**. However, only the *principal values*, expressed in radians, are returned by these functions.

EXAMPLE 24

```
ArcSin[1]
```
$$\frac{\pi}{2}$$

```
ArcCos[Cos[3π]]
```
π

> Cos[3π] = -1 but the principal value of ArcCos[-1] is π.

Hyperbolic functions are combinations of exponential functions which have interesting mathematical properties. There are six hyperbolic functions. The three basic ones are:

$$\sinh x = \frac{e^x - e^{-x}}{2} \qquad \cosh x = \frac{e^x + e^{-x}}{2} \qquad \tanh x = \frac{e^x - e^{-x}}{e^x + e^{-x}}$$

The other three, $\operatorname{sech} x$, $\operatorname{csch} x$, and $\coth x$, are reciprocals, respectively, of $\cosh x$, $\sinh x$, and $\tanh x$.

- The *Mathematica* representations of the six hyperbolic functions are **Sinh**, **Cosh**, **Tanh**, **Sech**, **Csch**, and **Coth**.

EXAMPLE 25

Compute a numerical approximation to $\sinh 2$.

```
Sinh[2]//N
3.62686
```

- The inverse hyperbolic functions are represented **ArcSinh**, **ArcCosh**, **ArcTanh**, **ArcSech**, **ArcCsch**, and **ArcCoth**.

Because **Cosh** and **Sech** are not one-to-one, **ArcCosh** and **ArcSech** return only positive values for real arguments.

EXAMPLE 26

```
ArcSinh[-2]//N
-1.44364
ArcCosh[2]//N
1.31696
```

One special command is worthy of mention at this time.

- **Print[*expression*]** prints *expression*, followed by a line feed.
- **Print[*expression1, expression2, . . .*]** prints *expression1, expression2, . . .* followed by a single line feed.

At first glance it may seem that **Print** is a redundant command, as simply typing the name of any object will reveal its value. However, it has a useful purpose (e.g., see Section 2.7, Loops, later in this chapter).

EXAMPLE 27

```
Print["This prints a line of text."]
This prints a line of text.
```

EXAMPLE 28

```
a = 1; b = 2; c = 3; d = 4; e = 5;
Print[a, b, c, d, e]
12345
```

Mathematica includes a class of functions ending in the letter Q:

ArgumentCountQ	AtomQ	DigitQ
EllipticNomeQ	EvenQ	ExactNumberQ
FreeQ	HypergeometricPFQ	InexactNumberQ
IntegerQ	IntervalMemberQ	InverseEllipticNomeQ
LegendreQ	LetterQ	LinkConnectedQ
LinkReadyQ	ListQ	LowerCaseQ
MachineNumberQ	MatchLocalNameQ	MatchQ
MatrixQ	MemberQ	NameQ
NumberQ	NumericQ	OddQ
OptionQ	OrderedQ	PartitionsQ
PolynomialQ	PrimeQ	SameQ
StringMatchQ	StringQ	SyntaxQ
TrueQ	UnsameQ	UpperCaseQ
ValueQ	VectorQ	

These functions are used to test for certain conditions and return a value of True or False. Their precise syntax can be determined from the Help menu or by using ? as illustrated in the next examples.

EXAMPLE 29

```
?PrimeQ
PrimeQ[expr] yields True if expr is a prime number, and yields
 False otherwise.

PrimeQ[5]
True
PrimeQ[6]
False
```

EXAMPLE 30

```
?PolynomialQ
PolynomialQ[expr, var] yields True if expr is a polynomial in
 var, and yields False otherwise. PolynomialQ[expr, {var1, ... }]
 tests whether expr is a polynomial in the vari.
```

$$\text{PolynomialQ}[x^2y + x + \sqrt{y}, x]$$
True
$$\text{PolynomialQ}[x^2y + x + \sqrt{y}, y]$$
False

SOLVED PROBLEMS

2.5   Compute numerical approximations to the square root and cube root of 10.

SOLUTION

$$\sqrt{10}\,//\text{N} \text{ or } \text{Sqrt}[10]\,//\text{N}$$
3.16228
$$\sqrt[3]{10}\,//\text{N} \text{ or } 10\,\verb|^|\,(1/3)\,//\text{N}$$
2.15443

2.6   The binomial coefficient $C(n,k) = \dfrac{n!}{k!(n-k)!}$.   Use this definition to compute $C(10,4)$.

SOLUTION

$\dfrac{10!}{4!(10-4)!}$ or `Factorial[10]/(Factorial[4] * Factorial[10-4])`

210

2.7   A fair die has six faces, numbered 1 through 6, and each occurs with equal probability.   Simulate four tosses of a fair die.

SOLUTION

```
Random[Integer, {1, 6}]
Random[Integer, {1, 6}]
Random[Integer, {1, 6}]
Random[Integer, {1, 6}]
6
1
5
3
```

2.8   Find a 15-significant-digit pseudorandom real number between π and 2π.

SOLUTION

```
Random[Real, {π, 2π}, 15]
4.13129131207734
```

2.9   What is the 27th Fibonacci number?

SOLUTION

```
Fibonacci[27]
196418
```

2.10   Show that there is no prime between 157 and 163.

SOLUTION

```
Prime[37]            ← We determine this by experimentation.
157
Prime[38]
163
```

Since 157 and 163 are consecutive primes, there is no prime between them.

2.11   What is the integer closest to $\sqrt{159}$?

SOLUTION

```
Round[Sqrt[159]] or √159//Round
13
```

2.12   Between what two consecutive integers does $(\pi^2 + 1)^5$ lie?

SOLUTION

```
Floor[(π² + 1)⁵]
151729
Ceiling[(π² + 1)⁵]
151730
```

The number $(\pi^2 + 1)^5$ lies between 151,729 and 151,730.

2.13   Compute the value of $\lceil x \rceil - \lfloor x \rfloor$ first using $x = 17$ and then using $x = \pi$.

SOLUTION

```
x = 17;
Ceiling[x] - Floor[x]
0
x = Pi;
Ceiling[x] - Floor[x]
1
```

> $\lceil x \rceil - \lfloor x \rfloor$ always equals 0 when x is an integer and 1 when x is noninteger.

2.14   What are the greatest common divisor and least common multiple of 5,355 and 40,425?

SOLUTION

```
GCD[5355, 40425]
105
LCM[5355, 40425]
2061675
```

2.15   Show that 15, 16, and 30 are relatively prime (numbers are relatively prime if they have no common factor other than 1).

SOLUTION

```
GCD[15, 16, 30]
1
```

Since their GCD = 1, their only common factor is 1, so they are relatively prime.

2.16   A theorem from number theory says that the product of the GCD and LCM of two numbers is always equal to the product of the numbers. Verify this using the numbers 74,613 and 85,085.

SOLUTION

```
a = 74613;
b = 85085;
GCD[a, b] * LCM[a, b]
6348447105
a * b
6348447105
```

Obviously the products are identical.

2.17   Show that 156,875,438,767 is not prime.

SOLUTION

```
FactorInteger[156875438767]
{{53,1},{2959913939,1}}
```

This shows that 156,875,438,767 is equal to the product of primes 53 and 2,959,913,939 and is therefore not prime.

2.18   How long did it take *Mathematica* to factor 156,875,438,767 in the previous problem?

SOLUTION

```
Timing[FactorInteger[156875438767]] or
FactorInteger[156875438767]//Timing
{0.06 Second,{{53, 1}, {2959913939, 1}}}
```

It took approximately 0.06 second (this time will vary from computer to computer).

2.19   Compute the natural logarithm of e^5.

SOLUTION

```
Log[e⁵] or Log[E^5] or Log[Exp[5]]
5
```

2.20   Compute the common logarithm (base 10) of e^5.   What is its numerical approximation?

SOLUTION

```
Log[10,e⁵] or Log[10, E^5] or Log[10, Exp[5]]
```
$$\frac{5}{\text{Log}[10]}$$
```
% //N
2.17147
```

2.21   Compute $\sqrt{3} + \sqrt{2}$ and $\sqrt{3} - \sqrt{2}$ to 50 significant digits.   Then compute their product.

SOLUTION

```
a = N[√3+√2, 50]
3.1462643699419723423291350657155704455124771291873
b = N[√3-√2, 50]
0.31783724519578224472575761729617428837313337843343
a * b
1.0000000000000000000000000000000000000000000000000
```

2.22   What is the *exact* value of sin 15°? Compute a 20-decimal-place approximation.

SOLUTION

```
Sin[15 Degree] or Sin[15°]
```
$$\frac{-1+\sqrt{3}}{2\sqrt{2}}$$
```
N[%, 20]
0.25881904510252076235
```

2.23   Select a random number x between 0 and 1 and compute $\sin^2 x + \cos^2 x$.

SOLUTION

```
x = Random[ ];
Sin[x]² + Cos[x]²
1
```

> Recall from trigonometry that
> $\sin^2 x + \cos^2 x = 1$ for all x.

2.24   Find a number between $-\pi/2$ and $\pi/2$ whose sin is 1/2.

SOLUTION

```
ArcSin[1/2]
```

$$\frac{\pi}{6}$$

2.25   Select a random number x between 0 and 1 and compute $\cosh^2 x - \sinh^2 x$.

SOLUTION

```
x = Random[];
Cosh[x]² - Sinh[x]²
1
```

> Hyperbolic functions have properties similar to
> trigonometric functions: $\cosh^2 x - \sinh^2 x = 1$ for all x.

2.26   Obtain an alternate representation of $\tanh(\ln x)$.

SOLUTION

```
Tanh[Log[x]]
```

$$\frac{-1 + x^2}{1 + x^2}$$

2.27   Approximately how many degrees are there in one radian?

SOLUTION

```
N[1/Degree]
57.2958
```

2.28   How much is $\infty + 100{,}000$?

SOLUTION

```
∞ + 100000
∞
```

2.29   What is a square root of the complex number $3 + 4i$?

SOLUTION

```
√3 + 4 I  or  Sqrt[3 + 4 I]
2 + i
```

2.30   The number of permutations of n objects taken k at a time is $P(n, k) = \dfrac{n!}{(n - k)!}$.   How many permutations of 20 objects taken 10 at a time are there?

SOLUTION

```
n = 20;
k = 10;
n!/(n-k)! or Factorial[n]/Factorial[n-k]
670442572800
```

2.31 Between what two consecutive integers does the natural logarithm of 100,000 lie?

SOLUTION

```
Floor[Log[100000]]
11
Ceiling[Log[100000]]
12
```

ln 100,000 lies between 11 and 12.

2.32 What is the quotient and remainder if 62,173,467 is divided by 9,542?

SOLUTION

```
Quotient[62173467, 9542]
6515
Mod[62173467, 9542]
7337
```

2.33 Find the greatest common divisor and least common multiple of 1,001 and 1,331.

SOLUTION

```
GCD[1001, 1331]
11
LCM[1001]
121121
```

2.34 How long does it take your computer to find the prime factorization of 10!?

SOLUTION

```
FactorInteger[10!]//Timing
{0.01 Second, {{2, 8}, {3, 4}, {5, 2}, {7, 1}}}
```

The factorization is $2^8 3^4 5^2 7^1$; times will vary depending on the speed of your CPU.

2.35 Find an algebraic expression for $\cos\left(\sin^{-1}\left(\dfrac{x^2}{x^2+1}\right)\right)$.

SOLUTION

```
Cos[ArcSin[x²/(x²+1)]]
```

$$\sqrt{1 - \frac{x^4}{(1+x^2)^2}}$$

2.36 Is 15,485,863 prime?

SOLUTION

```
PrimeQ[15485863]
True
```

2.3 BASIC ARITHMETIC OPERATIONS

As we have seen, basic arithmetic operations such as addition are performed by inserting an operation symbol between two numbers. Thus the sum of 3 and 5 would be obtained by typing $3 + 5$. However, in more advanced applications it is sometimes useful to represent these operations as functions. Towards this end *Mathematica* includes the following:

- `Plus[a, b,...]` computes the sum of a, b, ... `Plus[a, b]` is equivalent to **a + b**.
- `Times[a, b,...]` computes the product of a, b, ... `Times[a, b]` is equivalent to **a ∗ b**.
- `Subtract[a, b]` computes the difference of a and b. Only two arguments are permitted. `Subtract[a, b]` is equivalent to **a - b**.
- `Divide[a, b]` computes the quotient of a and b. Only two arguments are permitted. `Divide[a, b]` is equivalent to **a/b**.
- `Minus[a]` produces the additive inverse (negative) of a. `Minus[a]` is equivalent to **-a**.
- `Power[a, b]` computes a^b. `Power[a, b, c]` produces a^{b^c}, etc.

EXAMPLE 31

```
Plus[2, 3, 4]
9
Times[2, 3, 4]
24
Power[2, 3, 4]
2417851639229258349412352
```

In order to see the way in which *Mathematica* handles functions internally, the command `FullForm` is quite useful.

- `FullForm[expression]` exhibits the internal form of *expression*.

`FullForm` may be used for any *Mathematica* function, not only arithmetic operators.

EXAMPLE 32

```
FullForm[a + b + c]
Plus[a, b, c]
FullForm[a - b]
Plus[a, Times[-1, b]]
FullForm[(a ∗ b)^c]
Power[Times[a, b], c]
```

In addition to the standard operational symbols discussed above, there are a few additional commands which are useful in special situations.

- `Increment[x]` or **x + +** increases the value of x by 1 but returns the *old* value of x.
- `Decrement[x]` or **x - -** decreases the value of x by 1 but returns the *old* value of x.
- `PreIncrement[x]` or **+ + x** increases the value of x by 1 and returns the *new* value of x.
- `PreDecrement[x]` or **- - x** decreases the value of x by 1 and returns the *new* value of x.
- `AddTo[x, y]` or **x + = y** adds y to x and returns the new value of x.
- `SubtractFrom[x, y]` or **x - = y** subtracts y from x and returns the new value of x.
- `TimesBy[x, y]` or **x ∗ = y** multiplies y by x and returns the new value of x.
- `DivideBy[x, y]` or **x / = y** divides x by y and returns the new value of x.

In order for **x + +**, **+ + x**, **x - -**, **- - x**, **x + = y**, **x - = y**, **x ∗ = y** or **x / = y** to work, **x** must have a *numerical* value.

The next two examples illustrate the various addition commands. The commands for subtraction, multiplication, and division are similar.

EXAMPLE 33

```
x = 3;
x + +
3   ← the old value of x is returned
x
4   ← the actual value of x is 4
```

x + + is equivalent to the sequence
```
    x
    x = x + 1;
```

```
x = 3;
+ + x
4        ← the new value of x is returned
x
4        ← the actual value of x is 4
```

+ + x is equivalent to the statement x = x + 1

EXAMPLE 34

```
x = 3; y = 4;
x + y
7   ← the new sum is returned
x
4   ← x remains unchanged
```

```
x = 3; y = 4;
x + = y
7        ← the new sum is returned
x
7        ← the new value of x is 7
```

x + = y is equivalent to the statement x = x + y

SOLVED PROBLEMS

2.37 How does *Mathematica* evaluate the expression $a + bc/d$?

SOLUTION

```
FullForm[a + b * c/d]
Plus[a,Times[b,c,Power[d,-1]]]
```

2.38 How is the function Minus[x] treated internally in *Mathematica*?

SOLUTION

```
FullForm[Minus[x]]
Times[-1, x]
```

2.4 STRINGS

A string is an (ordered) sequence of characters. Strings have no numerical value and are often used as labels for tables, graphs, and other displays.

In *Mathematica*, a string is enclosed within quotation marks. Thus "abcde" is a string of five characters. Do not confuse "abcde" with abcde, as the latter is *not* a string.

Mathematica comes equipped with a number of string manipulation commands.

- **StringLength[*string*]** returns the number of characters in *string*.
- **Stringjoin[*string1*, *string2*, ...]** or *string1* <> *string2* <> ... concatenates two or more strings to form a new string whose length is equal to the sum of the individual string lengths.
- **StringReverse[*string*]** reverses the characters in *string*.

StringDrop eliminates characters from a string. There are five forms of this command.

- **StringDrop[*string*, n]** returns *string* with its first n characters dropped.
- **StringDrop[*string*, -n]** returns *string* with its last n characters dropped.
- **StringDrop[*string*, {n}]** returns *string* with its nth character dropped.
- **StringDrop[*string*, {-n}]** returns *string* with the nth character from the end dropped.
- **StringDrop[*string*, {m, n}]** returns *string* with characters m through n dropped.

`StringTake` takes characters from a string.   Its format is similar to `StringDrop`.

- `StringTake[`*string*`, n]` returns the first n characters of *string*.
- `StringTake[`*string*`, -n]` returns the last n characters of *string*.
- `StringTake[`*string*`, {n}]` returns the nth character of *string*.
- `StringTake[`*string*`, {-n}]` returns the nth character from the end of *string*.
- `StringTake[`*string*`, {m, n}]` returns characters m through n of *string*.

EXAMPLE 35

In this example we define `strng = "abcdefg"`.   The output is shown to the right of the command.   (To avoid conflict with the *Mathematica* symbol `'String'`, we do not use `'string'`.)

```
strng = "abcdefg"              abcdefg
strng <> "hijklmnop"           abcdefghijklmnop
StringLength[ strng]           7
StringReverse[ strng]          gfedcba
StringDrop[ strng, 2]          cdefg
StringDrop[ strng, -2]         abcde
StringDrop[ strng, {2}]        acdefg
StringDrop[ strng, {-2}]       abcdeg
StringDrop[ strng, {2, 5}]     afg
StringTake[ strng, 2]          ab
StringTake[ strng, -2]         fg
StringTake[ strng, {2}]        b
StringTake[ strng, {-2}]       f
StringTake[ strng, {2, 5}]     bcde
```

`StringInsert` allows you to insert characters within existing strings.

- `StringInsert[`*string1*`, `*string2*`, n]` yields a string with *string2* inserted starting at position n in *string1*.
- `StringInsert[`*string1*`, `*string2*`, -n]` yields a string with *string2* inserted starting at the nth position from the end of *string1*.
- `StringInsert[`*string1*`, `*string2*`, {n1, n2, ...}]` inserts a copy of *string2 at each of the positions* n1, n2, ... *of string1*.

`StringReplace` allows you to replace part of a string with another string.

- `StringReplace[`*string*`, `*string1* → *newstring1*`]` replaces *string1* by *newstring1* whenever it appears in *string*.
- `StringReplace[`*string*`, {`*string1* → *newstring1*`, `*string2* → *newstring2*`, ...}]` replaces *string1* by *newstring1*, *string2* by *newstring2*, ... whenever they appear in *string*.
- `StringPosition[`*string*`, `*substring*`]` returns a list of the start and end positions of all occurrences of *substring* with *string*.   (Lists are discussed in detail in Chapter 3.)

EXAMPLE 36

```
strng1 = "abcdefg";
strng2 = "123";
StringInsert[ strng1, strng2, 3]
ab123cdefg
StringInsert[ strng1, strng2, -3]
abcde123fg
StringInsert[ strng1, strng2, {1, 3, 5, 7}]
123ab123cd123ef123g
StringReplace[ strng1, "ab" → "AB"]
```

```
ABcdefg
StringReplace[ strng1, {"ab"→"AB", "fg"→"FG"}]
ABcdeFG
```

EXAMPLE 37

```
strng = "abcxabcxxabcxxxabc";
StringLength[strng]
18
StringPosition[strng, "abc"]
{{1, 3}, {5, 7}, {10, 12}, {16, 18}}
```

2.5 ASSIGNMENT, REPLACEMENT, AND LOGICAL RELATIONS

All programming languages must have the ability to make assignments in order to transfer the result of a calculation to a symbol which can be recalled for use later. *Mathematica* offers two types of assignment and there is often confusion as to which one to use in a given situation.

- **lhs = rhs** is an *immediate* assignment in which rhs is evaluated at the time the assignment is made.
- **lhs := rhs** is a *delayed* assignment in which rhs is evaluated <u>each time</u> the value of lhs is called.

In many situations both assignments produce identical results. There are, however, a few instances where one must be careful. The examples given below use ideas which are discussed in later chapters. They are self-explanatory, however, and will be easily understood.

EXAMPLE 38

When defining functions recursively (Section 2.9), the := operator *must* be used. For example,

```
f[0] = 1;
f[n_] := n f[n-1]
```

produces *n* factorial. Since *Mathematica* cannot compute f[n] until the value of n is specified, the delayed assignment, := , must be used. Using = causes recursion errors.

EXAMPLE 39

When defining piecewise functions (Section 2.9), one *must* use :=. For example,

```
g[x_] := x² /; x ≥ 0          ← /; is a conditional.  Assignment will be made only if x ≥ 0.
g[x_] := -x² /; x < 0
```

Using = causes trouble, as *Mathematica* cannot determine which branch will be taken until a value of x is supplied.

EXAMPLE 40

You may think that the := assignment is more general and can be safely used in any given situation. This is true to a certain extent, but there are times when one should use =. As an extreme, but reasonable, example, let us define

$$F[x\_] := \int_0^x t\, E^t\, Sin[t]\, dt$$

Each time a value of F is computed, *Mathematica* performs several "integration by parts" evaluations. Now imagine that many different values of F are needed, for example, in the instruction Plot[F[x], {x, 0, 5}].

This plots `F[x]` from 0 to 5 using many points. Each time the value of `F` is computed, the integral is reevaluated from scratch, applying integration by parts each time. The result is a lengthy delay in displaying the graph. Using = causes the graph to be displayed much more quickly.

Often, you will want to evaluate an expression *without* assigning a value to a symbol. This can be done with the `/.` replacement operator.

EXAMPLE 41

Suppose we want to evaluate $x^2 + 5x + 6$ when $x = 3$, but do not want to assign a value to x.

```
x² + 5 x + 6 / . x → 3
30
?x
Global'x          ← x is left undefined
```

`/.` can also be used to replace an expression by another expression. Several replacements can be made at the same time if braces are used. `/.` is equivalent to **ReplaceAll**.

EXAMPLE 42

```
√2 x + 3 + (2 x + 3)² / . 2 x + 3 → 3 y + 5
√3 y + 5 + (3 y + 5)²
```

EXAMPLE 43

```
2 x + 3 y / . {y → x, x → y}
3 x + 2 y
```

Do not confuse = with = =, a "logical" equality. `lhs == rhs` is `True` if and only if `lhs` and `rhs` have the same value; otherwise it is `False`. Logical equalities will be used extensively in connection with equation solving (Chapter 6).

Other logical relations are available. The following list summarizes them.

- `Equal[x, y]` or `x == y` is `True` if and only if `x` and `y` have the same value.
- `Unequal[x, y]` or `x != y` or `x ≠ y` is `True` if and only if `x` and `y` have different values.
- `Less[x, y]` or `x < y` is `True` if and only if `x` is numerically less than `y`.
- `Greater[x, y]` or `x > y` is `True` if and only if `x` is numerically greater than `y`.
- `LessEqual[x, y]` or `x <= y` or `x ≤ y` is `True` if and only if `x` is numerically less than `y` or equal to `y`.
- `GreaterEqual[x, y]` or `x >= y` or `x ≥ y` is `True` if and only if `x` is numerically greater than `y` or equal to `y`.

Note that `Equal` and `Unequal` can be used for comparing both numerical and certain nonnumerical quantities, while `Less`, `Greater`, `LessEqual`, and `GreaterEqual` are strictly numerical comparisons.

EXAMPLE 44

`1 == 2`	`1 != 2`	`1 <= 2`	`a + a == 2 a`
False	True	True	True
`2 == 2`	`2 != 2`	`2 <= 2`	`a < a`
True	False	True	a < a

Mathematica also includes the following logical operations.

- **And[p, q]** or **p && q** or **p∧q** is True if both p and q are True and False otherwise.
- **Or[p,q]** or **p || q** or **p∨q** is True if p or q (or both) are True and False otherwise.
- **Xor[p, q]** is True if p or q (but not both) are True and False otherwise.
- **Not[p]** or **!p** or **¬p** is True if p is False and False if p is True.
- **Implies[p, q]** or **p⇒q** is False if p is True and q is False, True otherwise.

Note: ⇒ can be obtained with the sequence [ESC] + [=] + [>] + [ESC].

Logical expressions can be compared using **LogicalExpand**. This command puts logical expressions into disjunctive normal form.

EXAMPLE 45

Use *Mathematica* to verify the distributive law: $p \wedge (q \vee r) = (p \wedge q) \vee (p \wedge r)$.

```
lhs = p && (q || r);
rhs = (p && q) || (p && r);
lhs == rhs
(p && (q || r)) == (p && q || p && r)
LogicalExpand[lhs] == LogicalExpand[rhs]
True
```

SOLVED PROBLEMS

2.39 The *Mathematica* command **Expand[*expression*]**, which will be discussed in Chapter 7, expands *expression* algebraically. Define two symbols a and b as Expand[(x+1)^3], using = and :=, respectively. Then let x = u + v and compute a and b.

SOLUTION

```
a = Expand[(x + 1) ^ 3]
1 + 3 x + 3 x² + x³                      ←expansion occurs immediately
b := Expand[(x + 1) ^ 3]                 ←expansion does not occur until b is called
x = u + v
u + v
a
1 + 3 (u + v) + 3 (u + v)² + (u + v)³    ←u + v replaces x after expansion
b
1 + 3u + 3u² + u³ + 3v + 6uv +
3u²v + 3v² + 3uv² + v³                   ←u + v replaces x before expansion
```

2.40 The command **Together**, which will be discussed in Chapter 7, combines two fractions into one with a common denominator. Define two symbols, y and z, as Together[a + b] using, respectively, = and := . Then let a = 1/x and b = 1/(x + 1) and compute y and z.

SOLUTION

```
y = Together[a + b]
a + b                    ←At this point a and b are not fractions so Together does nothing.
z := Together[a + b]
a = 1/x;
b = 1/(x + 1);
y
```

$$\frac{\dfrac{1}{x} + \dfrac{1}{1+x}}{z}$$

← Since `Together` was executed prior to the introduction
of the fractions, `Together[a+b] = a+b`.

$$\frac{1+2x}{x(1+x)}$$

← `Together` is executed *after* the
fractions are introduced.

2.41 The *Mathematica* command `Simplify[`*expression*`]` attempts to simplify the algebraic expression, *expression*. Type `a = Simplify[poly]` and `b := Simplify[poly]`. Then let `poly = x² + 2x + 1`. Compute `a` and `b` and explain the difference in output.

SOLUTION

```
a = Simplify[poly];
b := Simplify[poly];
poly = x² + 2x + 1;
a
1 + 2x + x²
b
(1 + x)²
```

Since `a` is computed immediately, its value is the simplified form of the symbol `poly`, which is just `poly`. Then `poly` is replaced by $x^2 + 2x + 1$. On the other hand, `b` is not evaluated until called in the next to last line, so *Mathematica* simplifies the polynomial.

2.42 Replace x by $x^2 + 2x + 3$ in the expression $x^2 + 5x + 6$.

SOLUTION

```
x² + 5x + 6 /. x → x² + 2x + 3
6 + 5(3 + 2x + x²) + (3 + 2x + x²)²
```

2.43 Replace y by $x + 1$ and z by $x + 2$ in the expression $(x + y + z)^2$.

SOLUTION

```
(x + y + z)² /. {y → x + 1, z → x + 2}
(3 + 3x)²
```

2.44 Use *Mathematica* to verify De Morgan's laws:

$$\neg(p \wedge q) = \neg p \vee \neg q \qquad \text{and} \qquad \neg(p \vee q) = \neg p \wedge \neg q$$

SOLUTION

```
LogicalExpand[!(p && q)] == LogicalExpand[!p || !q]
True
LogicalExpand[!(p || q)] == LogicalExpand[!p && !q]
True
```

2.6 SUMS AND PRODUCTS

Sums and products are of fundamental importance in mathematics and *Mathematica* makes their computation simple. Unlike many computer languages, initialization is automatic and the syntax is easy to apply, particularly if the BasicInput palette is used. Any symbol may be used as the index of summation (i is used in the description below). Negative increments are permitted wherever increment is used.

- **Sum[a[i],{i, imax}]** evaluates the sum $\displaystyle\sum_{i=1}^{imax} a[i]$.

- **Sum[a[i],{i, imin, imax}]** evaluates the sum $\displaystyle\sum_{i=imin}^{imax} a[i]$.

- **Sum[a[i],{i, imin, imax, increment}]** evaluates the sum $\displaystyle\sum_{i=imin}^{imax} a[i]$ in steps of increment.

EXAMPLE 46

To compute the sum of the squares of the first 20 consecutive integers, we can type

Sum[i^2, {i,1,20}] or $\displaystyle\sum_{i=1}^{20} i^2$

2870

> Note: Even though *Mathematica* allows the form **Sum[i^2, {i,20}]**, the use of the initial index, 1, is recommended for clarity.

EXAMPLE 47

Compute the sum $\dfrac{1}{15}+\dfrac{1}{17}+\dfrac{1}{19}+\ldots+\dfrac{1}{51}$.

Sum[1/i, {i,15,51,2}]

$$\dfrac{635013914758060441939}{684514075768739707575}$$

- Numerical approximations can be computed using **NSum**. Its syntax is identical to **Sum**.

EXAMPLE 48

Approximate the sum $\dfrac{1}{15}+\dfrac{1}{17}+\dfrac{1}{19}+\ldots+\dfrac{1}{51}$.

NSum[1/i, {i, 15, 51, 2}]
0.6557

The limits of a sum can be infinite. *Mathematica* uses sophisticated techniques to evaluate infinite summations.

EXAMPLE 49

Compute $\dfrac{1}{1}+\dfrac{1}{4}+\dfrac{1}{9}+\dfrac{1}{16}+\ldots$

Sum[1/i^2,{i, 1, Infinity}] or $\displaystyle\sum_{i=1}^{\infty}\dfrac{1}{i^2}$

$\dfrac{\pi^2}{6}$

Double sums can be computed using the following syntax or, more conveniently, by clicking twice on the $\sum$ symbol in the BasicInput palette. This also extends to triple sums, quadruple sums, and so forth.

- `Sum[a[i, j],{i, imax},{j, jmax}]` evaluates the sum $\displaystyle\sum_{i=1}^{imax}\sum_{j=1}^{jmax} a[i, j]$.

- `Sum[a[i, j],{i, imin, imax},{j, jmin, jmax}]` evaluates the sum

  $$\sum_{i=imin}^{imax}\sum_{j=jmin}^{jmax} a[i, j].$$

- `Sum[a[i,j],`
  `    {i, imin, imax, i_increment}, {j, jmin, jmax, j_increment}]` evaluates the

  sum $\displaystyle\sum_{i=imin}^{imax}\sum_{j=jmin}^{jmax} a[i, j]$ in steps of `i_increment` and `j_increment`.

- `NSum` has the same syntax as `Sum` and works in a similar manner to yield numerical approximations.

EXAMPLE 50

Compute the value of $\left(\dfrac{1}{1}+\dfrac{1}{2}+\dfrac{1}{3}+\dfrac{1}{4}\right)+\left(\dfrac{2}{1}+\dfrac{2}{2}+\dfrac{2}{3}+\dfrac{2}{4}\right)+\left(\dfrac{3}{1}+\dfrac{3}{2}+\dfrac{3}{3}+\dfrac{3}{4}\right)$.

`Sum[i/j,{i,1,3},{j,1,4}]` or $\displaystyle\sum_{i=1}^{3}\sum_{j=1}^{4}\dfrac{i}{j}$

$\dfrac{25}{2}$

Just as `Sum` computes sums, the *Mathematica* function `Product` computes products. Its syntax is much the same as `Sum`.

- `Product[a[i],{i, imax}]` evaluates the product $\displaystyle\prod_{i=1}^{imax} a[i]$.

- `Product[a[i],{i, imin, imax}]` evaluates the product $\displaystyle\prod_{i=imin}^{imax} a[i]$.

- `Product[a[i],{i, imin, imax, increment}]` evaluates the product $\displaystyle\prod_{i=imin}^{imax} a[i]$ in steps of `increment`.

- `NProduct`, with identical syntax, returns numerical approximations to each of the products described in `Product`.

Multiple products are also easily computed. The syntax for a double product is listed below, but the concept extends to triple products, and so forth.

- `Product[a[i, j],{i, imax},{j, jmax}]` evaluates the product $\displaystyle\prod_{i=1}^{imax}\prod_{j=1}^{jmax} a[i, j]$.

- `Product[a[i, j],{i, imin, imax},{j, jmin, jmax}]` evaluates the product

  $$\prod_{i=imin}^{imax}\prod_{j=jmin}^{jmax} a[i, j].$$

- ```
  Product[a[i,j],{i, imin, imax, i_increment},
                 {j, jmin, jmax, j_increment}]
  ```

  evaluates the product $\displaystyle\prod_{i\,=\,imin}^{imax}\prod_{j\,=\,jmin}^{jmax} a[i, j]$ in steps of `i_increment` and `j_increment`.

EXAMPLE 51

Compute the product of the consecutive integers 4 through 9.

`Product[i, {i, 4, 9}]` or $\displaystyle\prod_{i\,=\,4}^{9} i$

```
60480
```

EXAMPLE 52

The binomial coefficient $C(n,k) = \dfrac{n!}{k!(n-k)!}$ can be expressed as

$\left(\dfrac{n}{k}\right)\left(\dfrac{n-1}{k-1}\right)\left(\dfrac{n-2}{k-2}\right)\cdots\left(\dfrac{n-k+1}{1}\right)$ for more efficient computation.   Use this representation to compute $C(10,4)$.

```
n = 10;
k = 4;
Product[(n - i)/(k - i), {i, 0, k-1}]
```
or $\displaystyle\prod_{i\,=\,0}^{k-1}\dfrac{n-i}{k-1}$

```
210
```

SOLVED PROBLEMS

2.45   Compute the sum of the first 25 prime numbers.

SOLUTION

`Sum[Prime[k], {k, 1, 25}]` or $\displaystyle\sum_{k\,=\,1}^{25} Prime[k]$

```
1060
```

2.46   Compute the square root of the sum of the squares of the integers 15 through 30, inclusive.

SOLUTION

`Sqrt[Sum[k^2, {k, 15, 30}]]` or $\sqrt{\displaystyle\sum_{k\,=\,15}^{30} k^2}$

$2\sqrt{2110}$

2.47   Compute the infinite sum $1 + \dfrac{1}{2} + \dfrac{1}{4} + \dfrac{1}{8} + \dfrac{1}{16} + \cdots$

SOLUTION

`Sum[1/2^i, {i, 0, Infinity}]` or $\displaystyle\sum_{i\,=\,0}^{\infty} \dfrac{1}{2^i}$
```

2.48 Compute the sum $\frac{1}{2}+\frac{2}{3}+\frac{3}{4}+\cdots+\frac{99}{100}$.

SOLUTION

$$\sum_{i=1}^{99}\frac{i}{i+1}$$

$$\frac{2644148646393295574979137176981450 82779489}{2788815009188499086581352357412492142272}$$

2.49 Obtain a general formula for the sum of squares of the consecutive integers 1 through n.

SOLUTION

`Sum[k^2, {k, 1, n}]` or $\sum_{k=1}^{n}k^2$

$\frac{1}{6}(n)(1+n)(1+2n)$ ← *Mathematica* has "memorized" these standard formulas.

2.50 Compute the sum of the first 100 primes.

SOLUTION

$\sum_{i=1}^{100}$ `Prime[i]` or `Sum[Prime[i], {i, 1, 100}]`

24133

2.51 Compute the product of the natural logarithms of the integers 2 through 20. Obtain an approximation to 20 significant digits.

SOLUTION

`N[Product[Log[i], {i, 2,20}], 20]`

$1.3632878207490815857\times10^6$

2.52 Compute the sum $1+\left(1+\frac{1}{2}\right)+\left(1+\frac{1}{2}+\frac{1}{3}\right)+\cdots+\left(1+\frac{1}{2}+\frac{1}{3}+\cdots+\frac{1}{20}\right)$.

SOLUTION

`Sum[1/j, {i, 1, 20}, {j, 1, i}]`   or $\sum_{i=1}^{20}\sum_{j=1}^{i}\frac{1}{j}$

$\frac{41054655}{739024}$

2.53 Compute a numerical approximation of $\left(1+\frac{1}{2}\right)\left(1+\frac{1}{2}+\frac{1}{3}\right)\cdots\left(1+\frac{1}{2}+\frac{1}{3}+\cdots+\frac{1}{10}\right)$.

SOLUTION

`NProduct[Sum[1/j, {j, 1, i}], {i, 2, 10}]`   or $\prod_{i=2}^{10}\sum_{j=1}^{i}\frac{1}{j}$ `//N`

1871.44

2.7 LOOPS

Often you may need to repeat an operation or sequence of operations several times.   Although *Mathematica* offers the ability to compute sums and products conveniently using Sum and Product commands, there are times when your work may require the use of looping techniques.   *Mathematica* offers three basic looping functions: Do, While, and For.

- Do[*expression*,{k}] evaluates *expression* precisely k times.
- Do[*expression*,{i, imax}] evaluates *expression* imax times with the value of i changing from 1 to imax in increments of 1.
- Do[*expression*,{i, imin, imax}] evaluates *expression* with the value of i changing from imin to imax in increments of 1.
- Do[*expression*,{i, imin, imax, increment}] evaluates *expression* with the value of i changing from imin to imax in increments of increment.
- Do[*expression*,{i, imin, imax}, {j, jmin, jmax}] evaluates *expression* with the value of i changing from imin to imax and j changing from jmin to jmax in increments of 1. The variable i changes by 1 for each cycle of j. This is known as a nested Do loop.
- Do[*expression*,{i, imin, imax, i_increment}, {j, jmin, jmax, j_increment}] forms a nested Do loop allowing for incrementation.

The last two forms of the command may be extended to 3 or more variables.

EXAMPLE 53

```
Do[Print["This line will be repeated 5 times."], {5}]
This line will be repeated 5 times.
This line will be repeated 5 times.
This line will be repeated 5 times.
This line will be repeated 5 times.
This line will be repeated 5 times.
```

EXAMPLE 54

This example computes the sum of consecutive integers from 5 to 25.   (Of course, the Sum command is more convenient.)

```
mysum = 0;                          ←Initialization of mysum.   This step is important.
Do[mysum = mysum + k, {k, 5, 25}]
mysum
315
```

EXAMPLE 55

This example computes the sum of all fractions whose numerators and denominators are positive integers not exceeding 5.

```
fracsum = 0;
Do[fracsum = fracsum + i/j, {i,1,5}, {j,1,5}]
fracsum
```

$$\frac{137}{4}$$

- While[*condition*, *expression*] evaluates *condition*, then *expression*, repetitively, until *condition* is False.

If *expression* consists of multiple statements, they are separated by semicolons.

EXAMPLE 56

```
n = 1; While[n < 6, Print[n]; n = n + 1]
```
| n++ may be used in place of n = n+1 |

```
1
2
3
4
5
```

- **For[***initialization, test, increment, expression***]** executes *initialization*, then repeatedly evaluates *expression* and *increment* until *test* gives False.

The order of evaluation is *test*, *expression*, and then *increment*. The For loop terminates as soon as *test* gives False. If any of *initialization*, *test*, *increment*, or *expression* consists of multiple statements, they are separated by semicolons.

EXAMPLE 57

```
For[i = 1, i ≤ 5, i = i + 1, Print[i]]
1
2
3
4
5
```

Although it is not a loop, the If instruction is often used in conjunction with other loop commands.

- **If[***condition, true, false***]** evaluates *condition* and executes *true* if *condition* is True and executes *false* if condition is False.
- **If[***condition, true***]** evaluates *condition* and executes *true* if *condition* is True. If condition is False, no action is taken and Null is returned.
- **If[***condition, , false***]** evaluates *condition* and executes *false* if *condition* is False. If condition is True, no action is taken and Null is returned. (Note the double comma.)
- **If[***condition, true, false, neither***]** evaluates *condition* and executes *true* if *condition* is True, executes *false* if condition is False, and executes *neither* if *condition* is neither True or False.

EXAMPLE 58

```
If[2 == 2, Print ["a"], Print ["b"]]        ← 2 == 2 is True
a
If[2 == 3, Print ["a"], Print ["b"]]        ← 2 == 3 is False
b
If[7, Print ["a"], Print ["b"], Print ["c"]]   ← 7 is neither True or False
c
```

The next example, which separates primes from nonprimes, illustrates how the If instruction can be used in a Do loop.

EXAMPLE 59

```
Do[If[PrimeQ[k], Print[k], Print[" ", k]], {k, 1, 20}]
   1
 2
 3
   4
```

```
5
     6
7
        8
        9
        10
11
     12
13
        14
        15
        16
17
     18
19
     20
```

SOLVED PROBLEMS

2.54   Compute 10! using a `Do` loop.

SOLUTION

```
fact = 1;
n = 10;
Do[fact = fact*k, {k, n}]
fact
3628800
```

2.55   Compute 10! using a `While` loop.

SOLUTION

```
fact = 1;
n = 10;
While[n > 0, fact = fact*n; n --]
fact
3628800
```

2.56   Compute 10! using a `For` loop.

SOLUTION

```
For[fact = 1; n = 1, n ≤ 10, n++, fact = n * fact]
fact
3628800
```

2.57   Print all numbers from 1 to 20 which are *not* multiples of 2, 3, or 5.

SOLUTION

```
Do[If[Mod[k, 2] == 0 || Mod[k, 3] == 0 || Mod[k, 5] == 0, , Print[k]], {k, 1, 20}]
1
7
11
13
17
19
```

2.58   For each number k from 1 to 10, print half the number if k is even and twice the number if k is odd.

SOLUTION

```
Do[If[EvenQ[k], Print[k/2], Print[2 k]], {k, 1, 10}]
2
1
6
2
10
3
14
4
18
5
```

2.8   INTRODUCTION TO GRAPHING

The graph of a function offers tremendous insight into the function's behavior and can be of great value in the solution of problems.   *Mathematica* offers some very powerful graphics commands which are remarkably easy to implement.   Although there is a vast array of options available for customization of output, in this section we shall deal only with the most rudimentary forms using *Mathematica*'s defaults.   A more detailed discussion of graphics commands appears in Chapters 4 and 5.

The **Plot** command plots two-dimensional graphs.

- **Plot[f[x], {x, xmin, xmax}]** plots a two-dimensional graph of the function $f(x)$ on the interval xmin ≤ x ≤ xmax.
- **Plot[{f[x],{g[x]}, {x, xmin, xmax}]** plots two functions on one set of axes.   This extends in a natural way to three or more functions.

EXAMPLE 60

Plot the graph of $y = x^2$ on the interval $-5 \le x \le 5$.

```
Plot[x², {x, -5, 5}]
```

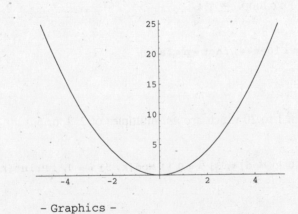

```
- Graphics -
```

The text "`- Graphics -`", which appears after the graph (and can be annoying) can be eliminated by placing a semicolon (;) at the end of the command.

SOLVED PROBLEMS

2.59 Plot the functions $y = x^2$ and $y = 2x + 10$, $-5 \leq x \leq 5$, on the same set of axes.

SOLUTION

`Plot[{x², 2 x + 10}, {x, -5, 5}];`   ← Note the use of the semicolon.

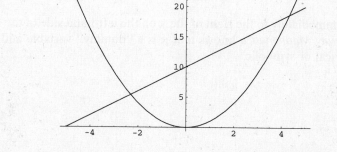

2.60 Sketch the graphs of $y = x^2$, $y = x^3$, and $y = x^4$, $0 \leq x \leq 1$, on the same set of axes.

SOLUTION

`Plot[{x², x³, x⁴}, {x, 0, 1}];`

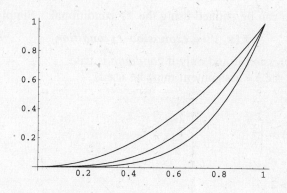

2.61 Sketch the graphs of the functions $y = x$, $y = -x$, and $y = x \sin x$ on the interval $-6\pi \leq x \leq 6\pi$ on one set of axes.

SOLUTION

`Plot[{x, -x, x Sin[x]}, {x, -6π, 6π}];`

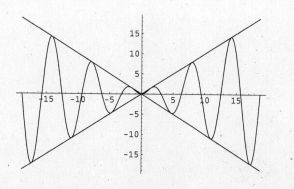

2.9 USER-DEFINED FUNCTIONS

Suppose we want to define a function f of a single variable.   If x is the independent variable, we write

```
f[x_] = . . . . .
```
or
```
f[x_] := . . . . .
```

where the right-hand side of the definition tells *Mathematica* how to compute the value of f for a given value of x. Any legitimate *Mathematica* operations, including references to built-in functions, are acceptable.

Note the underscore immediately to the right of the x on the left-hand side of the definition.   <u>This is crucial</u>.   It is the only way *Mathematica* knows that x is a "dummy" variable and can be replaced by any expression, numerical or symbolic.

EXAMPLE 61

```
f[x_] = x² + x³;
f[2]
12
f[2 x]
4 x² + 8 x³
f[Exp[x]]
e²ˣ + e³ˣ
f[λ]
λ² + λ³
```

A "piecewise" function can be defined using the `/;` conditional.   Simply put,

$$\texttt{f[x\_] := } \textit{expression} \texttt{ /; } \textit{condition}$$

assigns `f[x]` the value *expression* if and only if *condition* is true.
Note: In this application, the `:=` assignment <u>must</u> be used.

EXAMPLE 62

We define the function $f(x) = \begin{cases} x^2 & \text{if } x \le 2 \\ 8 - 2x & \text{if } x > 2 \end{cases}$.

```
f[x_] := x² /; x≤2
f[x_] := 8 - 2x /; x > 2
f[-4]
16

f[4]
0

Plot[f[x], {x, 0, 4}];
```

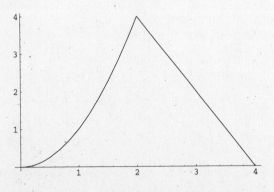

Functions are sometimes defined *recursively*. One or several values of the function are specified and later values are defined in terms of their predecessors.

EXAMPLE 63

The Fibonacci sequence can be defined recursively by defining $f(1) = 1$, $f(2) = 1$, and $f(n) = f(n-2) + f(n-1)$ for $n \geq 3$. We will compute the 25th Fibonacci number using this definition.

```
f[1] = 1;
f[2] = 1;
f[n_] := f[n - 2] + f[n - 1]
f[25]
75025
```

> Note the use of := here. This is important.
> Experiment and see what happens if = is used.

You may have noticed a long pause in the calculation of this number. To see this more precisely, we will time the operation. (Your times may be slightly different, depending upon your computer.)

```
f[25]//Timing
{18.18 Second,75025}
```

Intermediate calculations have not been stored. Each computation of f[n] necessitates the computation of f[n-2] and f[n-1], each of which causes all values of f down to f[3] to be computed. Since each intermediate value of f is computed recursively based upon the values of f[1] and f[2], the result is that it takes an extremely large number of iterations to compute f[25]. To eliminate this problem, we can store each value of f in memory as it is computed. They can then be recalled almost instantaneously.

EXAMPLE 64

```
f[1] = 1;
f[2] = 1;
f[n_] := f[n] = f[n - 2] + f[n - 1]    ← This causes Mathematica to store each f[n] value.
f[25]//Timing                           Type ?f to confirm this.
{0.Second,75025}
```

Functions of two or more variables can be defined in an analogous manner. The syntax is self-explanatory.

EXAMPLE 65

```
f[x_, y_] = x² + y³;
f[2, 3]
31
f[3, 2]
17
```

EXAMPLE 66

```
g[x_, y_, z_] = x + y * z;
g[2, 3, 4]
14
```

SOLVED PROBLEMS

2.62 Define $f(x)$ to be the polynomial $x^5 + 3x^4 - 7x^2 + 2$ and compute $f(2)$.

SOLUTION

```
f[x_] = x⁵ + 3x⁴ - 7x² + 2
2 - 7x² + 3x⁴ + x⁵
f[2]
54
```

2.63   Let $f(x) = \begin{cases} -x & \text{if } x \leq 0 \\ x^2 & \text{if } 0 < x \leq 3. \\ 18 - 3x & \text{if } x > 3 \end{cases}$

Sketch the graph of $f(x)$ for $-6 \leq x \leq 6$.

SOLUTION

```
f[x_] := -x /; x ≤ 0
f[x_] := x² /; 0 < x ≤ 3
f[x_] := 18 - 3x /; x > 3
Plot[f[x], {x, -6, 6}];
```

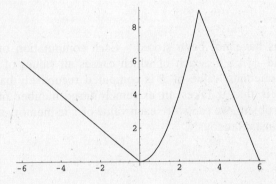

2.64   If $f(x)$ is defined on an interval $[a, b]$, the periodic extension of f with period $b - a$ is the function F such that

$$F(x) = \begin{cases} f(x) & \text{if } a \leq x \leq b \\ f(x - (b - a)) & \text{otherwise} \end{cases}$$

Let $f(x) = x^2$ if $-1 \leq x \leq 1$. Plot the periodic extension of f with period 2 from $x = 0$ to $x = 10$.

SOLUTION

```
f[x_] = x²;
extension[x_] := f[x] /; -1 ≤ x ≤ 1;
extension[x_] := extension[x - 2] /; x > 1;
Plot[extension[x], {x, 0, 10}];
```

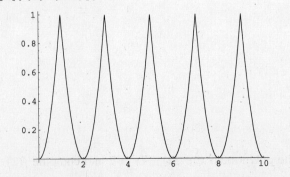

2.65   Define the function $f(n)$:

$$f(n) = \begin{cases} f(1) = 1 \\ f(2) = 2 \\ f(3) = 3 \\ f(n) = f(n-3) + f(n-2) + f(n-1) \text{ if } n \geq 4 \end{cases}$$

Compute $f(20)$.

SOLUTION

```
Clear[f]
f[1] = 1;
f[2] = 2;
f[3] = 3;
f[n_] := f[n] = f[n-3] + f[n-2] + f[n-1];
f[20]
101902
```

2.66   Define a function which represents the distance from the point (x, y) to $(3, 4)$ and compute the value of the function at the point $(5, -2)$.

SOLUTION

```
f[x_, y_] = √((x - 3)² + (y - 4)²); or f[x_, y_] = Sqrt[(x - 3)^2 + (y - 4)^2];
f[5, -2]
```
$2\sqrt{10}$

2.67   Define a function which represents the distance between the points $(x1, y1)$ and $(x2, y2)$ and use it to compute the distance from $(2, 3)$ to $(8, 11)$.

SOLUTION

```
d[x1_, y1_, x2_, y2_] = √((x2 - x1)² + (y2 - y1)²); or
d[x1_, y1_, x2_, y2_] = Sqrt[(x2 - x1)^2 + (y2 - y1)^2];
d[2, 3, 8, 11]
10
```

2.68   The area enclosed by a triangle whose sides have lengths a, b, and c is given by Heron's formula:

$$K = \sqrt{s(s-a)(s-b)(s-c)}$$

where $s = \dfrac{a + b + c}{2}$.   Express the area of a triangle as a function of a, b, and c and compute the area of the triangle whose sides are (a) 3, 4, and 5 and (b) 5, 9, and 12.

SOLUTION

```
s = a + b + c ;
    ──────────
        2

k[a_, b_, c_] = √(s(s - a)(s - b)(s - c));
k[3, 4, 5]
6
k[5, 9, 12]
```
$4\sqrt{26}$

2.10 OPERATIONS ON FUNCTIONS

If f and g are two functions with the same domain D, we define their sum, difference, product, and quotient pointwise; i.e.,

$$(f + g)(x) = f(x) + g(x) \qquad \text{for all } x \text{ in D}$$
$$(f - g)(x) = f(x) - g(x) \qquad \text{for all } x \text{ in D}$$
$$(fg)(x) = f(x)\,g(x) \qquad \text{for all } x \text{ in D}$$
$$(f/g)(x) = f(x)/g(x) \qquad \text{for all } x \text{ in D for which } g(x) \neq 0$$

If x is a number in the domain of g such that $g(x)$ is in the domain of f, we can define the composite function $f \circ g$:

$$(f \circ g)(x) = f(g(x))$$

The function $g \circ f$ can be defined in a similar manner.

The following example illustrates how to construct these functions.

EXAMPLE 67

```
f[x_] = √x;
g[x_] = x² + 2x + 3;
h1[x_] = f[x] + g[x]
```
$3 + \sqrt{x} + 2x + x^2$
```
h2[x_] = f[x] - g[x]
```
$-3 + \sqrt{x} - 2x - x^2$
```
h3[x_] = f[x]g[x]
```
$\sqrt{x}\,(3 + 2x + x^2)$
```
h4[x_] = f[x]/g[x]
```
$\dfrac{\sqrt{x}}{3 + 2x + x^2}$
```
h5[x_] = f[g[x]]
```
$\sqrt{3 + 2x + x^2}$
```
h6[x_] = g[f[x]]
```
$3 + 2\sqrt{x} + x$

The composition of two or more functions can be accomplished with the **Composition** command. Note that `Composition` is a *functional* operation and as such, its arguments are functions, `f`, not `f[x]`.

- **Composition[f1, f2, f3, ...]** constructs the composition `f1∘f2∘f3....`

EXAMPLE 68

```
f[x_] = √x;
g[x_] = x² + 2x + 3;
h1 = Composition[f, g];
h1[x]
```
$\sqrt{3 + 2x + x^2}$
```
h2 = Composition[g, f];
h2[x]
```
$3 + 2\sqrt{x} + x$

If we wish to compute the composition of a function with itself, we could, of course, use `Composition[f, f]`, `Composition[f, f, f]`, and so forth.   A more convenient tool is `Nest` or `NestList`.

- **`Nest[f, expression, n]`** applies f to *expression* successively n times.

- **`NestList[f, expression, n]`** applies f to *expression* successively n times and returns a list of all the intermediate calculations.   (Lists are discussed in detail in Chapter 3.)

EXAMPLE 69

```
f[x_] = x²;
Nest[f, x, 5]
x³²
NestList[f, x, 5]
{x, x², x⁴, x⁸, x¹⁶, x³²}
Nest[f, 2 x + 3, 5]
(3 + 2 x)³²
NestList[f, 2 x + 3, 5]
{3 + 2 x, (3 + 2 x)², (3 + 2 x)⁴, (3 + 2 x)⁸, (3 + 2 x)¹⁶, (3 + 2 x)³²}
```

SOLVED PROBLEMS

2.69   If $f(x) = \sin x + 2\cos x$ and $g(x) = 2\sin x - 3\cos x$, construct $(f+g)(x)$, $(f-g)(x)$, $(fg)(x)$, and $(f/g)(x)$ and evaluate them at $\pi/2$.

SOLUTION

```
f[x_] = Sin[x] + 2 Cos[x];
g[x_] = 2 Sin[x] - 3 Cos[x];
h1[x_] = f[x] + g[x]
-Cos[x] + 3 Sin[x]
h2[x_] = f[x] - g[x]
5 Cos[x] - Sin[x]
h3[x_] = f[x] g[x]
(2 Cos[x] + Sin[x]) (-3 Cos[x] + 2 Sin[x])
h4[x_] = f[x]/g[x]
```

$$\frac{2\,Cos[x] + Sin[x]}{-3\,Cos[x] + 2\,Sin[x]}$$

```
h1[π/2]
3
h2[π/2]
-1
h3[π/2]
2
h4[π/2]
```

$$\frac{1}{2}$$

2.70   Let $f(x) = \sqrt{1+x}$.   Compute $(f \circ f \circ f \circ f \circ f)(x)$.

SOLUTION

```
f[x_] = √(1+x);
Nest[f, x, 5]
```

$$\sqrt{1 + \sqrt{1 + \sqrt{1 + \sqrt{1 + \sqrt{1 + x}}}}}$$

2.71   Let $f(x) = \dfrac{1}{1+x}$.   Apply f to x five times and show the results of each application.   Then apply f to $x = 1$ twenty times.   What do you observe?

SOLUTION

```
f[x_] = 1/(1+x)
```

```
NestList[f, x, 5]
```

$$\left\{ x, \frac{1}{1+x}, \frac{1}{1+\frac{1}{1+x}}, \frac{1}{1+\frac{1}{1+\frac{1}{1+x}}}, \frac{1}{1+\frac{1}{1+\frac{1}{1+\frac{1}{1+x}}}}, \frac{1}{1+\frac{1}{1+\frac{1}{1+\frac{1}{1+\frac{1}{1+x}}}}} \right\}$$

```
NestList[f, 1, 20]
```

$$\left\{ 1, \frac{1}{2}, \frac{2}{3}, \frac{3}{5}, \frac{5}{8}, \frac{8}{13}, \frac{13}{21}, \frac{21}{34}, \frac{34}{55}, \frac{55}{89}, \frac{89}{144}, \frac{144}{233}, \frac{233}{377}, \right.$$

$$\left. \frac{377}{610}, \frac{610}{987}, \frac{987}{1597}, \frac{1597}{2584}, \frac{2584}{4181}, \frac{4181}{6765}, \frac{6765}{10946}, \frac{10946}{17711} \right\}$$

```
% //N
{1., 0.5, 0.666667, 0.6, 0.625, 0.615385,
 0.619048, 0.617647, 0.618182, 0.617978, 0.618056,
 0.618026, 0.618037, 0.618033, 0.618034, 0.618034,
 0.618034, 0.618034, 0.618034, 0.618034, 0.618034}
```

The numbers appear to be approaching a limit of approximately 0.618034.

2.72   If x is an approximation to $\sqrt{a}$, it can be shown that $\dfrac{1}{2}\left(x + \dfrac{a}{x}\right)$ is a better approximation.

(This is a special case of Newton's method.)   Use **NestList** to observe the first 10 approximations obtained in computing $\sqrt{3}$, starting with $x = 100$.

SOLUTION

```
a = 3;
```

$$f[x\_] = \frac{1}{2}\left(x + \frac{a}{x}\right);$$

```
NestList[f, 100, 10] // N
{100., 50.015, 25.0375, 12.5787, 6.40858, 3.43835, 2.15543, 1.77363, 1.73254,
 1.73205, 1.73205}
```

2.11 MODULES

Mathematica, by default, assumes that all objects are *global*. This means, for example, that if you define x to have a value of 3, x will remain 3 until its value is changed. In contrast, a *local* object has a limited scope valid only within a certain group of instructions.

Modules allow you to define local variables whose values are defined only within the module. Outside of the module, the object may either be undefined or may have a completely different value.

- **Module[{var1, var2,...}, body]** defines a module with local variables *var1*, *var2*, ...
- **Module[{var1 = v1, var2 = v2, ...}, body]** defines a module with local variables *var1*, *var2*, ... initialized to *v1*, *v2*, ..., respectively.

EXAMPLE 70

```
x = 3;                      ← global variable x is set to 3
Module[{x = 8}, x + 1]      ← module is defined with local variable x initialized to 8
9                           ← x is incremented
x                           ← global x is called
3                           ← original value of x is returned
```

It is often useful to group several commands into one unit to be executed as a group. This is especially true if complicated structures involving loops are involved. Several commands may be incorporated within *body* if they are separated by semicolons.

EXAMPLE 71

```
Module[{x = 1, y = 2}, x = x + 3 ; y = y + 4; Print[x y]]
24                  ← x and y are initialized to 1 and 2, respectively, x is incremented
                      by 3, y is incremented by 4, and the two are multiplied.
```

It is often convenient to define a function whose value is a module. This allows considerably more flexibility when dealing with functions whose definitions are complicated. When defining a function in this manner, it is important that the delayed assignment, **:=**, be used.

EXAMPLE 72

The following defines the factorial function. The value of x0 is assumed to be a nonnegative integer. The variables fact and x, which are initialized to be 1 and x0, respectively, are local so there is no conflict with any variables of the same name elsewhere in the program. x0 is a "dummy" variable.

```
f[x0_] := Module[{fact = 1, x = x0},
             While[x > 1, fact = x*fact; x = x - 1];
             Print[fact]]
f[0]
1
f[5]
120
f[10]
3628800
```

To clarify how a module works, consider the next example. Although the same module is executed three times, the variable, which appears as x, is actually assigned three different local names. Because of this clever "bookkeeping," all three are independent and none will conflict with global variable x.

EXAMPLE 73

```
x = 3
3
Module[{x}, Print[x]]
x$5
Module[{x}, Print[x]]
x$6
Module[{x}, Print[x]]
x$7
x
3
```

> Since all three values of x are given different internal names, there can be no conflict.

SOLVED PROBLEMS

2.73 Write a module which will take an integer and return all its factors.

SOLUTION

```
factorlist[x0_] :=
 Module[{x = 1}, While[x ≤ x0, If[Mod[x0, x] == 0, Print[x]]; x++]]
```

```
factorlist[1]              factorlist[90]
1                          1
                           2
factorlist[10]             3
1                          5
2                          6
5                          9
10                         10
                           15
factorlist[11]             18
1                          30
11                         45
                           90
```

2.74 A very crude way of determining the position of a prime within the sequence of primes is to examine the list of all primes up to and including the prime in question and determine its position in the list. If the number is not in the list, then the number is not prime. Construct a module which will determine whether a number is prime, and if so, determine its position. If not, return a message indicating that it is not prime.

SOLUTION

```
pos[x0_] := Module[{x = 1, prm},
  prm = False;
  While[Prime[x] ≤ x0 && Not[prm], If[Prime[x] == x0, prm = True]; x ++];
  If[prm, Print[x - 1], Print["Not a Prime"]]]
```

```
pos[1]
Not a Prime
pos[2]
1
pos[3]
2
pos[101]
26
pos[1001]
Not a Prime
```

2.75 A famous conjecture in mathematics asserts that if you start with a positive integer n and replace it by $n/2$ if n is even and by $3n + 1$ if n is odd, and repeat the process over and over in an iterative manner, that you will *always* wind up with 1. (This conjecture has never been proved or disproved.) Construct a module which simulates this iterative process.

SOLUTION

We first define a function, successor, which will define one iteration step.

```
successor[n_] := If[EvenQ[n], n/2, 3n + 1]
```

Next we introduce a module allvalues, which will produce a list of all successors, starting with the successor of n.

```
allvalues[n_] := Module[{m = n}, While[m ≠ 1, m = successor[m]; Print[m]]]

allvalues[6]
    3
    10
    5
    16
    8
    4
    2
    1
```

Since this list might be long if n is large, and all we are really interested in is the final value and the number of iterations it takes to get there, another module might be more appropriate.

```
finalvalue[n_] :=
 Module[{m = n, k = 0}, While[m != 1, m = successor[m]; k++];
  Print["final value = ", m, ", # iterations = ", k]]
```

finalvalue lists the final value of the process, together with the number of iterations needed to reach the final value.

```
finalvalue[6]
final value = 1, # iterations = 8
finalvalue[100]
final value = 1, # iterations = 25
finalvalue[1000]
final value = 1, # iterations = 111
```

CHAPTER 3

Lists

3.1 INTRODUCTION

Lists are general objects that contain collections of other objects. In reading this chapter you will see that lists are used for a variety of applications. Therefore, *Mathematica* offers an extensive collection of list manipulation commands.

The objects within a list are contained within braces, {}. Alternatively, the **List** command may be used to define a list.

- **List[*elements*]**, where *elements* represents the members of the list separated by commas, is equivalent to **{*elements*}**.

EXAMPLE 1

{1, 2, 3, 4} is a list of numbers. List[a, b, c, d] is a list of undefined symbols.

```
List[a, b, c, d]
{a, b, c d}                      ←List[a, b, c, d] is equivalent to {a, b, c, d}.
```

Lists can be given symbolic names so that they can be easily referenced. Any operation performed on a list will be performed on *each* element of the list.

EXAMPLE 2

```
lst = {1, 2, 3, 4, 5, 6, 7, 8, 9, 10}
{1, 2, 3, 4, 5, 6, 7, 8, 9, 10}
1/lst
```
$\{1, \frac{1}{2}, \frac{1}{3}, \frac{1}{4}, \frac{1}{5}, \frac{1}{6}, \frac{1}{7}, \frac{1}{8}, \frac{1}{9}, \frac{1}{10}\}$
```
lst²
```
{1, 4, 9, 16, 25, 36, 49, 64, 81, 100}
$\sqrt{\text{lst}}$
$\{1, \sqrt{2}, \sqrt{3}, 2, \sqrt{5}, \sqrt{6}, \sqrt{7}, 2\sqrt{2}, 3, \sqrt{10}\}$

> We use **lst** rather than **list** in the examples to avoid conflict with the *Mathematica* symbol **List**.

If two or more lists contain the same number of elements, new lists can be created using standard operations.

EXAMPLE 3

```
lst1 = {1, 2, 3, 4, 5};
lst2 = {2, 3, 2, 3, 2};
```

```
lst1 + lst2
{3, 5, 5, 7, 7}
lst1 * lst2
{2, 6, 6, 12, 10}
lst1/lst2
```
$\{\frac{1}{2}, \frac{2}{3}, \frac{3}{2}, \frac{4}{3}, \frac{5}{2}\}$
```
lst1^lst2
{1, 8, 9, 64, 25}
```

SOLVED PROBLEMS

3.1   Construct a list of the factorials of the integers 1 through 10.

SOLUTION

```
lst = {1, 2, 3, 4, 5, 6, 7, 8, 9, 10};
lst!
{1, 2, 6, 24, 120, 720, 5040, 40320, 362880, 3628800}
```

3.2   Construct a list of the first 10 positive integer powers of 2 .

SOLUTION

```
lst = {1, 2, 3, 4, 5, 6, 7, 8, 9, 10};
2^lst
{2, 4, 8, 16, 32, 64, 128, 256, 512, 1024}
```

3.3   Construct a list whose elements are the sum of the squares and cubes of the first five positive integers.

SOLUTION

```
lst = (1, 2, 3, 4, 5};
lst² + lst³ or lst ^ 2 + lst ^ 3
{2, 12, 36, 80, 150}
```

3.4   Define $lst1 = \{1, 3, 5, 7, 9\}$ and $lst2 = \{2, 4, 6, 8, 10\}$. Construct a list whose five elements are the products of the corresponding entries of the two lists.

SOLUTION

```
lst1 = {1, 3, 5, 7, 9};
lst2 = {2, 4, 6, 8, 10};
lst1 * lst2
{2, 12, 30, 56, 90}
```

3.2 GENERATING LISTS

The most common lists are lists of equally spaced numbers. The **Range** command allows convenient construction.   The values of m, n, and d in the following description need not be integer-valued.   Negative values are acceptable as well.

- **Range[n]** generates a list of the first n consecutive integers.
- **Range[m, n]** generates a list of consecutive integers from m to n.
- **Range[m, n, d]** generates a list of integers from m through n in increments of d.

EXAMPLE 4

```
Range[10]
{1, 2, 3, 4, 5, 6, 7, 8, 9, 10}
Range[5, 10]
{5, 6, 7, 8, 9, 10}
Range[25, 5, -2]
{25, 23, 21, 19, 17, 15, 13, 11, 9, 7, 5}
Range[1/3, 1, 1/12]
```

$$\left\{\frac{1}{3}, \frac{5}{12}, \frac{1}{2}, \frac{7}{12}, \frac{2}{3}, \frac{3}{4}, \frac{5}{6}, \frac{11}{12}, 1\right\}$$

```
Range[1, 2, .1]
{1, 1.1, 1.2, 1.3, 1.4, 1.5, 1.6, 1.7, 1.8, 1.9, 2.}
```

Lists with more complicated structure can be constructed with the **Table** and **Array** commands. There are several different forms.

- **Table[***expression***, {n}]** generates a list containing n copies of *expression*.
- **Table[***expression***, {k, n}]** generates a list of the values of *expression* as k varies from 1 to n.
- **Table[***expression***, {k, m, n}]** generates a list of the values of *expression* as k varies from m to n.
- **Table[***expression***, {k, m, n, d}]** generates a list of the values of *expression* as k varies from m to n in steps of d.

EXAMPLE 5

```
Table[x, {10}]
{x, x, x, x, x, x, x, x, x, x}
Table[k², {k, 10}]
{1, 4, 9, 16, 25, 36, 49, 64, 81, 100}
Table[1/k, {k, 5, 13}]
```

$$\left\{\frac{1}{5}, \frac{1}{6}, \frac{1}{7}, \frac{1}{8}, \frac{1}{9}, \frac{1}{10}, \frac{1}{11}, \frac{1}{12}, \frac{1}{13}\right\}$$

```
Table[√k, {k, 5, 13, 2}]
```

$$\left\{\sqrt{5}, \sqrt{7}, 3, \sqrt{11}, \sqrt{13}\right\}$$

- **Array[f, n]** generates a list consisting of n values, f[1], f[2], ..., f[n].
- **Array[f, n, r]** generates a list consisting of n values, f[i], starting with f[r].

EXAMPLE 6

```
f[x_] = x² + x + 1;
Array[f, 10]
{3, 7, 13, 21, 31, 43, 57, 73, 91, 111}

Array[f, 10, 0]
{1, 3, 7, 13, 21, 31, 43, 57, 73, 91}          ← The first element is f[0].
```

Nested lists are lists which contain other lists. For example,

$$\{\{1, 2, 3, 4\}, \{2, 3, 4, 5\}, \{3, 4, 5, 6\}\}$$

is a nested list of three objects, each of which is a list of four integers. Nested lists can be generated using the Table and Array commands.

- `Table[`*expression*`, {i, m`$_i$`, n`$_i$`}, {j, m`$_j$`, n`$_j$`}]` generates a nested list whose values are *expression*, computed as `j` goes from m$_j$ to n$_j$ and as `i` goes from m$_i$ to n$_i$.
- `Array[f, {m, n}]` generates a nested list consisting of an array of `m` elements, each of which is an array of `n` elements, whose values are `f[i,j]` as `j` goes from 1 to `n` and `i` goes from 1 to `m`. Here `f` is a function of two variables.
- `Array[f, {m, n}, {r, s}]` generates a nested list consisting of an array of `m` elements, each of which is an array of `n` elements. The first element of the first sublist is `f[r, s]`.

In each of the above descriptions, the rightmost index varies most rapidly. The lists corresponding to this index are innermost. All indices have unit increments.

Each of the above descriptions extends in a natural way to lists of greater depth.

EXAMPLE 7

```
Table[i + j, {i, 1, 3}, {j, 1, 5}]
{{2, 3, 4, 5, 6}, {3, 4, 5, 6, 7}, {4, 5, 6, 7, 8}}
```

Each value of `i` causes a list `{i+1,i+2,i+3,i+4,i+5}` to be generated. The three lists are enclosed within an outer set of list braces.

```
Table[i + j, {i, 1, 5}, {j, 1, 3}]
{{2, 3, 4}, {3, 4, 5}, {4, 5, 6}, {5, 6, 7}, {6, 7, 8}}
```

EXAMPLE 8

```
g[x_, y_] = x² + 3 y;
Array[g, {3, 4}]
{{4, 7, 10, 13}, {7, 10, 13, 16}, {12, 15, 18, 21}}
Array[g, {4, 3}]
{{4, 7, 10}, {7, 10, 13}, {12, 15, 18}, {19, 22, 25}}
Array[g, {4, 3}, {0, 0}]
{{0, 3, 6}, {1, 4, 7}, {4, 7, 10}, {9, 12, 15}}
```

Often it will be convenient to construct lists of letters and other characters.

- `Characters[`*string*`]` produces a list of characters in *string*.
- `CharacterRange["`*char1*`", "`*char2*`"]` produces a list of characters from *char1* to *char2*, based upon their standard ASCII values (assuming an American English alphabet).

EXAMPLE 9

```
Characters["Mathematica"]
{M, a, t, h, e, m, a, t, i, c, a}
```

EXAMPLE 10

```
CharacterRange["a", "e"]
{a, b, c, d, e}
CharacterRange[" ", "~"]
{ , !, ", #, $, %, &, ', (, ), *, +, ,, -, ., /, 0, 1,
 2, 3, 4, 5, 6, 7, 8, 9, :, ;, <, =, >, ?, @, A, B, C, D,
 E, F, G, H, I, J, K, L, M, N, O, P, Q, R, S, T, U, V, W, X,
 Y, Z, [, \, ], ^, _, `, a, b, c, d, e, f, g, h, i, j, k,
 l, m, n, o, p, q, r, s, t, u, v, w, x, y, z, {, |, }, ~}
```

Even though the output of `Characters` and `CharacterRange` appear to be individual characters, in actuality they are strings of length 1. By *Mathematica*'s convention, the quotation marks are not printed.

EXAMPLE 11

```
digits = CharacterRange["0", "9"]
{0, 1, 2, 3, 4, 5, 6, 7, 8, 9}          ←These are not numbers but strings of characters of length 1.
FullForm[digits]
List["0", "1", "2", "3", "4", "5", "6", "7", "8", "9"]
```

SOLVED PROBLEMS

3.5 Construct a list of the positive multiples of 7 which do not exceed 100.

SOLUTION

```
Range[7, 100, 7]  or  Table[7k, {k, 1, 14}]
{7, 14, 21, 28, 35, 42, 49, 56, 63, 70, 77, 84, 91, 98}
```

3.6 Construct a list of the first 10 prime numbers.

SOLUTION 1

```
Table[Prime[k], {k, 1, 10}]
{2, 3, 5, 7, 11, 13, 17, 19, 23, 29}
```

SOLUTION 2

```
Array[Prime, 10]
{2, 3, 5, 7, 11, 13, 17, 19, 23, 29}
```

SOLUTION 3

```
Prime[Range[10]]
{2, 3, 5, 7, 11, 13, 17, 19, 23, 29}
```

3.7 Construct a list of the reciprocals of the first 10 even integers.

SOLUTION

```
Table[1/k, {k, 2, 20, 2}]  or  1/Range[2, 20, 2]
```

$$\left\{\frac{1}{2}, \frac{1}{4}, \frac{1}{6}, \frac{1}{8}, \frac{1}{10}, \frac{1}{12}, \frac{1}{14}, \frac{1}{16}, \frac{1}{18}, \frac{1}{20}\right\}$$

3.8 Construct a list of five objects, each of which is a list consisting of six integers.   The first list is to contain the first six multiples of 2, the second, multiples of 3, the third, multiples of 4, and so forth.

SOLUTION

```
Table[i * j, {i, 2, 6}, {j, 1, 6}]
{{2, 4, 6, 8, 10, 12}, {3, 6, 9, 12, 15, 18}, {4, 8, 12, 16, 20, 24},
 {5, 10, 15, 20, 25, 30}, {6, 12, 18, 24, 30, 36}}
```

3.9 Let $p(x) = x^2 - 8x + 10$.   Compute the values of $p(x)$ for $x = 1, 2, 3, \ldots, 10$.

SOLUTION

```
p[x_] = x² - 8x + 10;
Array[p, 10]  or  p[Range[10]]
{3, -2, -5, -6, -5, -2, 3, 10, 19, 30}
```

3.3   LIST MANIPULATION

- **Length[lst]** returns the length of lst, i.e., the number of elements in lst.
- **First[lst]** returns the element of lst in the first position.
- **Last[lst]** returns the element of lst in the last position.

EXAMPLE 12

```
lst = {a, b, c, d, e, f, g};
Length[lst]
7
First[lst]
a
Last[lst]
g
```

The function **Part** returns individual elements of a list.

- **Part[lst, k]** or **lst[[k]]** returns the kth element of lst.
- **Part[lst, -k]** or **lst[[-k]]** returns the kth element from the end of lst.

Part[lst, 1] and **Part[lst, -1]** are equivalent to **First[lst]** and **Last[lst]**, respectively.

EXAMPLE 13

```
lst = {a, b, c, d, e, f, g};
Part[lst, 1] or lst[[1]]
a
Part[lst, 3] or lst[[3]]
c
Part[lst, -3] or lst[[-3]]
e
Part[lst, -1] or lst[[-1]]
g
```

EXAMPLE 14

```
lst = {{a, b, c, d}, {e, f, g, h}, {i, j, k, l}};
First[lst]
{a, b, c, d}
Last[lst]
{i, j, k, l}
lst[[2]]
{e, f, g, h}
```

Since lst[[2]] is itself a list, its third entry, for example, can be obtained as lst[[2]][[3]] (the third entry of the second list). For convenience, this can be represented lst[[2, 3]] or Part[lst, 2, 3]. Part[Part[lst, 2], 3] can also be used, but is somewhat clumsy.

EXAMPLE 15

```
lst = {{a, b, c, d}, {e, f, g, h}, {i, j, k, l}};
lst[[2]][[3]]
g
lst[[2, 3]]
```

```
g
Part[lst, 2, 3]
g
Part[Part[lst, 2], 3]
g
```

Lists can be modified several different ways.   If lst is any list of objects,

- **Rest[lst]** returns lst with its *first* object deleted.
- **Take[lst, n]** returns a list consisting of the first n objects of lst.
- **Take[lst, {n}]** returns a list consisting of the nth object of lst.
- **Take[lst, -n]** returns a list consisting of the last n objects of lst.
- **Take[lst, {-n}]** returns a list consisting of the nth object from the end of lst.
- **Take[lst, {m, n}]** returns a list consisting of the objects of lst in positions m through n inclusive.

EXAMPLE 16

```
lst = {a, b, c, d, e, f, g};
Rest[lst]
{b, c, d, e, f, g}
Take[lst, 3]
{a, b, c}
Take[lst, -3]
{e, f, g}
Take[lst, {3}]
{c}
Take[lst, {-3}]
{e}
Take[lst, {2, 5}]
{b, c, d, e}
```

Elements can be deleted from a list by using the **Delete** command.

- **Delete[lst, n]** deletes the element in the nth position of lst.
- **Delete[lst, -n]** deletes the element in the nth position from the end of lst.

EXAMPLE 17

```
lst = {a, b, c, d, e, f, g};
Delete[lst, 3]
{a, b, d, e, f, g}
Delete[lst,-3]
{a, b, c, d, f, g}
```

The function **Drop** allows a little more flexibility.

- **Drop[lst, n]** returns lst with its first n objects deleted.
- **Drop[lst, -n]** returns lst with its last n objects deleted.
- **Drop[lst, {n}]** returns lst with its nth object deleted.
- **Drop[lst, {-n}]** returns lst with the nth object from the end deleted.
- **Drop[lst, {m, n}]** returns lst with objects m through n deleted.

Note: Drop[lst, {n}] is equivalent to Delete[lst, n] and Drop[lst, {-n}] is equivalent to Delete[lst, -n].

EXAMPLE 18

```
lst = {a, b, c, d, e, f, g};
Drop[lst, 2]
{c, d, e, f, g}
Drop[lst, -2]
{a, b, c, d, e}
Drop[lst, {2}]
{a, c, d, e, f, g}
Drop[lst, {-2}]
{a, b, c, d, e, g}
Drop[lst, {2, 4}]
{a, e, f, g}
```

There are a variety of list functions which allow elements to be inserted into a list.

- **Append[lst, x]** returns lst with x inserted to the right of its last element.
- **Prepend[lst, x]** returns lst with x inserted to the left of its first element.
- **Insert[lst, x, n]** returns lst with x inserted in position n.
- **Insert[lst, x, -n]** returns lst with x inserted in the nth position from the end.

EXAMPLE 19

```
lst = {1, 2, 3, 4, 5, 6, 7, 8, 9, 10};
Append[lst, x]
{1, 2, 3, 4, 5, 6, 7, 8, 9, 10, x}
Prepend[lst, x]
{x, 1, 2, 3, 4, 5, 6, 7, 8, 9, 10}
Insert[lst, x, 4]
{1, 2, 3, x, 4, 5, 6, 7, 8, 9, 10}
Insert[lst, x, -4]
{1, 2, 3, 4, 5, 6, 7, x, 8, 9, 10}
```

Objects in a list can be replaced by other objects using **ReplacePart**.

- **ReplacePart[lst, x, n]** replaces the object in the nth position of lst by x.
- **ReplacePart[lst, x, -n]** replaces the object in the nth position from the end by x.

EXAMPLE 20

```
lst = {1, 2, 3, 4, 5, 6, 7, 8, 9, 10};
ReplacePart[lst, x, 7]
{1, 2, 3, 4, 5, 6, x, 8, 9, 10}
ReplacePart[lst, x, -7]
{1, 2, 3, x, 5, 6, 7, 8, 9, 10}
```

Lists can be rearranged using **Sort** and **Reverse**.

- **Sort[lst]** sorts the list lst in increasing order. Real numbers are ordered according to their numerical value. Letters are arranged lexicographically, with capital letters coming after lowercase letters.
- **Reverse[lst]** reverses the order of the elements of lst.

EXAMPLE 21

```
lst = {1, 5, -3, 0, 2.5};
Sort[lst]
{-3, 0, 1, 2.5, 5}
```

EXAMPLE 22

```
lst = {z, x, Y, w, X, y, Z, W};
Sort[lst]
{w, W, x, X, y, Y, z, Z}
```

EXAMPLE 23

```
lst = {a, b, c, d, e, f, g};
Reverse[lst]
{g, f, e, d, c, b, a}
```

Cycling of lists is made possible by use of the functions **RotateLeft** and **RotateRight**.

- **RotateLeft[lst]** cycles each element of lst one position to the left. The leftmost element is moved to the extreme right of the list.
- **RotateLeft[lst, n]** cycles the elements of lst precisely n positions to the left. The leftmost n elements are moved to the extreme right of the list in their same relative positions. If n is negative, rotation occurs to the right.
- **RotateRight[lst]** cycles each element of the list lst one position to the right. The rightmost element is moved to the extreme left of the list.
- **RotateRight[lst, n]** cycles the elements of lst precisely n positions to the right. The rightmost n elements are moved to the extreme left of the list in their same relative positions. If n is negative, rotation occurs to the left.

EXAMPLE 24

```
lst = {1, 2, 3, 4, 5, 6, 7, 8, 9, 10};
RotateLeft[lst]
{2, 3, 4, 5, 6, 7, 8, 9, 10, 1}
RotateLeft[lst, 3]
{4, 5, 6, 7, 8, 9, 10, 1, 2, 3}
RotateLeft[lst, -3]
{8, 9, 10, 1, 2, 3, 4, 5, 6, 7}
RotateRight[lst]
{10, 1, 2, 3, 4, 5, 6, 7, 8, 9}
RotateRight[lst, 3]
{8, 9, 10, 1, 2, 3, 4, 5, 6, 7}
RotateRight[lst, -3]
{4, 5, 6, 7, 8, 9, 10, 1, 2, 3}
```

Lists can be concatenated using **Join**.

- **Join[lst1, lst2]** combines the two lists lst1 and lst2 into one list consisting of the elements from lst1 and from lst2.

Join makes no attempt to eliminate repetitive elements. However, repetition can be conveniently eliminated with the Union command (see Section 5.4).

Join can be generalized in a natural way to combine more than two lists.

EXAMPLE 25

```
lst1 = {1, 2, 3, 4, 5};
lst2 = {3, 4, 5, 6, 7};
Join[lst1, lst2]
{1, 2, 3, 4, 5, 3, 4, 5, 6, 7}
```

Nested lists, which are very common, can have a complicated structure. There are a few *Mathematica* commands which can help you understand and manipulate them.

- **Depth[1st]** returns *one more* than the number of levels in the list structure.
- **Level[1st, {levelspec}]** returns a list consisting of those objects which are at level *levelspec* of 1st.
- **Level[1st, levelspec]** returns a list consisting of those objects which are at or below level *levelspec* of 1st.

EXAMPLE 26

```
1st = {1, {2, {3, 4, 5}}};
Depth[1st]
4
Level[1st, {1}]
{1, {2, {3, 4, 5}}}
Level[1st, {2}]
{2, {3, 4, 5}}
Level[1st, {3}]
{3, 4, 5}
Level[1st, 3]
{1, 2, 3, 4, 5, {3, 4, 5}, {2, {3, 4, 5}}}
```

$4 - 1 = 3$. This tells us that 1st contains lists within lists within itself. Note that Depth always returns one more than the actual number of levels in the list. This is for technical reasons dealing with the structure of *Mathematica* commands. For now, just remember that the number of levels is 1 less than Depth.

- **Flatten[1st]** converts a nested list to a simple list containing the innermost objects of 1st.
- **Flatten[1st, n]** flattens a nested list n times, each time removing the outermost level.
- **FlattenAt[1st, n]** flattens the sublist which is at the nth position of the list by one level. If n is negative, *Mathematica* counts backward, starting at the end of the list.

EXAMPLE 27

```
1st = {1, {2, 3}, {4, 5, {6}}, {7, {8, {9, 10}}}};
Flatten[1st]
{1, 2, 3, 4, 5, 6, 7, 8, 9, 10}
Flatten[1st, 1]
{1, 2, 3, 4, 5, {6}, 7, {8, {9, 10}}}
Flatten[1st, 2]
{1, 2, 3, 4, 5, 6, 7, 8, {9, 10}}
FlattenAt[1st, 3]
{1, {2, 3}, 4, 5, {6}, {7, {8, {9, 10}}}}    ← Only the *third* sublist of 1st is
FlattenAt[1st, -1]                              flattened one level.
{1, {2, 3}, {4, 5, {6}}, 7, {8, {9, 10}}}
```

Flatten converts a nested list into a simpler list. **Partition** takes simple lists and converts them into nested lists in a very organized and convenient way.

- **Partition[1st, k]** converts 1st into sublists of length k. If 1st contains $kn + m$ elements, where $m < k$, Partition will create n sublists and the remaining m elements will be dropped.

Partition is a very convenient command for generating tables and matrices. Only the simplest form of the command has been described. The reader, if interested, is urged to investigate other forms in the help browser.

EXAMPLE 28

```
lst = Range[12]
{1, 2, 3, 4, 5, 6, 7, 8, 9, 10, 11, 12}
Partition[lst, 4]
{{1, 2, 3, 4}, {5, 6, 7, 8}, {9, 10, 11, 12}}
Partition[lst, 5]
{{1, 2, 3, 4, 5}, {6, 7, 8, 9, 10}}
Partition[lst, 6]
{{1, 2, 3, 4, 5, 6}, {7, 8, 9, 10, 11, 12}}
```

SOLVED PROBLEMS

3.10 The *Mathematica* function **IntegerDigits** returns a list containing the digits of an integer.   How many digits are there in 100! and what is the 50th digit from the left and from the right?

SOLUTION

```
lst = IntegerDigits[100!]
{9, 3, 3, 2, 6, 2, 1, 5, 4, 4, 3, 9, 4, 4, 1, 5, 2, 6, 8, 1, 6, 9, 9, 2,
 3, 8, 8, 5, 6, 2, 6, 6, 7, 0, 0, 4, 9, 0, 7, 1, 5, 9, 6, 8, 2, 6, 4, 3,
 8, 1, 6, 2, 1, 4, 6, 8, 5, 9, 2, 9, 6, 3, 8, 9, 5, 2, 1, 7, 5, 9, 9, 9,
 9, 3, 2, 2, 9, 9, 1, 5, 6, 0, 8, 9, 4, 1, 4, 6, 3, 9, 7, 6, 1, 5, 6, 5,
 1, 8, 2, 8, 6, 2, 5, 3, 6, 9, 7, 9, 2, 0, 8, 2, 7, 2, 2, 3, 7, 5, 8, 2,
 5, 1, 1, 8, 5, 2, 1, 0, 9, 1, 6, 8, 6, 4, 0, 0, 0, 0, 0, 0, 0, 0, 0, 0,
 0, 0, 0, 0, 0, 0, 0, 0, 0, 0, 0, 0, 0, 0, 0}
```

```
Length[lst]
158
Part[lst, 50] or lst[[50]]
1
Part[lst, -50] or lst[[-50]]
2
```

3.11 Compute the sum of the digits of the 100th Fibonacci number.

SOLUTION

We use **IntegerDigits** (see Problem 3.10).

```
lst = IntegerDigits[Fibonacci[100]]
{3, 5, 4, 2, 2, 4, 8, 4, 8, 1, 7, 9, 2, 6, 1, 9, 1, 5, 0, 7, 5}
```

$$\texttt{Sum[lst[[k]], \{k, 1, Length[lst]\}]} \quad \text{or} \quad \sum_{k=1}^{\text{Length[lst]}} \texttt{lst[[k]]}$$

93

3.12 The command **Table[i * j, {i, 3, 10}, {j, 2, 7}]** generates a nested list of numbers. Add the 4th number in the 5th sublist to the 3rd number in the 6th sublist.

SOLUTION

```
lst = Table[i * j, {i, 3, 10}, {j, 2, 7}]
{{6, 9, 12, 15, 18, 21}, {8, 12, 16, 20, 24, 28},
 {10, 15, 20, 25, 30, 35}, {12, 18, 24, 30, 36, 42},
 {14, 21, 28, 35, 42, 49}, {16, 24, 32, 40, 48, 56},
 {18, 27, 36, 45, 54, 63}, {20, 30, 40, 50, 60, 70}}
lst[[5, 4]] + lst[[6, 3]]
67
```

3.13 The *Mathematica* function `RealDigits` returns a list containing the digits of an approximate real number. This list contains two sublists—the first contains the digits of the approximation and the second is the number of digits to the left of the decimal point. Compute a 20-significant-digit approximation of π and determine the next to the last decimal digit.

SOLUTION

```
approx = N[Pi,20]
3.1415926535897932385
lst = RealDigits[approx]
{{3, 1, 4, 1, 5, 9, 2, 6, 5, 3, 5, 8, 9, 7, 9, 3, 2, 3, 8, 5}, 1}
lst[[1, -2]]
8
```

3.14 Construct a list consisting of the consecutive integers 1 to 10 followed by 20 to 30.

SOLUTION

```
lst = Drop[Range[30], {11, 19}]
{1, 2, 3, 4, 5, 6, 7, 8, 9, 10, 20, 21, 22, 23, 24, 25, 26, 27, 28, 29, 30}
```

3.15 Construct a list consisting of the consecutive integers 1 to 10, followed by 99 followed by 11 to 20.

SOLUTION

```
lst = Insert[Range[20], 99, 11]
{1, 2, 3, 4, 5, 6, 7, 8, 9, 10, 99, 11, 12, 13, 14, 15, 16, 17, 18, 19, 20}
```

3.16 Construct a list of the integers 1 to 20 in *descending* order.

SOLUTION 1

```
lst = Range[20, 1, -1]
{20, 19, 18, 17, 16, 15, 14, 13, 12, 11, 10, 9, 8, 7, 6, 5, 4, 3, 2, 1}
```

SOLUTION 2

```
lst = Range[20] // Reverse       ← This is equivalent to Reverse[Range[20]].
{20, 19, 18, 17, 16, 15, 14, 13, 12, 11, 10, 9, 8, 7, 6, 5, 4, 3, 2, 1}
```

3.17 Sort the letters of the word "MISSISSIPPI" alphabetically.

SOLUTION

```
lst = Characters["MISSISSIPPI"]
{M, I, S, S, I, S, S, I, P, P, I}
Sort[lst]
{I, I, I, I, M, P, P, S, S, S, S}
```

3.18 Construct a list of numbers from 0 to 2π in increments of $\pi/6$.

SOLUTION

```
Range[0, 2Pi, Pi/6]  or  Range[0, 2π, π/6]
```

$$\left\{0, \frac{\pi}{6}, \frac{\pi}{3}, \frac{\pi}{2}, \frac{2\pi}{3}, \frac{5\pi}{6}, \pi, \frac{7\pi}{6}, \frac{4\pi}{3}, \frac{3\pi}{2}, \frac{5\pi}{3}, \frac{11\pi}{6}, 2\pi\right\}$$

3.19   Flavius Joseph was a Jewish historian in the first century.   He writes about a group of 10 Jews in a cave who, rather than surrender to the Romans, chose to commit suicide, one by one.   They formed a circle and every other one was killed.   Who is the lone survivor?

SOLUTION

We number the people 1 through 10 and define a list consisting of these 10 integers.

```
1st = Range[10]
{1, 2, 3, 4, 5, 6, 7, 8, 9, 10}
```

The first person to go is number 2.   We eliminate him by rotating the list one position to the left and dropping his number from the list.

```
1st = Rest[RotateLeft[1st]]
{3, 4, 5, 6, 7, 8, 9, 10, 1}
```

The new list begins with 3 and omits the number 2.   To determine the survivor, we repeat the process until only one number remains.

```
1st = Rest[RotateLeft[1st]]
{5, 6, 7, 8, 9, 10, 1, 3}
1st = Rest[RotateLeft[1st]]
{7, 8, 9, 10, 1, 3, 5}
1st = Rest[RotateLeft[1st]]
{9, 10, 1, 3, 5, 7}
1st = Rest[RotateLeft[1st]]
{1, 3, 5, 7, 9}
1st = Rest[RotateLeft[1st]]
{5, 7, 9, 1}
1st = Rest[RotateLeft[1st]]
{9, 1, 5}
1st = Rest[RotateLeft[1st]]
{5, 9}
1st = Rest[RotateLeft[1st]]
{5}
```

> Number 5 is the survivor.

Although it is interesting to see how our list progresses from step to step, the above technique would not appropriate for a long list.   A more efficient procedure would involve a simple `While` loop.

```
1st = Range[10];
While[Length[1st] > 1, 1st = Rest[RotateLeft[1st]]]
1st
{5}
```

3.20   Determine which elements are in the highest level of the list
{a, {b, c}, {{d, e}, {f, g}, {{h, i}}, {j, {k, l, m}}}}.

SOLUTION

```
1st = {a, {b, c}, {{d, e}, {f, g}, {{h, i}}, {j, {k, l, m}}}};
Depth[1st]
5                    ←remember to subtract 1 to determine the highest level
Level[1st, {4}]
{h, i, k, l, m}
```

3.21   Reduce the depth of {a, {b, c}, {{d, e}, {f, g}, {{h, i}}, {j, {k, l, m}}}} by one level; by two levels.

SOLUTION

```
lst = {a, {b, c}, {{d, e}, {f, g}, {{h, i}}, {j, {k, l, m}}}};
Flatten[lst, 1]
{a, b, c, {d, e}, {f, g}, {{h, i}}, {j, {k, l, m}}}
Flatten[lst, 2]
{a, b, c, d, e, f, g, {h, i}, j, {k, l, m}}
```

3.22 Take the list of characters A through X and construct a list with six sublists, each containing four distinct letters.

SOLUTION

```
lst = CharacterRange["A", "X"]
{A, B, C, D, E, F, G, H, I, J, K, L, M, N, O, P, Q, R, S, T, U, V, W, X}
Partition[lst, 4]
{{A, B, C, D}, {E, F, G, H}, {I, J, K, L}, {M, N, O, P}, {Q, R, S, T}, {U, V, W, X}}
```

3.4 SET THEORY

Sets are represented as lists in *Mathematica*. Sets are manipulated using the basic list functions **Union**, **Intersection**, and **Complement**.

- **Union[lst1, lst2]** combines lists lst1 and lst2 into one sorted list, eliminating any duplicate elements. Although only two lists are presented in this description, any number of lists may be used. As a special case, **Union[lst]** will eliminate duplicate elements in lst.
- **Intersection[lst1, lst2]** returns a sorted list of elements common to lst1 and lst2. If lst1 and lst2 are disjoint, i.e., they have no common elements, the command **Intersection[lst1, lst2]** returns the empty list, {}.
- **Complement[universe, lst1]** returns a sorted list consisting of those elements of universe which are not in lst1. In this context, universe represents the universal set.
- **Complement[universe, lst1, lst2]** returns sorted list consisting of those elements of universe which are not in lst1 or lst2. This command extends in a natural way to more than two lists.

EXAMPLE 29

```
lst = {a, b, c, a, c, c, c, b, b}
{a, b, c, a, c, c, c, b, b}
Union[lst]
{a, b, c}
```

EXAMPLE 30

```
universe = {1, 2, 3, 4, 5, 6, 7, 8, 9, 10};
lst1 = {1, 3, 5, 7};
lst2 = {5, 7, 8, 10};
Union[lst1, lst2]
{1, 3, 5, 7, 8, 10}
Intersection[lst1, lst2]
{5, 7}
Complement[universe, lst1]
{2, 4, 6, 8, 9, 10}
Complement[universe, lst1, lst2]
{2, 4, 6, 9}
```

Using the BasicInput palette, the symbols ∪ and ∩ may be used to represent union and intersection, respectively.

- `lst1 ∪ lst2` is equivalent to `Union[lst1, lst2]`.
- `lst1 ∩ lst2` is equivalent to `Intersection[lst1, lst2]`.

EXAMPLE 31

```
lst1 = {1, 2, 3, 4, 5};
lst2 = {3, 4, 5, 6, 7};
lst1 ∪ lst2
{1, 2, 3, 4, 5, 6, 7}
lst1 ∩ lst2
{3, 4, 5}
```

There are a number of useful set commands available in the package `DiscreteMath`Combinatorica``. Among them are `CartesianProduct`, `Subsets`, and `KSubsets`.

By definition, the Cartesian product of two sets, A and B, is the set of ordered pairs of elements, the first taken from A and the second from B.

- `CartesianProduct[lst1, lst2]` returns the Cartesian product of `lst1` and `lst2`.

EXAMPLE 32

```
<<DiscreteMath`Combinatorica`       ←This loads the package.  See Chapter 1.
lst1 = {a, b, c, d};
lst2 = {x, y, z};
CartesianProduct[lst1, lst2]
{{a, x}, {a, y}, {a, z}, {b, x}, {b, y}, {b, z},
 {c, x}, {c, y}, {c, z}, {d, x}, {d, y}, {d, z}}
```

A subset of A is any set, each of whose elements is a member of A. The empty set is a subset of every set. Including the empty set, a set of *n* elements has 2^n subsets. The set of *all* subsets of A is called the *power set* of A.

- `Subsets[lst]` returns a list containing all subsets of `lst`, including the empty set, i.e., the power set of `lst`.
- `KSubsets[lst, k]` returns a list containing all subsets of `lst` of size k.

EXAMPLE 33

```
<<DiscreteMath`Combinatorica`       ←Omit if you have already loaded the package.
lst = {a, b, c, d};
Subsets[lst]
{{}, {a}, {a, b}, {b}, {b, c}, {a, b, c}, {a, c}, {c}, {c, d},
 {a, c, d}, {a, b, c, d}, {b, c, d}, {b, d}, {a, b, d}, {a, d}, {d}}
Subsets[lst]//Sort        ←//Sort presents the subsets in a more organized way.
{{}, {a}, {b}, {c}, {d}, {a, b}, {a, c}, {a, d}, {b, c}, {b, d},
 {c, d}, {a, b, c}, {a, b, d}, {a, c, d}, {b, c, d}, {a, b, c, d}}
KSubsets[lst, 3]
{{a, b, c}, {a, b, d}, {a, c, d}, {b, c, d}}
```

SOLVED PROBLEMS

3.23   Which different letters are contained in the word "MISSISSIPPI"?

SOLUTION

```
Union[Characters["MISSISSIPPI"]]
{I, M, P, S}
```

3.24 Find the union and intersection of the sets {a, b, c, d, e, f, g}, {c, d, e, f, g, h, i}, and {e, f, g, h, i, j, k}.

SOLUTION

```
set1 = {a, b, c, d, e, f, g};
set2 = {c, d, e, f, g, h, i};
set3 = {e, f, g, h, i, j, k};
Union[set1, set2, set3]  or  set1 ∪ set2 ∪ set3
{a, b, c, d, e, f, g, h, i, j, k}
Intersection[set1, set2, set3]  or  set1 ∩ set2 ∩ set3
{e, f, g}
```

3.25 Find all the elements of the set {a, b, c, d, e, f, g} which are *not* in {a, c, d, e}.

SOLUTION

```
set1 = {a, b, c, d, e, f, g};
set2 = {a, c, d, e};
Complement[set1, set2]
{b, f, g}
```

3.26 The 20th prime is 71.   Find all the numbers less than 71 which are *not* prime.

SOLUTION

```
universe = Range[71];
primenumbers = Table[Prime[k], {k, 1, 20}];
Complement[universe, primenumbers]
{1, 4, 6, 8, 9, 10, 12, 14, 15, 16, 18, 20, 21, 22, 24, 25, 26, 27, 28,
 30, 32, 33, 34, 35, 36, 38, 39, 40, 42, 44, 45, 46, 48, 49, 50, 51,
 52, 54, 55, 56, 57, 58, 60, 62, 63, 64, 65, 66, 68, 69, 70}
```

3.27 Construct a list consisting of the consonants of the alphabet.

SOLUTION

```
letters = CharacterRange["a", "z"];
vowels = Characters["aeiou"];
consonants = Complement[letters, vowels]
{b, c, d, f, g, h, j, k, l, m, n, p, q, r, s, t, v, w, x, y, z}
```

3.28 Find all the numbers less than 1,000 which are *both* prime and Fibonacci.

SOLUTION

```
k = 1; lst1 = {};
While[Fibonacci[k] ≤ 1000, lst1 = Append[lst1, Fibonacci[k]]; k++]
k = 1; lst2 = {};
```

```
While[Prime[k] ≤ 1000, lst2 = Append[lst2, Prime[k]]; k++]
lst1 ∩ lst2
{2, 3, 5, 13, 89, 233}
```

3.29   Create a list which contains all the subsets of {a, b, c, d, e}.   How many subsets are there?

SOLUTION

```
<<DiscreteMath`Combinatorica`        ← The package must be loaded first.
letters = {a, b, c, d, e};
Subsets[letters] //Sort
{{}, {a}, {b}, {c}, {d}, {e}, {a, b}, {a, c}, {a, d}, {a, e}, {b, c},
 {b, d}, {b, e}, {c, d}, {c, e}, {d, e}, {a, b, c}, {a, b, d},
 {a, b, e}, {a, c, d}, {a, c, e}, {a, d, e}, {b, c, d}, {b, c, e},
 {b, d, e}, {c, d, e}, {a, b, c, d}, {a, b, c, e}, {a, b, d, e},
 {a, c, d, e}, {b, c, d, e}, {a, b, c, d, e}}
Length[%]
32
```

3.30   Create a list which contains all the subsets of {a, b, c, d, e} which contain precisely three elements.   How many are there?

SOLUTION

```
<<DiscreteMath`Combinatorica`
letters = {a, b, c, d, e};
KSubsets[letters, 3]
{{a,b,c}, {a,b,d}, {a,b,e}, {a,c,d}, {a,c,e}, {a,d,e}, {b,c,d},
 {b,c,e}, {b,d,e}, {c,d,e}}
Length[%]
10
```

3.5 TABLES AND MATRICES

Mathematica represents tables and matrices as nested lists.   Internally, there is no distinction made in the way they are stored, but they can be represented differently using the functions **MatrixForm** and **TableForm**.   It is often more convenient to use **//MatrixForm** or **//TableForm** to the right of the matrix or table name.

Matrices and tables can be entered directly as nested lists.   A matrix or table having m rows and n columns would be a list of m sublists, each containing n entries.

Matrices and tables can also be conveniently entered by going to Input⇒Create Table/ Matrix /Palette . . . .   This also allows some convenient options.   Clicking OK yields an empty grid—use the Tab key to cycle from entry to entry.

$$\begin{pmatrix} \square & \square & \square \\ \square & \square & \square \\ \square & \square & \square \end{pmatrix}$$

Create Table/Matrix/Palette

Make:
- ○ Table (plain GridBox)
- ● Matrix
- ○ Palette

Number of rows: `5`

Number of columns: `5`

- ☐ Draw lines between rows
- ☐ Draw lines between columns
- ☐ Draw frame

- ☐ Fill with: `0`
- ☐ Fill diagonal: `1`

[OK] [Cancel] [Help]

EXAMPLE 34

```
lst = {{1, 2, 3, 4}, {5, 6, 7, 8}, {9, 10, 11, 12}}
{{1, 2, 3, 4}, {5, 6, 7, 8}, {9, 10, 11, 12}}
MatrixForm[lst] or lst //MatrixForm
```

$$\begin{pmatrix} 1 & 2 & 3 & 4 \\ 5 & 6 & 7 & 8 \\ 9 & 10 & 11 & 12 \end{pmatrix}$$

Two special matrix-generating commands are worth remembering because of their frequency in applications.

- `IdentityMatrix[n]` produces an $n \times n$ matrix with 1's on the main diagonal and 0's elsewhere.
- `DiagonalMatrix[lst]`, where `lst` is a list of objects, produces a square matrix of order `Length[lst]` with the elements of `lst` on the main diagonal and 0's elsewhere.

EXAMPLE 35

```
IdentityMatrix[3]                          DiagonalMatrix[{1, 2, 3}]
```

$$\begin{pmatrix} 1 & 0 & 0 \\ 0 & 1 & 0 \\ 0 & 0 & 1 \end{pmatrix} \qquad \begin{pmatrix} 1 & 0 & 0 \\ 0 & 2 & 0 \\ 0 & 0 & 3 \end{pmatrix}$$

Once defined, matrices can be combined using the operations of addition, subtraction, and scalar and matrix multiplication. The operation of matrix multiplication is represented by a period (.). Matrices are discussed in greater detail in Chapter 12, Linear Algebra.

EXAMPLE 36

$$A = \begin{pmatrix} 1 & 2 & 3 \\ 4 & 5 & 6 \\ 7 & 8 & 9 \end{pmatrix}$$

> The matrix is created using Input⇒Create Table/ Matrix /Palette . . . .
> *Mathematica* outputs the matrix as a nested list.

```
{{1, 2, 3}, {4, 5, 6}, {7, 8, 9}}
```

$$B = \begin{pmatrix} 2 & 1 & 5 \\ 4 & 7 & 2 \\ 1 & 3 & 2 \end{pmatrix}$$

```
{{2, 1, 5}, {4, 7, 2}, {1, 3, 2}}
```

A + B // MatrixForm

$$\begin{pmatrix} 3 & 3 & 8 \\ 8 & 12 & 8 \\ 8 & 11 & 11 \end{pmatrix}$$

A - B // MatrixForm

$$\begin{pmatrix} -1 & 1 & -2 \\ 0 & -2 & 4 \\ 6 & 5 & 7 \end{pmatrix}$$

3 A // MatrixForm

$$\begin{pmatrix} 3 & 6 & 9 \\ 12 & 15 & 18 \\ 21 & 24 & 27 \end{pmatrix}$$

A . B // MatrixForm

$$\begin{pmatrix} 13 & 24 & 15 \\ 34 & 57 & 42 \\ 55 & 90 & 69 \end{pmatrix}$$

It is useful to remember that if lst is a simple list of numbers, lst.lst yields the sum of their squares. The result is printed as a single number without braces.

EXAMPLE 37

```
lst = {1,2,3,4,5};
lst.lst
55
```

Tables are also stored as nested lists, but are represented as tables with **TableForm**. Although this command allows representation of tables of any dimension, we shall discuss only one- and two-dimensional tables in this book.

■ **TableForm[lst:** *options***]** prints the elements of lst in a rectangular array.

lst//TableForm is equivalent to **TableForm[lst]** if no options are needed.

EXAMPLE 38

```
lst = {{12, 7, 10}, {105, 205, 7}, {3, 30, 300}};
TableForm[lst]  or  lst//TableForm
```

```
12      7       10
105     205     7
3       30      300
```

Observe that the numbers in a table are, by default, left-justified.   This can sometimes make the table confusing to read.   Justification can be controlled with the `TableAlignments` option.

- `TableAlignments → Left` justifies the columns to the left (default).
- `TableAlignments → Right` justifies the columns to the right.
- `TableAlignments → Center` centers the columns.

EXAMPLE 39

```
lst = {{12,7,10}, {105,205,7}, {3,30,300}};
TableForm[lst, TableAlignments → Right]
```

```
 12      7      10
105    205       7
  3     30     300
```

```
TableForm[lst, TableAlignments → Center]
```

```
 12      7      10
105    205       7
  3     30     300
```

Row and column headings can be inserted by using the option `TableHeadings` within the `TableForm` command.

- `TableHeadings → None` is *Mathematica*'s default.
- `TableHeadings → Automatic` produces consecutive integer labels for both rows and columns.

Each row and column of a table can be labeled separately using strings (characters enclosed within double quotes) or *Mathematica* expressions.   The general form of this option is

- `TableHeadings → {rowlist, columnlist}`

where *rowlist* is a list of row labels and *columnlist* is a list of column labels.   If you desire to have row labels but not column labels, or column labels but not row labels, simply replace *rowlist* or *columnlist* by `None`.

EXAMPLE 40

```
lst = {{a, b, c}, {d, e, f}, {g, h, i}};
TableForm[lst, TableHeadings → Automatic]
```

```
        1       2       3
1       a       b       c
2       d       e       f
3       g       h       i
```

EXAMPLE 41

```
lst = {{a, b, c}, {d, e, f}, {g, h, i}};
TableForm[lst,
          TableHeadings → {{"row1", "row2", "row3"},
          {"column1", "column2", "column3"}},
          TableAlignments → Center]
```

	column1	column2	column3
row1	1	b	c
row2	d	e	f
row3	g	h	i

```
TableForm[lst,
          TableHeadings → {None, {"column1", "column2", "column3"}},
          TableAlignments → Center]
```

column1	column2	column3
a	b	c
d	e	f
g	h	i

```
TableForm[lst,
          TableHeadings → {{"row1", "row2", "row3"}, None},
          TableAlignments → Center]
```

row1	a	b	c
row2	d	e	f
row3	g	h	i

TableDirections is an option which determines how the entries of the table should be placed.

- **TableDirections → Column** (default) prints the table with the first element of each inner list in the first column, the second element of each inner list in the second column, and so forth.
- **TableDirections → Row** interchanges the positions of the columns with the rows.

EXAMPLE 42

```
Clear[a]
lst = Array[a, {3, 4}]
{{a[1, 1], a[1, 2], a[1, 3], a[1, 4]}, {a[2, 1], a[2, 2], a[2, 3],
  a[2, 4]}, {a[3, 1], a[3, 2], a[3, 3], a[3, 4]}}
TableForm[lst, TableDirections → Column]
a[1, 1] a[1, 2] a[1, 3] a[1, 4]
a[2, 1] a[2, 2] a[2, 3] a[2, 4]          The elements a[1, 1], a[2, 1], and
a[3, 1] a[3, 2] a[3, 3] a[3, 4]          a[3, 1] form the first column.
TableForm[lst, TableDirections → Row]
a[1, 1] a[2, 1] a[3, 1]
a[1, 2] a[2, 2] a[3, 2]                  The elements a[1, 1], a[2, 1], and
a[1, 3] a[2, 3] a[3, 3]                  a[3, 1] form the first row.
a[1, 4] a[2, 4] a[3, 4]
```

By default, *Mathematica* suppresses trailing 0's to the right of decimal points. This often leads to tables which are not uniformly displayed. However, the default can be overridden and the output can be formatted to your specification.

PaddedForm allows the output of a command or calculation to be formatted.

- **PaddedForm[*expression*, n]** prints the value of *expression*, leaving space for a total of n digits. This form of the command can be used for integers or real number approximations. **Note:** The decimal point is not counted as a position.
- **PaddedForm[*expression*, {n, f}]** prints the value of *expression*, leaving space for a total of n digits, f of which are to the right of the decimal point. The fractional portion of the number is rounded if any digits are deleted.

EXAMPLE 43

```
a = 123.456789;
PaddedForm[a, 12]
     123.456789                    ←3 spaces to the left of the number
PaddedForm[a, 20]
             123.456789            ←11 spaces to the left of the number
PaddedForm[a, {20, 3}]
                123.457            ←14 spaces to the left of the number, the third
                                     decimal is rounded to 7
```

As with any function, PaddedForm can be applied to an entire table.

EXAMPLE 44

The following command defines a table of square and cube roots of the first 10 positive integers:

```
1st = Table[{n, N[√n], N[∛n]}, {n, 1, 10}];
```

First we will print the table using the standard TableForm command.

```
TableForm[1st]
```

1	1.	1.
2	1.41421	1.25992
3	1.73205	1.44225
4	2.	1.5874
5	2.23607	1.70998
6	2.44949	1.81712
7	2.64575	1.91293
8	2.828423	2.
9	3.	2.08008
10	3.16228	2.15443

Now we use PaddedForm to pad the entire table.

```
PaddedForm[TableForm[1st],{7,5}]
```

1.00000	1.00000	1.00000
2.00000	1.41421	1.25992
3.00000	1.73205	1.44225
4.00000	2.00000	1.58740
5.00000	2.23607	1.70998
6.00000	2.44949	1.81712
7.00000	2.64575	1.91293
8.00000	2.82843	2.00000
9.00000	3.00000	2.08008
10.00000	3.16228	2.15443

If we wish to format the individual columns differently, this can be accomplished by padding the individual entries of the list, rather than the whole table.

```
paddedn := PaddedForm[n, 2]
paddedsquaroot := PaddedForm[N[√n], {11, 7}]
paddedcuberoot := PaddedForm[N[∛n], {7, 3}]
lst = Table[{paddedn, paddedsquaroot, paddedcuberoot}, {n, 1, 10}];
TableForm[lst]
```

```
    1        1.0000000        1.000
    2        1.4142136        1.260
    3        1.7320508        1.442
    4        2.0000000        1.587
    5        2.2360680        1.710
    6        2.4494897        1.817
    7        2.6457513        1.913
    8        2.8284271        2.000
    9        3.0000000        2.080
   10        3.1622777        2.154
```

Spacing between rows and columns can be controlled with **TableSpacing**. This option specifies the number of spaces to put between entries in each dimension.

- **TableSpacing → {*rowspaces, columnspaces*}**. *rowspaces* specifies the number of blank lines between successive lines of the table; *columnspaces* specifies the number of blank characters between columns.

EXAMPLE 45

```
lst = {{a, b, c}, {d, e, f}, {g, h, i}};
TableForm[lst, TableSpacing → {0, 0}]
abc                    ←no spacing between rows or columns
def
ghi
```

```
TableForm[lst, TableSpacing → {1, 3}]
a    b    c            ←1 line between rows, 3 spaces between columns


d    e    f


g    h    i
```

```
TableForm[lst, TableSpacing → {3, 1}]
a b c                  ←3 lines between rows, 1 space between columns


d e f


g h i
```

Lists can be expressed as single columns with **ColumnForm**.

- **ColumnForm[lst]** presents lst as a single column of objects.
- **ColumnForm[lst, *horizontal*]** specifies the horizontal alignment of each row. Acceptable values of *horizontal* are **Left** (default), **Center**, and **Right**.
- **ColumnForm[lst, *horizontal, vertical*]** allows vertical alignment of the column. Acceptable values are **Above**, **Below** (default), and **Center**.

EXAMPLE 46

```
lst = {a, bb, ccc};
ColumnForm[lst]
a
bb
ccc
ColumnForm[lst,Right]
  a
 bb
ccc
```

SOLVED PROBLEMS

3.31 Construct a 3×3 matrix whose entries are consecutive integers, increasing as we go to the right and down.

SOLUTION

```
lst = Table[3i + j, {i, 0, 2}, {j, 1, 3}]
{{1, 2, 3}, {4, 5, 6}, {7, 8, 9}}
lst // MatrixForm
```

$$\begin{pmatrix} 1 & 2 & 3 \\ 4 & 5 & 6 \\ 7 & 8 & 9 \end{pmatrix}$$

3.32 The Hilbert matrix is a square matrix whose element in position (i, j) is $\dfrac{1}{i+j-1}$. Construct the Hilbert matrix of order 5.

SOLUTION

```
a[i_, j_] = 1/(i + j - 1);
hilbert = Array[a, {5, 5}]
```

$$\left\{\left\{1, \frac{1}{2}, \frac{1}{3}, \frac{1}{4}, \frac{1}{5}\right\}, \left\{\frac{1}{2}, \frac{1}{3}, \frac{1}{4}, \frac{1}{5}, \frac{1}{6}\right\}, \left\{\frac{1}{3}, \frac{1}{4}, \frac{1}{5}, \frac{1}{6}, \frac{1}{7}\right\}, \left\{\frac{1}{4}, \frac{1}{5}, \frac{1}{6}, \frac{1}{7}, \frac{1}{8}\right\}, \left\{\frac{1}{5}, \frac{1}{6}, \frac{1}{7}, \frac{1}{8}, \frac{1}{9}\right\}\right\}$$

```
hilbert // MatrixForm
```

$$\begin{pmatrix} 1 & \frac{1}{2} & \frac{1}{3} & \frac{1}{4} & \frac{1}{5} \\ \frac{1}{2} & \frac{1}{3} & \frac{1}{4} & \frac{1}{5} & \frac{1}{6} \\ \frac{1}{3} & \frac{1}{4} & \frac{1}{5} & \frac{1}{6} & \frac{1}{7} \\ \frac{1}{4} & \frac{1}{5} & \frac{1}{6} & \frac{1}{7} & \frac{1}{8} \\ \frac{1}{5} & \frac{1}{6} & \frac{1}{7} & \frac{1}{8} & \frac{1}{9} \end{pmatrix}$$

3.33 Construct the 5×5 identity matrix.

SOLUTION

```
IdentityMatrix[5]
```

$$\begin{pmatrix} 1 & 0 & 0 & 0 & 0 \\ 0 & 1 & 0 & 0 & 0 \\ 0 & 0 & 1 & 0 & 0 \\ 0 & 0 & 0 & 1 & 0 \\ 0 & 0 & 0 & 0 & 1 \end{pmatrix}$$

3.34 Construct a 5×5 matrix having the first five primes as diagonal entries and 0's elsewhere.

SOLUTION

```
diag = Table[Prime[k], {k, 1, 5}]
{2, 3, 5, 7, 11}
DiagonalMatrix[diag] //MatrixForm
```

$$\begin{pmatrix} 2 & 0 & 0 & 0 & 0 \\ 0 & 3 & 0 & 0 & 0 \\ 0 & 0 & 5 & 0 & 0 \\ 0 & 0 & 0 & 7 & 0 \\ 0 & 0 & 0 & 0 & 11 \end{pmatrix}$$

3.35 Construct a table having three columns. The first column lists the consecutive integers 1 through 10 and the second and third columns are their squares and cubes. Label the three columns "integers," "squares," and "cubes."

SOLUTION

```
1st = Table[{k, k², k³}, {k, 1, 10}];

TableForm[1st,
    TableHeadings → {None, {"integers", "squares", "cubes"}},
    TableAlignments → Right]
```

```
integers       squares        cubes
       1             1             1
       2             4             8
       3             9            27
       4            16            64
       5            25           125
       6            36           216
       7            49           343
       8            64           512
       9            81           729
      10           100          1000
```

3.36 If c represents the temperature in degrees Celsius, its corresponding Fahrenheit temperature is $f = \frac{9}{5}c + 32$ degrees. Construct a labeled table showing, horizontally, the Fahrenheit equivalents of Celsius temperatures from 1 to 10 degrees in increments of 1 degree.

SOLUTION

```
f = 9/5 c + 32;

1st = Table[{c, PaddedForm[N[f], {3, 1}]}, {c, 1, 10}];

TableForm[1st, TableDirections → Row,
        TableHeadings → {None, {"Celsius", "Fahrenheit"}},
        TableAlignments → Center]
```

```
Celsius      1    2    3    4    5    6    7    8    9   10
Fahrenheit 33.8 35.6 37.4 39.2 41.0 42.8 44.6 46.4 48.2 50.0
```

3.37 Construct a table showing the radian equivalents of angles from 0° to 30° in increments of 5°.

SOLUTION

```
lst = Table[{deg, N[deg Degree]}, {deg, 0, 30, 5}];
TableForm[lst, TableDirections→Row,
    TableHeadings→{None, {"Degrees", "Radians"}},
    TableAlignments→Center]
```

> Degree is a *Mathematica* constant (see Chapter 2).

Degrees	0	5	10	15	20	25	30
Radians	0	0.0872665	0.174533	0.261799	0.349066	0.436332	0.523599

3.38 If p dollars is invested for t years in a bank account paying an annual interest rate of r compounded n times a year, the amount of money after k periods is $p\left(1+\dfrac{r}{n}\right)^k$ dollars. If $1000 is invested in an account paying 6% compounded quarterly, make a table showing how much money has accumulated during a three-year period.

SOLUTION

```
p = 1000;
r = .06;
n = 4;
t = 3;
a = p(1 + r/n)^k;
lst = Table[{k, a}, {k, 1, nt}];
TableForm[lst, TableHeadings→{None, {"period", "amount"}}]
```

period	amount
1	1015.
2	1030.22
3	1045.68
4	1061.36
5	1077.28
6	1093.44
7	1109.84
8	1126.49
9	1143.39
10	1160.54
11	1177.95
12	1195.62

3.39 If p dollars is invested in a bank account paying a rate of r compounded n times a year, the amount of money after t years is $p\left(1+\dfrac{r}{n}\right)^{nt}$ dollars. If interest is compounded continuously, the amount after t years is pe^{rt}. If $1000 is invested in an account paying 6% annually, make a table showing how much money is in the account at the end of each year for 10 years if interest is compounded quarterly, monthly, daily, and continuously.

SOLUTION

```
p = 1000.;
r = 0.06;
a = PaddedForm[p (1 + r/4)^4t, {7, 2}];
b = PaddedForm[p (1 + r/12)^12t, {7, 2}];
```

```
c = PaddedForm[p (1 + r/365)^365t, {7, 2}];
d = PaddedForm[p Exp [rt], {7, 2}];
tt = PaddedForm[t, 2];
1st = Table[{tt, a, b, c, d}, {t, 1, 10}]
TableForm[1st, TableHeadings →
      {None, {"year", " quarterly", " monthly", " daily", "continuously"}}]
```

year	quarterly	monthly	daily	continuously
1	1061.36	1061.68	1061.83	1061.84
2	1126.49	1127.16	1127.49	1127.50
3	1195.62	1196.68	1197.20	1197.22
4	1268.99	1270.49	1271.22	1271.25
5	1346.86	1348.85	1349.83	1349.86
6	1429.50	1432.04	1433.29	1433.33
7	1517.22	1520.37	1521.91	1521.96
8	1610.32	1614.14	1616.01	1616.07
9	1709.14	1713.70	1715.93	1716.01
10	1814.02	1819.40	1822.03	1822.12

3.40 The payment on a monthly mortgage of a dollars is $\dfrac{a \times \dfrac{r}{12}}{1 - \left(1 + \dfrac{r}{12}\right)^{-12n}}$ where n is the number

of years the money is borrowed and r is the annual rate of interest.   Construct a table showing the monthly payments on a 30-year mortgage of $250,000 at rates of 6% to 8% in increments of $\frac{1}{4}$%.

SOLUTION

```
a = 250000;
n = 30;
```

$$payment := \dfrac{a \dfrac{r}{12}}{1 - \left(1 + \dfrac{r}{12}\right)^{-12 n}};$$

```
1st = Table[{PaddedForm[r, {5, 4}],
          PaddedForm[payment, {6, 2}]}, {r, .06, .08, .0025}];
TableForm[1st, TableHeadings → {None, {" rate", "payment"}}]
```

rate	payment
0.0600	1498.88
0.0625	1539.29
0.0650	1580.17
0.0675	1621.50
0.0700	1663.26
0.0725	1705.44
0.0750	1748.04
0.0775	1791.03
0.0800	1834.41

CHAPTER 4

Two-Dimensional Graphics

4.1 PLOTTING FUNCTIONS OF A SINGLE VARIABLE

Anyone who has ever tried to plot a graph using one of the standard programming languages will appreciate the ease with which graphs can be produced in *Mathematica*.   In many instances, only one instruction is all that is needed to produce a pictorial representation of a function or a more general relationship between two variables.

Although *Mathematica*'s defaults work well in most instances, there are many options available to control subtleties.   We shall describe the more common ones in this section and present a variety of examples which illustrate the ease with which graphs may be constructed.

The basic command for drawing the graph of a function is **Plot**.   Although x is used as the independent variable in the description below, any symbol may be used in its place.

- **Plot[f[x], {x, xmin, xmax}]** plots the graph of $f(x)$ on the interval $xmin \le x \le xmax$.

EXAMPLE 1

Plot the parabola $f(x) = x^2$ from -3 to 3.

```
Plot[x², {x, -3, 3}]
```

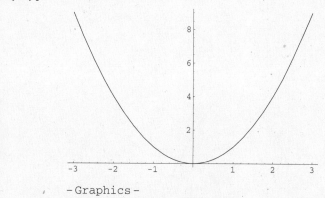

```
-Graphics-
```

The description, - Graphics -, which follows the graph, indicates that the graph is a "graphics" object.   It may be suppressed by using a semicolon (;) to the right of the **Plot** command.

Two functions can be plotted on the same set of axes.

- `Plot[{f[x], g[x]}, {x, xmin, xmax}]` plots the graphs of $f(x)$ and $g(x)$ from `xmin` to `xmax` on the same set of axes. This command can be generalized in a natural way to plot three or more functions.

EXAMPLE 2

Plot $f(x) = x^2$ and $g(x) = 9 - x^2$ from -3 to 3.

`Plot[{x², 9 - x²}, {x, -3, 3}];`   ← note the use of a semicolon here

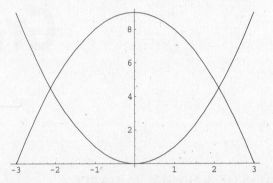

The `Show` command is useful for plotting several graphics objects simultaneously, particularly when their domains are different intervals.

- `Show[g1, g2, ...]` plots several graphs on a common set of axes.

EXAMPLE 3

Suppose we wish to plot the graph of $y = x^2$ on the interval $[-4, 4]$ and the graph of $y = \sin x$ on the interval $[0, 2\pi]$, but wish to plot them on one set of axes. We define two graphics objects, `g1` and `g2`:

`g1 = Plot[x² - 9, {x, -4, 4}];`

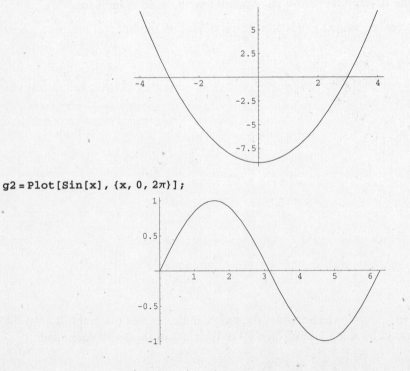

`g2 = Plot[Sin[x], {x, 0, 2π}];`

Now we apply the `Show` command.   Note how the axes are adjusted to exhibit both graphs.

`Show[g1, g2];`

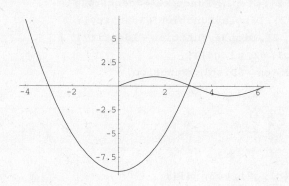

You will notice that in constructing `g1` and `g2` in the previous example, each curve was first drawn individually on its own axis.   To suppress the graphics output, the option **DisplayFunction** can be used. `DisplayFunction` specifies which function will control the graphics output.

- **DisplayFunction → Identity** suppresses graphics output.
- **DisplayFunction → $DisplayFunction** restores the ability to display graphics.

EXAMPLE 4

`g1 = Plot[x² - 9, {x, -4, 4}, DisplayFunction → Identity];`
`g2 = Plot[Sin[x], {x, 0, 2π}, DisplayFunction → Identity];`
`Show[g1, g2, DisplayFunction → $DisplayFunction];`

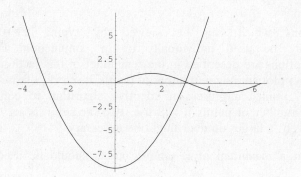

Only the combined graph is drawn.

Another useful command for drawing multiple graphs is **GraphicsArray**.

- **GraphicsArray[{g1, g2, ...}]** plots a row of graphics objects.
- **GraphicsArray[{g11, g12, ...}, {g21, g22, ...}}]** plots a two-dimensional array of graphics objects.

Note: Unlike `Plot`, `GraphicsArray` does not produce a graphics object.   Therefore, the graph must be viewed with `Show`.

EXAMPLE 5

```
g1 = Plot[x, {x, -2, 2}, DisplayFunction→Identity];
g2 = Plot[-x, {x, -2, 2}, DisplayFunction→Identity];
g3 = Plot[x², {x, -2, 2}, DisplayFunction→Identity];
g4 = Plot[-x², {x, -2, 2}, DisplayFunction→Identity];
g = GraphicsArray[{g1, g2, g3, g4}];
Show[g, DisplayFunction→$DisplayFunction];
```

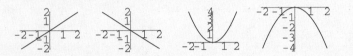

```
gg = GraphicsArray[{g1, g2}, {g3, g4}];
Show[gg, DisplayFunction→$DisplayFunction];
```

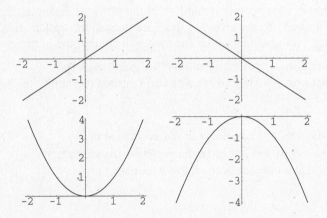

Plot has a variety of options which can be viewed by typing **?? Plot** or **Options[Plot]**. These options may be used individually or in conjunction with one another. Some of the more common options are described in the remainder of this section.

Since *Mathematica* cannot (obviously) plot an infinite number of points, it selects a finite number of equally spaced points as "sample" points and uses an adaptive algorithm to construct a smooth-looking curve. The maximum number of points it will use, **PlotPoints**, is set to 25 by default. If the curve "wiggles" excessively, a larger number might be necessary.

- **PlotPoints → n** specifies that a maximum of n sample points should be used in the construction of the graph.
- **PlotRange** is an option which specifies which points should be included in the plot.

  - **PlotRange → Automatic** is *Mathematica*'s default. Any points whose vertical coordinates appear to be too large to produce a pleasing graph (by *Mathematica*'s standards) are omitted from the graph.
  - **PlotRange → All** forces *Mathematica* to plot all points.
  - **PlotRange → {ymin, ymax}** plots only those points whose vertical coordinates fall between ymin and ymax.
  - **PlotRange → {{xmin, xmax}, {ymin, ymax}}** plots those points whose horizontal coordinates fall between xmin and xmax and whose vertical coordinates fall between ymin and ymax.

EXAMPLE 6

```
Plot[x³, {x, -10, 10}];                    Plot[x³, {x, -10, 10}, PlotRange → All];
```

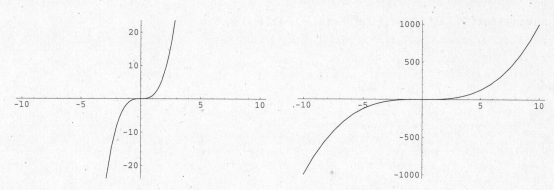

When you plot a graph, you will notice that the horizontal and vertical axes are usually not the same length.   By default, the ratio of vertical axis length to horizontal axis length is `1/GoldenRatio`, where `GoldenRatio` = $(1 + \sqrt{5})/2$.   The designers of *Mathematica* felt that this ratio is the most comfortable and pleasing to the eye.   It can be changed with the option **AspectRatio**, a graphics option which determines the height-to-width ratio of a graph.

- **AspectRatio → Automatic** computes the aspect ratio from the actual coordinate values of the plot.
- **AspectRatio → *ratio*** sets the ratio of vertical to horizontal axis length to the value *ratio*.

EXAMPLE 7

```
Plot[x², {x, -5, 5}];                      Plot[x², {x, -5, 5}, AspectRatio → Automatic];
```

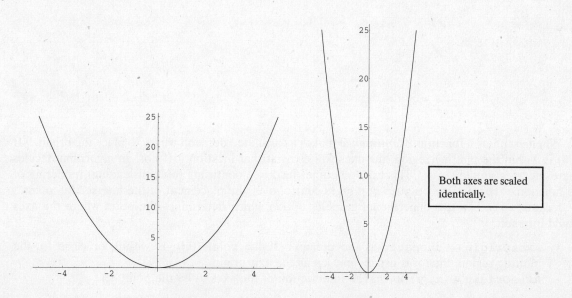

Both axes are scaled identically.

EXAMPLE 8

The following command should produce a circle of radius 3 centered at the origin. However, because of unequal axis scaling, the graph appears as an ellipse.

```
Plot[{-Sqrt[9 - x²], Sqrt[9 - x²]}, {x, -3, 3}];
```

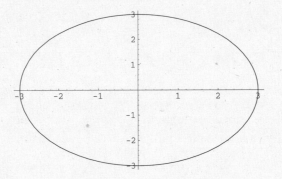

We can make the circle appear round by setting AspectRatio → Automatic.

```
Plot[{-Sqrt[9 - x²], Sqrt[9 - x²]}, {x, -3, 3}, AspectRatio→Automatic];
```

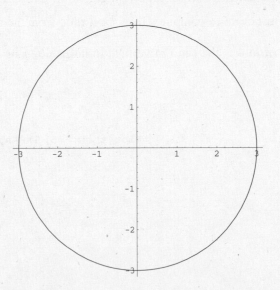

When graphing a function, *Mathematica* makes a calculated decision where to place the origin. If $(0,0)$ is within the plotting region, the axes will cross at that location. If not, an algorithm decides where the axes should cross. This can sometimes lead to a confusing (and misleading) rendering of the function. The option AxesOrigin gives control over the placement of the intersection point.

AxesOrigin is a two-dimensional graphics option which determines the point where the axes should intersect.

- **AxesOrigin → Automatic** is the default. If the point $(0,0)$ is within, or close to, the plotting region, then it is usually chosen as the axis origin.
- **AxesOrigin → {x, y}** forces the intersection of the axes to be the point (x, y).

EXAMPLE 9

```
Plot[5 + x⁴, {x, 1, 2}];
```

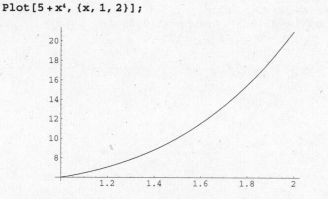

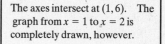

The axes intersect at $(1,6)$.   The graph from $x = 1$ to $x = 2$ is completely drawn, however.

```
Plot[5 + x⁴, {x, 1, 2}, AxesOrigin → {0, 0}];
```

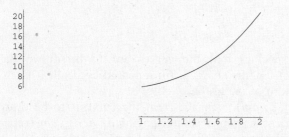

Now the axes intersect at $(0,0)$, but only the portion of the axes corresponding to the specified domain is actually drawn.   The problem is easily remedied using `PlotRange`.

```
Plot[5 + x⁴, {x, 1, 2}, AxesOrigin → {0, 0}, PlotRange → {{0, 2}, {0, 25}}];
```

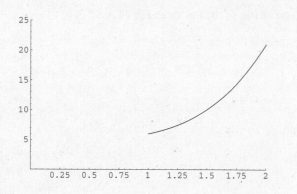

`PlotStyle` allows alteration to the appearance of a graph.   This is particularly useful when several graphs are drawn simultaneously on the same set of axes.   Its form is

- `PlotStyle → style` if only one style option is used.
- `PlotStyle → {style1, style2, ...}` if several style options are desired.   If more than one graph is to be modified, the styles are applied cyclically.

Listed below are some of the more common plot styles:

- `GrayLevel[x]` for $0 \le x \le 1$ allows lightening of the image.   The closer x is to 1 the lighter the image will appear.

EXAMPLE 10

```
Plot[{Sin[x], Sin[2x], Sin[3x]}, {x, -π, π},
   PlotStyle → {GrayLevel[0.0], GrayLevel[0.5], GrayLevel[0.8]}];
```

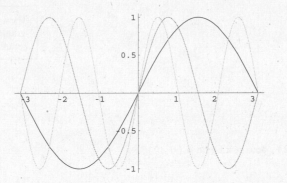

- **Dashing[{$r1$, $r2$, ..., rm}]** specifies that the curves are to be drawn dashed with successive segments and spaces of lengths $r1, r2, ..., rm$ repeated cyclically. Each r value is given as a fraction of the total width of the graph.
- **AbsoluteDashing[{$a1$, $a2$, ..., am}]** specifies that the curve is to be drawn dashed, with successive segments and spaces having absolute lengths $a1, a2, ..., am$ (repeated cyclically). The absolute lengths are measured in printer's points, approximately 1/72 of an inch.

EXAMPLE 11

```
Plot[{x², 2x², 3x²}, {x, -3, 3},
   PlotStyle → {Dashing[{.01}], Dashing[{.03}], Dashing[{.03, .1}]}];
```

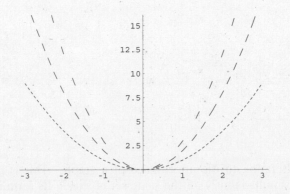

- **Thickness[r]** specifies that the graph is to be drawn with a thickness r. The thickness r is given as a fraction of the total width of the graph. The default value for two-dimensional graphs is 0.004.
- **AbsoluteThickness[d]** specifies that the graph is to be drawn with absolute thickness d. The absolute thickness is measured in printer's points, approximately equal to 1/72 of an inch.

EXAMPLE 12

```
Plot[{x², 2x², 3x²}, {x, -3, 3}, PlotStyle →
        {Thickness[.005], Thickness[.01], Thickness[.02]}];
```

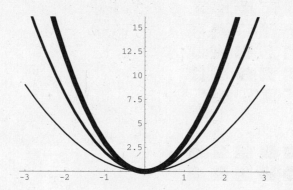

There are several graphics options which allow graphs to be drawn in color.

- **Hue[*hue*]** is a color specification. As *hue* varies from 0 to 1, the corresponding color runs through red, yellow, green, cyan, blue, magenta, and back to red again.
- **Hue[*hue, saturation, brightness*]** specifies colors in terms of hue, saturation, and brightness levels. The values of *saturation* and *brightness* must be between 0 and 1.
- **RGBColor[*red, green, blue*]** specifies the mixture of red, green, and blue to produce a certain color. The values of *red*, *green*, and *blue* must be between 0 and 1. RGBColor[1, 0, 0] produces a pure red display, RGBColor[0, 1, 0] produces green, and RGBColor[0, 0, 1] produces blue.
- **CMYKColor[*cyan, magenta, yellow, black*]** specifies the mixture of cyan, magenta, yellow, and black to produce a certain color. The values of *cyan*, *magenta*, *yellow*, and *black* must be between 0 and 1. CMYKColor is used for printing colored graphs on paper.

EXAMPLE 13

```
Plot[{x², 2x², 3x²}, {x, -3, 3},
 PlotStyle → {RGBColor[1, 0, 0], RGBColor[0, 1, 0],
        RGBColor[0, 0, 1]}];
```

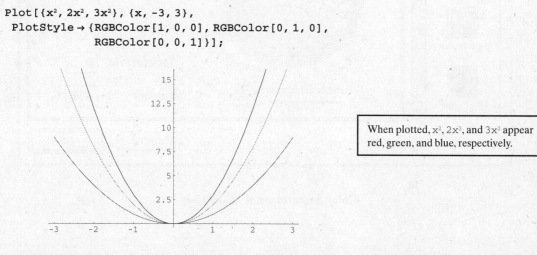

> When plotted, x^2, $2x^2$, and $3x^2$ appear red, green, and blue, respectively.

Mathematica makes it easy to compute the RGB "formula" for custom colors. Simply go to Input⇒Color Selector and click on the color of your choice.   The exact RGB combination for the color selected will be placed into your *Mathematica* notebook at the cursor position.

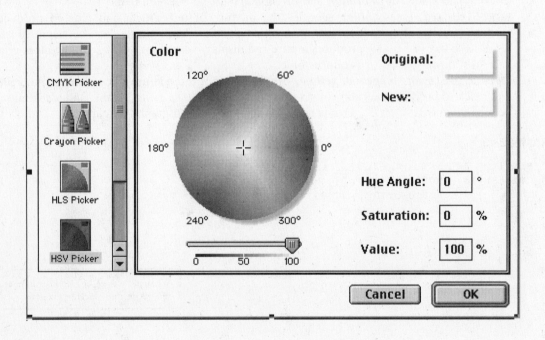

Color Selector on a PC

Color Selector on a Macintosh

The package **Graphics`Colors`** contains a list of predefined colors. Type **AllColors** after loading the package to see the entire list by name. Once a color is selected, it may be used by name in place of the RGBColor specification. To see the RGB formula, just type the name of the color.

EXAMPLE 14

```
<<Graphics`Colors`
AliceBlue
RGBColor[0.941206, 0.972503, 1.]
```

There are two graphics options which can be used to label graphs. **PlotLabel** specifies an overall label for the graph. **AxesLabel** allows one or both axes to be labeled with an appropriate description.

- **PlotLabel → *description*** labels the graph with a title.
- **AxesLabel → *label*** specifies a label for the *y* axis only.
- **AxesLabel → {*x-label*, *y-label*}** specifies labels for both the *x* and *y* axes.
- **AxesLabel → None** specifies that neither axis should be labeled. This is *Mathematica*'s default.

description, *label*, *x-label* and *y-label* may be any *Mathematica* symbol or string (text enclosed within double quotes). Care should be taken when using both PlotLabel and AxesLabel to ensure that the labels do not overlap.

EXAMPLE 15

```
Plot[Sin[x], {x, 0, 2π},
    AxesLabel → {"Values of x", "Values of sin x"}];
```

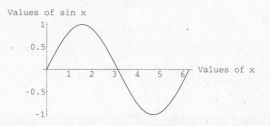

PlotLegend is a useful option that can be used to label the graphs. However, it is contained within the package **Graphics`Legend`**, which must be loaded prior to its use.

- **PlotLegend[{*text1*, *text2*, ...}]** attaches *text1*, *text2*, ... to each description specified in PlotStyle. If there are more PlotStyle descriptions than text descriptions, the text descriptions are repeated in a cyclic manner.

EXAMPLE 16

```
<<Graphics`Legend`
Plot[{x², 2x², 3x²}, {x, -3, 3}, PlotStyle → {Dashing[{.01}],
  Dashing[{.03}], Dashing[{.03, .08}]}, PlotLegend → {x², 2x², 3x²}];
```

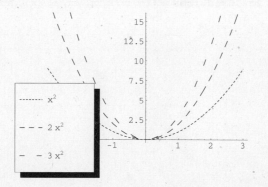

If desired, graphs can be enclosed within a rectangular frame.   Additionally, one or both axes of a graph can be suppressed.

Frame specifies whether a frame should be drawn around the graph.

- **Frame → True** specifies that a rectangular frame is to be drawn around the graph.
- **Frame → False** specifies that no frame is to be drawn (default).

Axes specifies whether the axes should be drawn.

- **Axes → True** specifies that both axes will be drawn (default).
- **Axes → False** draws no axes.
- **Axes → {False, True}** draws a y axis but no x axis.
- **Axes → {True, False}** draws an x axis but no y axis.

EXAMPLE 17

Plot[$\dfrac{1}{x^2+1}$, {x, -3, 3}, Frame → True, Axes → False];

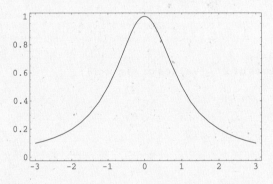

Gridlines specifies that a rectangular grid be drawn in the graph.

- **GridLines → None** specifies that no gridlines are to be drawn (default).
- **GridLines → Automatic** specifies that gridline positions are to be chosen by *Mathematica*.
- **GridLines → {*xlist*, *ylist*}** specifies that gridline positions are to be drawn at the specified locations.   *xlist* and *ylist* are lists of numbers or may (individually) be specified as Automatic, in which case *Mathematica* will choose their location.

EXAMPLE 18

When plotting trigonometric graphs, it is convenient to have vertical grid lines placed at multiples of $\pi/2$.

Plot[Sin[x], {x, 0, 2π}, GridLines → {{0, π/2, π, 3π/2, 2π}, Automatic}];

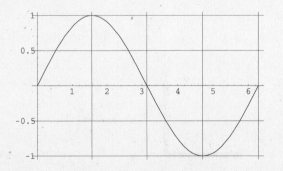

Tick marks and corresponding labeling along the axes can be controlled with the option **Ticks**. **FrameTicks** offers similar options along the edges of a frame when Frame → True is set.

- **Ticks → None** specifies that no tick marks are to be drawn. The numerical labeling of the axes is suppressed.
- **Ticks → Automatic** specifies that tick marks will be drawn. This is the default.
- **Ticks → {*xlist, ylist*}** specifies that tick marks will be drawn at the specified locations. *xlist* and *ylist* are lists of numbers or may be specified as Automatic.

EXAMPLE 19

Here are three ways to plot the graph $y = \dfrac{x^2}{x^2+1}$.

```
Plot [ x²/(x²+1) ,{x, -3, 3}];
```

```
Plot [ x²/(x²+1) ,{x, -3, 3}, Ticks → None];
```

```
Plot [ x²/(x²+1) ,{x, -3, 3}, Ticks → {{-3, 3}, Automatic}];
```

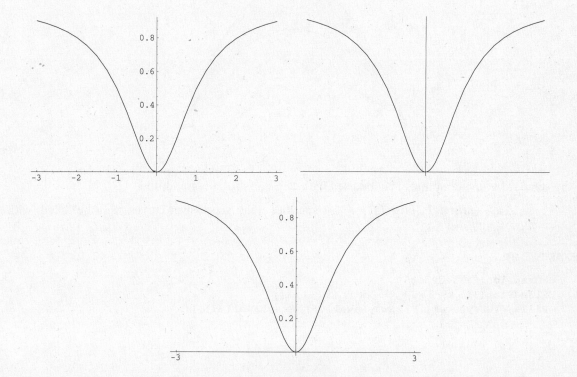

The command **FilledPlot** will plot a shaded graph. The command comes in the package **Graphics`FilledPlot`**, which must be loaded prior to its use.

- **FilledPlot[f[x], {x, xmin, xmax}]** plots f[x] and shades the region bounded by the function and the x axis.
- **FilledPlot[{f1[x], f2[x], ...}, {x, xmin, xmax}]** plots f1[x], f2[x],... and shades the regions bounded by successive graphs in different colors.

EXAMPLE 20

```
<<Graphics`FilledPlot`
FilledPlot[1 - x², {x, -1, 1}];
FilledPlot[{1 - x², 2 - 2x²}, {x, -1, 1}];
FilledPlot[{1 - x², 2 - 2x², 3 - 3x²}, {x, -1, 1}];
```

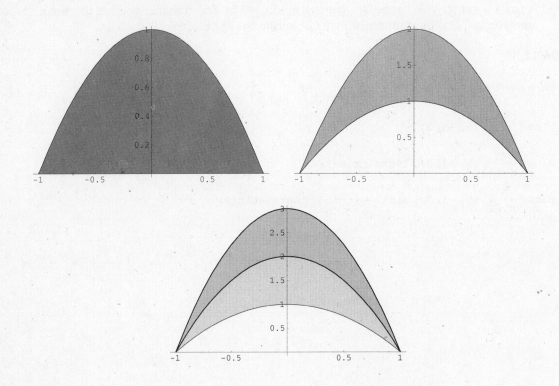

Any standard graphics options may be used with `FilledPlot`. In addition,

- **Fills** → {*fillstyle1*, *fillstyle2*, ...} specifies that successive curves will be filled with *fillstyle1*, *fillstyle2*, ...

EXAMPLE 21

```
<<Graphics`FilledPlot`
FilledPlot[{1 - x², 2 - 2x², 3 - 3x²}, {x, -1, 1},
 Fills → {GrayLevel[.8], GrayLevel[.6], GrayLevel[.3]}];
```

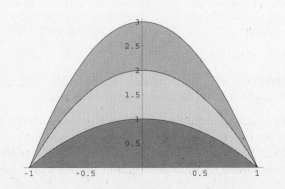

SOLVED PROBLEMS

4.1   Plot the graph of $y = xe^{-x}$ from $x = 0$ to $x = 5$.

SOLUTION

```
Plot[x Exp[-x], {x, 0, 5}];
```

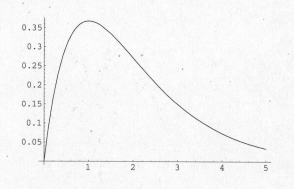

4.2   Plot $f(x) = |1 - |x||$ on the interval $[-3, 3]$.

SOLUTION

```
Plot[Abs[1 - Abs[x]], {x, -3, 3}];
```

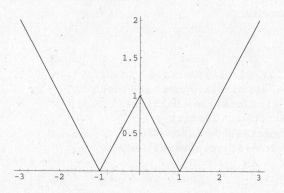

4.3   The standard normal curve used in probability and statistics is the graph of the function

$$f(x) = \frac{1}{\sqrt{2\pi}} e^{-\frac{1}{2}x^2}$$

Sketch the graph for $-3 \le x \le 3$.

SOLUTION

```
f[x_] = 1/(Sqrt[2*Pi])*Exp[(-1/2)x^2];
```

or $f[x\_] = \dfrac{1}{\sqrt{2\pi}} \, Exp[-\dfrac{1}{2} \, x^2];$

```
Plot[f[x], {x, -3, 3}];
```

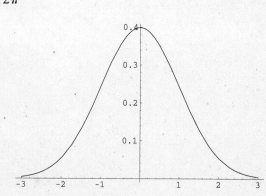

4.4   Plot the graphs $y = \sin x$, $y = 2 \sin x$, and $y = 3 \sin x$ from -2π to 2π on the same set of axes.

SOLUTION

```
Plot[{Sin[x], 2Sin[x], 3Sin[x]}, {x, -2π, 2π}];
```

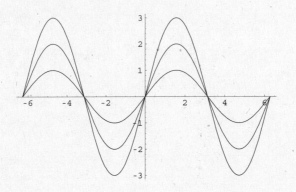

4.5   The graphs of inverse functions are symmetric with respect to the line $y = x$.   Plot the inverse functions $f(x) = x^2$ and $f^{-1}(x) = \sqrt{x}$, $0 \le x \le 4$, as solid curves and the line $y = x$ as a dotted line and observe the symmetry.

SOLUTION

```
g1 = Plot[x², {x, 0, 2}, DisplayFunction → Identity];
g2 = Plot[√x, {x, 0, 4}, DisplayFunction → Identity];
g3 = Plot[x, {x, 0, 4}, PlotStyle → Dashing[{0.01}],
        DisplayFunction → Identity];
Show[g1, g2, g3, AspectRatio → Automatic,
    DisplayFunction → $DisplayFunction];
```

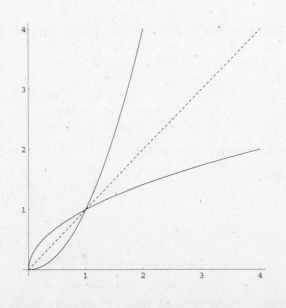

4.6   Sketch the graphs of $y = x^2$, $y = -x^2$, and $y = x^2 \sin 10x$, $-2\pi \leq x \leq 2\pi$, on a single set of axes enclosed by a frame.

SOLUTION

```
Plot[{x², -x², x²Sin[10x]}, {x, -2π, 2π}, Frame → True];
```

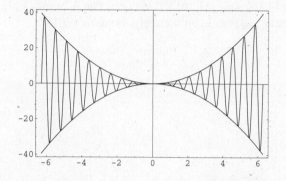

4.7   The family of Chebyshev polynomials is used in approximation theory and numerical analysis. *Mathematica* represents these polynomials as `ChebyshevT[n, x]`. On a single set of axes, using some device to distinguish the curves, plot a labeled graph showing the Chebyshev polynomials of degree 2, 3, and 4.

SOLUTION

```
<<Graphics`Legend`
Plot[{ChebyshevT[2, x], ChebyshevT[3, x], ChebyshevT[4, x]},
     {x, -2, 2}, PlotStyle → {GrayLevel[0], GrayLevel[.4], GrayLevel[.7]},
     PlotLegend → {T2, T3, T4}];
```

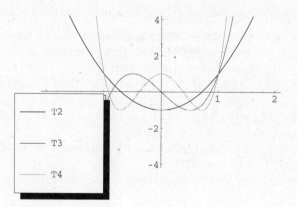

ALTERNATE SOLUTION

```
<<Graphics`Legend`
Plot[{ChebyshevT[2, x], ChebyshevT[3, x], ChebyshevT[4, x]},
     {x, -2, 2}, PlotStyle →
          {RGBColor[1, 0, 0], RGBColor[0, 1, 0], RGBColor[0, 0, 1]},
     PlotLegend → {T2, T3, T4}];
```

(color graph not shown)

4.8   Sketch the graphs of $y = 1 + \sin x$, $0 \le x \le 2\pi$, $y = 2 + \sin x$, $2\pi \le x \le 4\pi$, and $y = 3 + \sin x$, $4\pi \le x \le 6\pi$, on one set of axes.

SOLUTION

```
g1 = Plot[1 + Sin[x], {x, 0, 2π}, DisplayFunction → Identity];
g2 = Plot[2 + Sin[x], {x, 2π, 4π}, DisplayFunction → Identity];
g3 = Plot[3 + Sin[x], {x, 4π, 6π}, DisplayFunction → Identity];
Show[g1, g2, g3, DisplayFunction → $DisplayFunction];
```

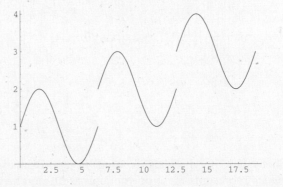

4.2   ADDITIONAL GRAPHICS COMMANDS

Standard geometric shapes can be constructed with the **Graphics** command and viewed with the **Show** command.

- **Graphics[*primitive*]** constructs a graphics object using a graphics primitive.

The following are a few of the more common graphics primitives available in *Mathematica*:

- **Circle[{x, y}, r]** creates a circle centered at (x, y) having radius r.
- **Disk[{x, y}, r]** creates a disk (filled circle) centered at (x, y) having radius r.
- **Point[{x, y}]** plots a point at coordinate (x, y).
- **Line[{{x1, y1}, {x2, y2}, ...}]** draws lines connecting points $(x1, y1)$, $(x2, y2), \ldots$.
- **Rectangle[{x1, y1}, {x2, y2}]** creates a filled rectangle having $(x1, y1)$ and $(x2, y2)$ as opposite ends of a diagonal.
- **Polygon[{{x1, y1}, {x2, y2}, ...}]** constructs a filled polygon having points $(x1, y1)$, $(x2, y2), \ldots$ as vertices.
- **Text[*textstring*, {x, y}]** prints a string of text centered at position (x, y).

When viewing graphics objects using Show, the default for Axes is Axes → False. If desired, Axes → True may be included as an option.

EXAMPLE 22

```
g1 = Graphics[Circle[{0, 0}, 1]];
g2 = Graphics[Line[{{-1, -1}, {-1, 1}, {1, 1}, {1, -1}, {-1, -1}}]];
g3 = Graphics[Polygon[{{-1, 0}, {0, 1}, {1, 0}, {0, -1}}]];
Show[g1, g2, g3, AspectRatio → Automatic];
```

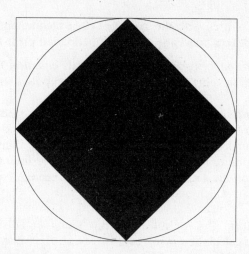

Curves are sometimes defined parametrically; i.e., the *x* and *y* coordinates of points are defined as two independent functions of a third variable. Parametric curves, which are usually more complex in their behavior, can be viewed using `ParametricPlot`.

- `ParametricPlot[{x[t],y[t]}, {t, tmin, tmax}]` plots the parametric equations $x = x(t)$, $y = y(t)$ over the interval `tmin ≤ t ≤ tmax`.
- `ParametricPlot[{{x1[t],y1[t]}, {x2[t],y2[t]}, ...}, {t, tmin, tmax}]` plots several sets of parametric equations over `tmin ≤ t ≤ tmax`

EXAMPLE 23

```
ParametricPlot[{t³ - 2t, t² - t}, {t, -2, 2}];
```

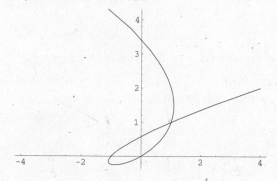

EXAMPLE 24

```
x[t_] = Cos[t] - Cos[100t] Sin[t];
y[t_] = 2 Sin[t] - Sin[100t];
ParametricPlot[{x[t], y[t]}, {t, 0, 2π}, PlotPoints → 50];
```

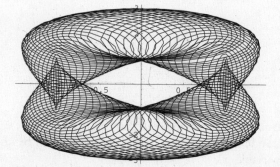

Implicit functions can be plotted with the **ImplicitPlot** command.   However, it is contained in the **Graphics`ImplicitPlot`** package and must be loaded prior to its use.

- **ImplicitPlot[*equation*, {x, xmin, xmax}]** plots *equation* for values of x ranging from xmin to xmax.   For each value of x, values of y are computed and the set of generated points (x, y) is plotted.
- **ImplicitPlot[*equation*, {x, xmin, xmax}, {y, ymin, ymax}]** plots *equation* by treating the equation as a function in three-dimensional space, and generates a contour of the equation cutting through the plane where z equals zero.   This method is faster but may generate rougher graphs, especially around singularities or intersections of the curve.
- **ImplicitPlot[*equation*, {x, xmin, x1, x2, ..., xmax}]** plots the equation avoiding the points x1, x2, ... which may cause difficulty.

equation must be of the form lhs==rhs.   Note the double equal sign in the middle.

Because of the amount of calculation involved, ImplicitPlot may take some time to complete a graph.   The second form of the command (using contour methods) is somewhat faster.

EXAMPLE 25

Plot the equation $x^2 y^2 = (y+1)^2(4-y^2)$ for $-10 \le x \le 10$.   (Conchoid of Nicomedes.)

```
<<Graphics`ImplicitPlot`
ImplicitPlot[x² y² = = (y + 1)² (4 - y²), {x, -10, 10}, PlotRange → All];
```

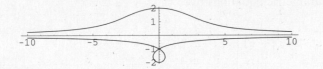

EXAMPLE 26

Plot the equation $x^3 + y^3 = 6xy$ for $-4 \le x \le 4$.   (Folium of Descartes.)

```
<<Graphics`ImplicitPlot`
ImplicitPlot[x³ + y³ = = 6 x y, {x, -4, 4}];
```

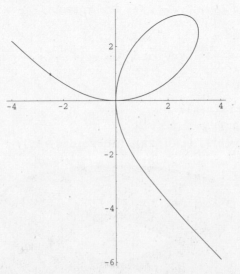

In the next example we must use the contour method since ImplicitPlot cannot solve equations involving transcendental functions.   This is accomplished by specifying an interval for y as well as the independent variable, x.

EXAMPLE 27

Plot $\cos(x - y) = y \sin x$, $-2 \le x \le 2$.

```
<<Graphics`ImplicitPlot`
ImplicitPlot[Cos[x-y]==y Sin[x], {x, -2, 2}, AxesOrigin→{0, 0}];
```

Solve::tdep : The equations appear to involve transcendental
　　functions of the variables in an essentially non-algebraic way.
ImplicitPlot::epfail:
　　Cos[x-y]==y Sin[x] could not be solved for points to plot.

```
ImplicitPlot[Cos[x-y]==y Sin[x], {x, -2, 2}, {y, -2, 2}, AxesOrigin→{0, 0}];
```

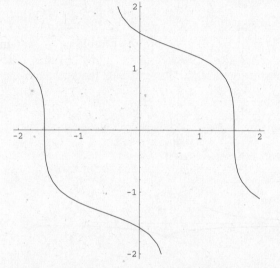

For curves defined in polar coordinates, the command **PolarPlot**, available in the package Graphics`Graphics` is available.

- **PolarPlot[f[θ], {θ, θmin, θmax}]** generates a plot of the polar equation $r = f(\theta)$ as θ varies from θmin to θmax.
- **PolarPlot[{f1[θ], f2[θ], ...}, {θ, θmin, θmax}]** plots several polar graphs on one set of axes.

Note: The default aspect ratio for PolarPlot is AspectRatio→Automatic.

EXAMPLE 28

```
<<Graphics`Graphics`
PolarPlot[3 (1-Cos[θ]), {θ, 0, 2π}];
```

(This curve is called a cardioid.)

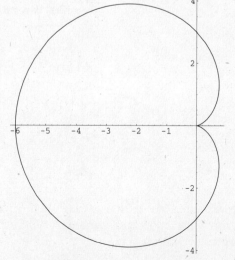

EXAMPLE 29

Plot the three-leaf rose $r = \sin 3\theta$ inside the unit circle $r = 1$.

```
<<Graphics`Graphics`
PolarPlot[{1, Sin[3θ]}, {θ, 0, 2π}];
```

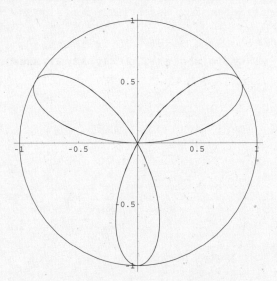

SOLVED PROBLEMS

4.9   Sketch the parabola $y = x^2 - 9$ and a circle of radius 3 centered at the origin.

SOLUTION

```
g1 = Plot[x² - 9, {x, -4, 4}, DisplayFunction → Identity];
g2 = Graphics[Circle[{0, 0}, 3]];
g3 = Graphics[Text["CIRCLE IN A PARABOLA", {0, 6}]];
Show[g1, g2, g3, AspectRatio → Automatic, DisplayFunction → $DisplayFunction];
```

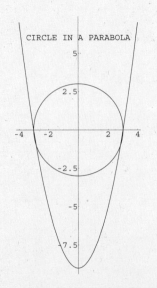

4.10 The curve traced by a point on a circle as the circle rolls along a straight line is called a *cycloid* and has parametric equations $x = r(\theta - \sin\theta)$, $y = r(1 - \cos\theta)$ where r represents the radius of the circle.   Plot the cycloid formed as a circle of radius 1 makes four complete revolutions.

SOLUTION

```
ParametricPlot[{θ - Sin[θ], 1 - Cos[θ]}, {θ, 0, 8π}, AspectRatio→Automatic];
```

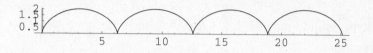

4.11 Let P be a point at a distance a from the center of a circle of radius r.   (Imagine the point being placed on a spoke of a bicycle wheel.) The curved traced by P as the circle rolls along a straight line is called a *trochoid*.   Its parametric equations are $x = r\theta - a\sin\theta$, $y = r - a\cos\theta$.   Sketch the trochoid with $r = 1$, $a = 1/2$ as the circle makes four revolutions.   What would the graph look like if $r = 1$, $a = 2$ so that the point is outside the circle?

SOLUTION

```
r = 1;
a = 1/2;
ParametricPlot[{r θ - a Sin[θ], r - a Cos[θ]}, {θ, 0, 8π},
    AxesOrigin → {0, 0}, PlotRange → {{0, 8π}, {0, 1.5}}];
```

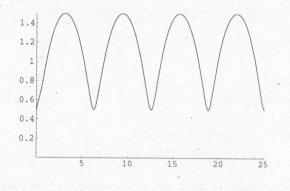

```
a = 2;
ParametricPlot[{r θ - a Sin[θ], r - a Cos[θ]}, {θ, 0, 8π},
    AxesOrigin → {0, 0}];
```

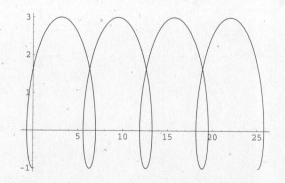

4.12 A circle of radius b rolls on the inside of a larger circle of radius a. The curve traced out by a fixed point initially at $(a, 0)$ is called a hypocycloid and has equations

$$x = (a - b)\cos\theta + b\cos\left(\frac{a - b}{b}\,\theta\right)$$

$$y = (a - b)\sin\theta - b\sin\left(\frac{a - b}{b}\,\theta\right)$$

Sketch the hypocycloid for $a = 1$, $b = 4$ and then again for $a = 8$, $b = 5$.

SOLUTION

```
x[θ_]:=(a-b)Cos[θ]+bCos[(a-b)θ/b]
y[θ_]:=(a-b)Sin[θ]-bSin[(a-b)θ/b]
a=4;
b=1;
ParametricPlot[{x[θ],y[θ]},{θ,0,2π},AspectRatio→Automatic];
```

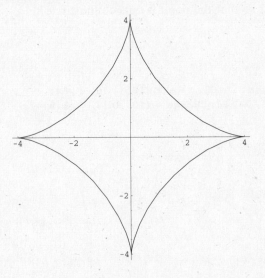

```
a=8;
b=5;
ParametricPlot[{x[θ],y[θ]},{θ,0,10π},AspectRatio→Automatic];
```

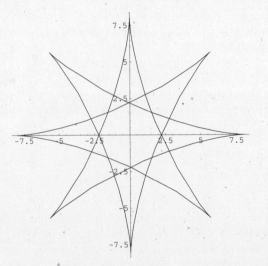

4.13   Sketch the graph defined by the equation $y^2 = x^3(2 - x)$, $0 \leq x \leq 2$.

SOLUTION

```
<<Graphics`ImplicitPlot`
ImplicitPlot[y² == x³(2 - x), {x, 0, 2}];
```

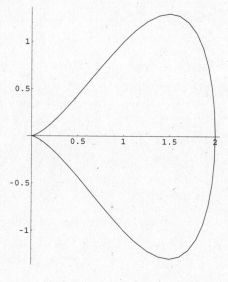

4.14   Sketch the graph of Tschirnhauser's cubic: $y^2 = x^3 + 3x^2$, $-3 \leq x \leq 3$.

SOLUTION

```
<<Graphics`ImplicitPlot`
ImplicitPlot[y² == x³ + 3x², {x, -3, 3}];
```

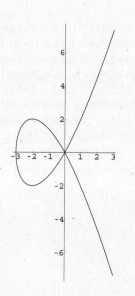

4.15   The polar graph $r = \theta$ is called the spiral of Archimedes.   Sketch the graph for $0 \leq \theta \leq 10\pi$ and then for $-10\pi \leq \theta \leq 10\pi$.

SOLUTION

```
<<Graphics`Graphics`
PolarPlot[θ, {θ, 0, 10π}];
PolarPlot[θ, {θ, -10π, 10π}];
```

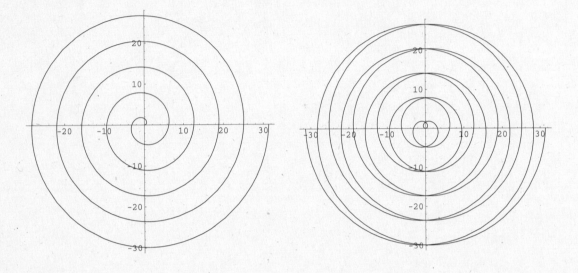

4.16   The equations $r = \sin n\theta$, where n is a positive integer, represent a family of polar curves called roses.   Investigate the behavior of this family and form a conjecture about how the number of loops is related to n.

SOLUTION

```
<<Graphics`Graphics`
g1 = PolarPlot[Sin[2θ], {θ, 0, 2π}, Ticks → False,
     PlotLabel → "n = 2", DisplayFunction → Identity];
g2 = PolarPlot[Sin[3θ], {θ, 0, 2π}, Ticks → False,
     PlotLabel → "n = 3", DisplayFunction → Identity];
g3 = PolarPlot[Sin[4θ], {θ, 0, 2π}, Ticks → False,
     PlotLabel → "n = 4", DisplayFunction → Identity];
g4 = PolarPlot[Sin[5θ], {θ, 0, 2π}, Ticks → False,
     PlotLabel → "n = 5", DisplayFunction → Identity];
g = GraphicsArray[{{g1, g2}, {g3, g4}}];
Show[g, DisplayFunction → $DisplayFunction];
```

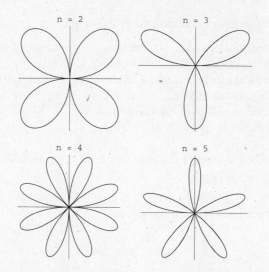

Conclusion: If n is odd, the rose will have n leaves.   If n is even, there will be $2n$ leaves.

4.17   Sketch the cardioid $r = 1 - \cos \theta$ and the circle $r = 1$ on the same set of axes.

SOLUTION

```
<<Graphics`Graphics`
PolarPlot[{1-Cos[θ],1}, {θ, 0, 2π}];
```

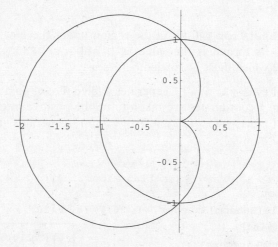

4.3   SPECIAL TWO-DIMENSIONAL PLOTS

Discrete functions, i.e., functions defined on a discrete set, can be visualized using the special plotting function `ListPlot`.

- `ListPlot[{y1, y2, ...}]` plots points whose y coordinates are `y1, y2, ...`. The x coordinates are taken to be the positive integers, $1, 2, ...$
- `ListPlot[{{x1, y1}, {x2, y2}, ..., }]` plots the points $(x1, y1)$, $(x2, y2)$, ...

Standard graphics options are permitted.   The form of the command would then be

- `ListPlot[{y1, y2, ...}, options]`
- `ListPlot[{{x1, y1}, {x2, y2}, ..., }, options]`

The most useful graphics options used with **ListPlot** are

- **PlotStyle → PointSize[d]**.   d specifies the diameter of the point as a fraction of the overall width of the graph.   The default value is 0.008.
- **PlotStyle → AbsolutePointSize[d]**.   Here d represents printer's points, which are approximately 1/72 of an inch.
- **PlotJoined → True** will connect the points.

EXAMPLE 30

The following plots a list of the squares of the positive integers 1 through 20.

```
squares = Table[k², {k, 1, 20}];
ListPlot[squares];
ListPlot[squares, PlotStyle → PointSize[.02]];
```

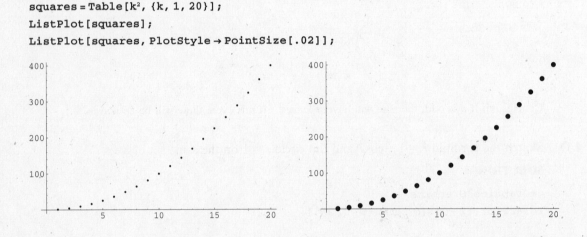

As an alternative, **FilledListPlot** will fill in the region under the plot, creating a filled polygon which describes the data.   Its format is the same as **ListPlot**.   FilledListPlot is found in **Graphics`FilledPlot`**, which must be loaded before use.

- **FilledListPlot[{y1, y2, ...}]** generates a filled plot of points whose y coordinates are y1, y2, ...   The x coordinates are taken to be the positive integers, 1, 2, ...
- **FilledListPlot[{{x1, y1}, {x2, y2}, ..., }]** generates a filled plot of points (x1, y1), (x2, y2), ...

EXAMPLE 31

```
<<Graphics`FilledPlot`
randomintegers = Table[Random[Integer, {1, 20}], {k, 1, 30}];
ListPlot[randomintegers];
FilledListPlot[randomintegers];
```

The **Table** command generates a list of 30 random integers, each between 1 and 20.

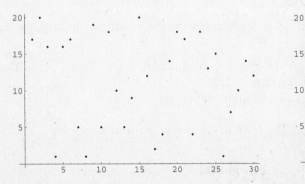

Several lists can be plotted simultaneously using **MultipleListPlot**. Its syntax is similar to `ListPlot` except that several lists may be specified. `MultipleListPlot` is found in **Graphics`MultipleListPlot`**, which must be loaded before use.

- **MultipleListPlot[*list1*, *list2*, ...]**
- **MultipleListPlot[*list1*, *list2*, ..., *options*]**

The list may be of the form `{{x1, y1}, {x2, y2}, ... }` or `{y1, y2, ...}`, in which case the *x* coordinates are taken to be the positive integers 1, 2, ...

Each of the lists is plotted using a different plotting symbol. The option **PlotJoined** controls whether to connect the points.

- **PlotJoined → True** causes the points to be connected with distinguishing lines.

EXAMPLE 32

```
<<Graphics`MultipleListPlot`
lst1 = {{0, -4}, {1, 0}, {1, 3}, {5, 2}, {7, 7}};
lst2 = {{0, -1}, {2, -1}, {3, -5}, {4, 0}, {7, 5}};
lst3 = {{0, 0}, {3, 6}, {5, -4}, {6, 1}, {7, 1}};
MultipleListPlot[lst1, lst2, lst3];
MultipleListPlot[lst1, lst2, lst3, PlotJoined → True];
```

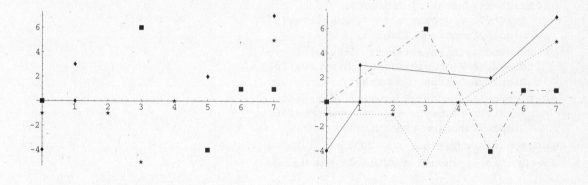

Different types of bar graphs can be drawn with *Mathematica*, using commands which are located in the package `Graphics`Graphics`.

- **BarChart[*datalist1*, *datalist2*, ..., *options*]** draws a simple bar graph. If more than one data list is specified, the bars are grouped side by side according to their position within the list. Each list is of the form `{y1, y2, ...}` or `{{x1, y1}, {x2, y2}, ...}` and the lists may contain different quantities of numbers.
- **StackedBarChart[*datalist1*, *datalist2*, ..., *options*]** is used when more than one data list is present. The bars of the graph are stacked rather than placed side by side.
- **PercentileBarChart[*datalist1*, *datalist2*, ..., *options*]** generates a stacked bar chart except that the total height of each bar is constant at 100%.
- **GeneralizedBarChart[*datalist1*, *datalist2*, ..., *options*]** generates a bar chart of the data lists, where each data list specifies position, height, and width of the bars.

Some of the more useful options available for use with `BarChart` are

- **BarOrientation → Horizontal** causes the bars to be placed horizontally rather than vertically. **BarOrientation → Vertical** is the default.

- **BarEdges → False** eliminates the outlines surrounding each bar.
- **BarValues → True** labels each bar with its numerical value.
- **BarLabels → *stringlist*** labels each bar according to the position of the string in *stringlist*.
- **TextStyle** allows you to change the default font and size used in the graph's text. **TextStyle → {FontFamily → *fontname*, FontSize → *size*}** is a simple, but useful, application.
- **BarSpacing → *value*** specifies how much space to place between each bar, as a fraction of the width of one bar.
- **BarGroupSpacing → *value*** specifies how much space to place between each group of bars, as a fraction of the width of one bar.
- **BarStyle** specifies the style (**RGBColor**, **Hue**, **GrayLevel**, etc.) of each bar.
- **BarEdgeStyle** specifies the style of the bar edges.

EXAMPLE 33

```
<<Graphics`Graphics`
dataset1 = {5, 1, 2, 8, 5};
dataset2 = {1, 6, 3, 5, 2};
g1 = BarChart[dataset1,
    BarStyle → {GrayLevel[.7]},
    DisplayFunction → Identity];
g2 = BarChart[dataset1, dataset2,
    BarStyle → {GrayLevel[.7], GrayLevel[0]},
    DisplayFunction → Identity];
g3 = StackedBarChart[dataset1, dataset2,
    BarStyle → {GrayLevel[.7], GrayLevel[0]},
    DisplayFunction → Identity];
g4 = PercentileBarChart[dataset1, dataset2,
    BarStyle → {GrayLevel[.7], GrayLevel[0]},
    DisplayFunction → Identity];
g = GraphicsArray[{{g1, g2}, {g3, g4}}];
Show[g, DisplayFunction → $DisplayFunction];
```

> The default **BarStyle** is red and blue bars. **GrayLevel** is used for presentation in this book in this and subsequent examples.

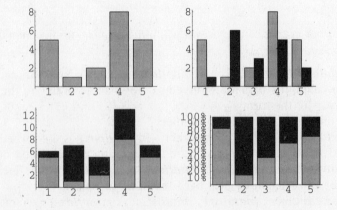

EXAMPLE 34

```
<<Graphics`Graphics`
dataset1 = {5, 1, 2, 8, 5};
dataset2 = {1, 6, 3, 5, 2};
```

```
g1 = BarChart[dataset1, dataset2,
    BarLabels → {"Mon.", "Tues.", "Wed.", "Thurs.", "Fri."},
    BarSpacing → .1, BarGroupSpacing → .5,
    BarStyle → {GrayLevel[.7], GrayLevel[0]},
    DisplayFunction → Identity];
g2 = BarChart[dataset1, dataset2,
    BarLabels → {"Mon.", "Tues.", "Wed.", "Thurs.", "Fri."},
    BarSpacing → .1, BarGroupSpacing → .5,
    BarOrientation → Horizontal,
    BarStyle → {GrayLevel[.7], GrayLevel[0]},
    DisplayFunction → Identity];
g = GraphicsArray[{g1, g2}];
Show[g, DisplayFunction → $DisplayFunction];
```

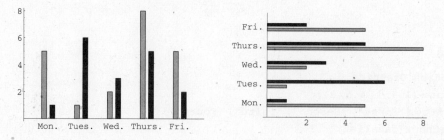

Pie charts may be constructed using the **PieChart** command, also located in **Graphics`Graphics`**.

- **Piechart[{x1, x2, ... }]** converts the list of numbers {x1, x2, ... } into a "pie" display of circular sectors whose sizes are proportional to the values x1, x2, ...
- **Piechart[{x1, x2, ... }, *options*]** constructs the pie chart allowing various options.

Some of the available options associated with PieChart are

- **PieLabels → *labellist*** where *labellist* is a list of text strings corresponding to the labeling of the wedges of the chart. The default labeling of the wedges is the consecutive integers 1,2,.... **PieLabels → None** forces no labeling.
- **PieExploded → All** yields an "exploded" pie chart. If not all of the wedges are to be exploded, **PieExploded → {*wedgenumber1, wedgenumber2, ...*}** will cause only wedges *wedgenumber1, wedgenumber2, ...* to be exploded.
- **PieStyle → *stylelist***, where *stylelist* is a list of graphic styles (RGBColor, Hue, GrayLevel, etc.), allows various styles to be applied to the wedges of the pie. If there are more wedges than styles, PieStyle applies the list cyclically.
- **PieLineStyle → *linestylelist***, where *linestylelist* is a list of line styles, allows the lines bordering individual wedges to be styled. If there are more wedges than styles, PieLineStyle applies the list cyclically.

EXAMPLE 35

```
<< Graphics`Graphics`
datalist = {1.5, 3, 4.5, 9};
PieChart[datalist, PieStyle →
    {GrayLevel[.6], GrayLevel[.7], GrayLevel[.8], GrayLevel[.9]}];
```

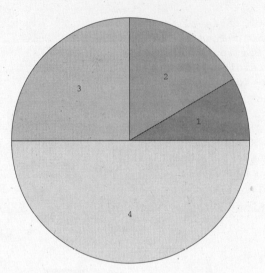

EXAMPLE 36

```
<< Graphics`Graphics`
datalist = {1.5, 3, 4.5, 9};
labellist = {"First", "Second", "Third", "Fourth"};
g1 = PieChart[datalist,
    PieLabels → labellist,
    PieExploded → All,
    PieStyle → {GrayLevel[.6], GrayLevel[.7], GrayLevel[.8], GrayLevel[.9]},
    DisplayFunction → Identity];
g2 = PieChart[datalist,
    PieLabels → labellist,
    PieExploded → {2, 3},
    PieStyle →
      {GrayLevel[.6], GrayLevel[.7], GrayLevel[.8], GrayLevel[.9]},
    DisplayFunction → Identity];
g = GraphicsArray[{g1, g2}];
Show[g, DisplayFunction → $DisplayFunction];
```

> Only the second and third wedges will be exploded.

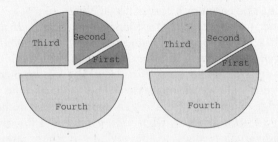

SOLVED PROBLEMS

4.18 Plot the first 50 prime numbers.

SOLUTION

```
primelist = Table[Prime[k], {k, 1, 50}];
ListPlot[primelist];
```

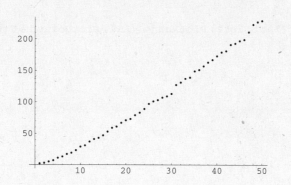

4.19 Plot the points $(0,0)$, $(2,7)$, $(3,5)$, and $(4,11)$ and connect them with straight line segments.

SOLUTION

```
lst = {{0, 0}, {2, 7}, {3, 5}, {4, 11}};
ListPlot[lst, PlotJoined → True];
```

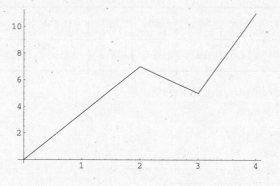

4.20 Plot the set of points corresponding to the values of $n!$ and 2^n for $n = 1, 2, 3, 4,$ and 5. First plot individual points and then connect them with line segments.

SOLUTION

```
<<Graphics`MultipleListPlot`
lst1 = Table[n!, {n, 1, 5}];
lst2 = Table[2ⁿ, {n, 1, 5}];
MultipleListPlot[lst1, lst2, PlotRange → All];
```

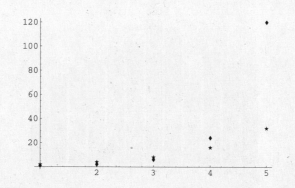

```
MultipleListPlot[lst1, lst2, PlotRange → All, PlotJoined → True];
```

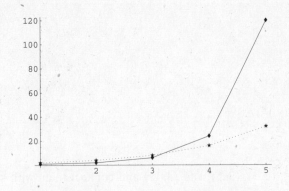

4.21 The monthly sales for XYZ Corp. (in thousands of dollars) were

Jan	Feb	Mar	Apr	May	June	July	Aug	Sept	Oct	Nov	Dec
13.2	15.7	17.4	12.6	19.7	22.6	20.2	18.3	16.2	15.0	12.1	8.6

Construct a bar graph illustrating this data.

SOLUTION

```
<<Graphics`Graphics`
months = {"Jan", "Feb", "Mar", "Apr", "May", "Jun", "Jul", "Aug",
          "Sep", "Oct", "Nov", "Dec"};
salesdata = {13.2, 15.7, 17.4, 12.6, 19.7, 22.6, 20.2, 18.3,
             16.2, 15.0, 12.1, 8.6};
BarChart[salesdata,
  BarLabels → months,
  BarValues → True,
  PlotRange → All,
  TextStyle → {FontFamily → Helvetica, FontSize → 9}];
```

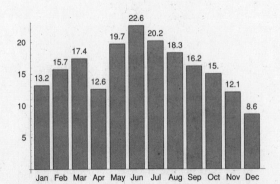

4.22   Construct a monochromatic pie chart illustrating the data in the previous problem.

SOLUTION

```
<<Graphics`Graphics`
months = {"Jan", "Feb", "Mar", "Apr", "May", "Jun", "Jul", "Aug",
          "Sep", "Oct", "Nov", "Dec"};
salesdata = {13.2, 15.7, 17.4, 12.6, 19.7, 22.6, 20.2, 18.3,
             16.2, 15.0, 12.1, 8.6};
PieChart[salesdata, PieLabels → months, PieStyle → {GrayLevel[.8]}];
```

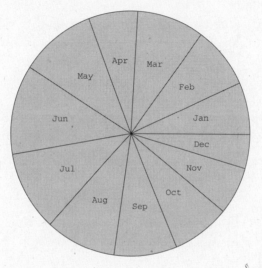

4.4   ANIMATION

Animation effects can be produced quickly and easily through the use of the option Animate Selected Graphics, which can be found in the Cell menu. This option displays several different graphics images rapidly in succession, producing the illusion of movement.

The first step in animation begins with the construction of a sequence of images, each slightly different from the other. (The more images you have and the less difference between them, the smoother your "movie" will look.) Then use your mouse to select the image cell brackets. Finally, go to Cell⇒Animate Selected Graphics and *Mathematica* does the rest.

The following example gives an interesting animated description of the behavior of the odd powers of x^n as n gets larger. (*Mathematica* places the graphs vertically under one another.)

EXAMPLE 37

```
Do[Plot[xᵏ, {x, -1, 1},
    PlotRange → {-1, 1}, Ticks → False], {k, 1, 19, 2}]
```

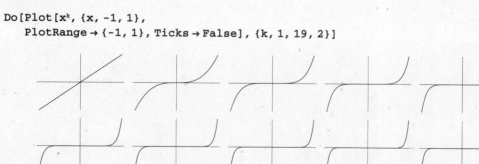

Select the cell brackets for the images, go to Cell⇒Animate Selected Graphics, and watch the animation.

SOLVED PROBLEMS

4.23   Construct an animation of the spiral of Archimedes, $r = \theta$ as θ varies from 8π to 10π.

SOLUTION

```
<<Graphics`Graphics`
Do[PolarPlot[θ, {θ, 0, 8π+φ}, Ticks→False,
    PlotRange→{{-10π, 10π}, {-10π, 10π}}], {φ, 0, 2π, π/6}]
```

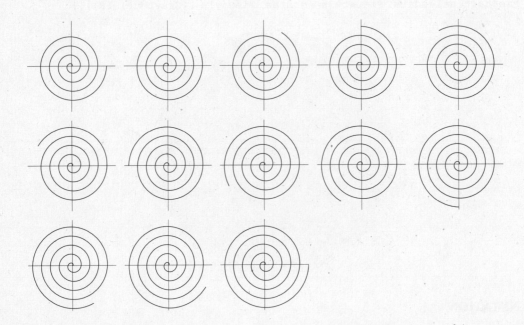

Select the cell brackets for the images, go to Cell⇒Animate Selected Graphics, and the spiral will appear to rotate.

4.24   Construct an animation of a point "rolling" along a sine curve from 0 to 2π.

SOLUTION

First we construct the sine curve.

```
sincurve = Plot[Sin[x], {x, 0, 2π}, Ticks→False];
```

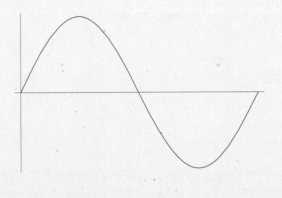

Now we construct the sequence of points as disks with radii 0.05.

```
Do[Show[sincurve, Graphics[Disk[{x, Sin[x]}, 0.05]],
        PlotRange → {{0, 2π}, {-1, 1}},
        AspectRatio → Automatic], {x, 0, 2π, π/6}]
```

This results in the following sequence of pictures:

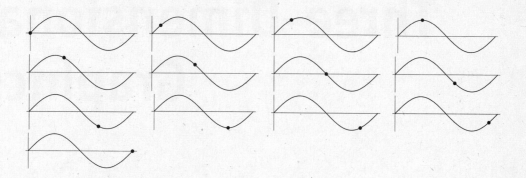

Select the cell brackets for the images, go to Cell⇒Animate Selected Graphics, and the point will appear to move along the curve.

CHAPTER 5

Three-Dimensional Graphics

5.1 PLOTTING FUNCTIONS OF TWO VARIABLES

A function of two variables may be viewed as a surface in three-dimensional space. The simplest command for plotting a surface is **Plot3D**.

- **Plot3D[f[x, y], {x, xmin, xmax}, {y, ymin, ymax}]** plots a three-dimensional graph of the function f[x, y] above the rectangle xmin ≤ x ≤ xmax, ymin ≤ y ≤ ymax.

EXAMPLE 1

```
Plot3D[Sin[x-y],{x, -π, π}, {y, -π, π}];
```

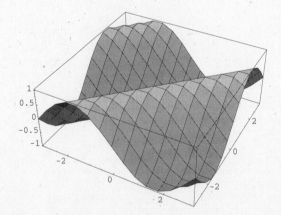

The option **PlotPoints** specifies the number of points to be used, in each direction, to produce the graph. Unlike two-dimensional graphics, the default for a three-dimensional plot is **PlotPoints→15**. This often leads to graphs with ragged surfaces. Increasing the value of PlotPoints will alleviate this condition.

- **PlotPoints→n** specifies that n sample points should be used in each direction.
- **PlotPoints→{nx, ny}** specifies that nx and ny sample points are to be used along the *x* axis and *y* axis, respectively.

The next example shows how an increase in the value of `PlotPoints` affects the "smoothness" of the resulting graph.

EXAMPLE 2

```
f[x_, y_] = x²y² Exp[-(x² + y²)];
Plot3D[f[x, y], {x, -2, 2}, {y, -2, 2}];
Plot3D[f[x, y], {x, -2, 2}, {y, -2, 2}, PlotPoints→25];
Plot3D[f[x, y], {x, -2, 2}, {y, -2, 2}, PlotPoints→40];
```

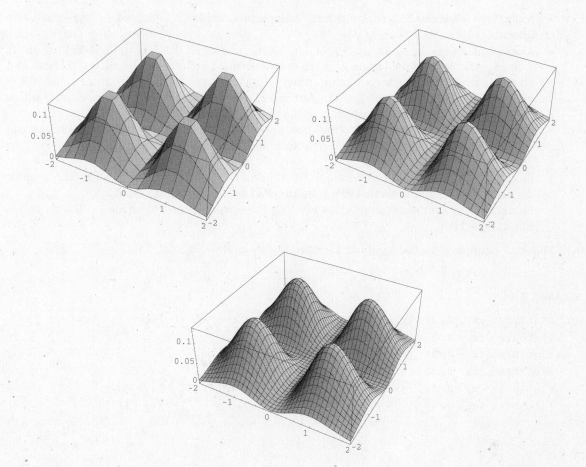

Most two-dimensional graphics options will work with **Plot3D**. There are a few extra options as well. The most popular are:

- **Axes→False** will suppress the axes from being drawn.
- **Axes→{*true_or_false, true_or_false, true_or_false*}**, where *true_or_false* is either `True` or `False`, will determine which axes will be drawn.
- **Boxed→False** will suppress the bounding box containing the graph.
- **BoxRatios→{sx, sy, sz}** specifies the ratios of side lengths for the bounding box of the three dimensional picture.
- **Ticks→False** will eliminate tick marks and corresponding labeling along all axes.
  Ticks→{*true_or_false, true_or_false, true_or_false*}, where *true_or_false* is either `True` or `False`, will control ticks on individual axes.

`FaceGrids` is an option that draws grid lines on the faces of the bounding box.

- `FaceGrids`→`All` draws grid lines on all six faces of the bounding box.
- `FaceGrids`→`None` (default) draws no grid lines.
- `FaceGrids`→`{{x1, y1, z1 }, {x2, y2, z2 }, . . . , {x6, y6, z6 }}` allows grids to be drawn on individual faces. Two of the three numbers in each sublist must be 0 and the third, ± 1, to specify one of the six possible faces.

`AxesEdge` is an option that specifies on which edges of the bounding box axes should be drawn.

- `AxesEdge`→`Automatic` (default) lets *Mathematica* decide on which edges axes should be drawn.
- `AxesEdge`→`{{$y_1$, $z_1$ }, {$x_2$, $z_2$ }, {$x_3$, $y_3$ }}` where each of the x, y, and z values are either 1 or -1, indicates on which edges axes are to be drawn. 1 indicates that the axes will be drawn on the edge with the larger coordinate value; -1 indicates the smaller coordinate value. Any of the three sublists can be replaced by `Automatic`, in which case *Mathematica* decides where to place the axis, or `None`, in which case the axis is not drawn.
- `BoxStyle` is an option which specifies how the bounding box is to be drawn. `BoxStyle` can be set to a list of graphics directives such as `Dashing`, `Thickness`, `GrayLevel`, or `RGBColor`.
- `Mesh` is an option which determines whether a mesh should be drawn on the graphic surface. The default is `Mesh`→`True`; `Mesh`→`False` eliminates the mesh.
- `Shading` is an option which specifies whether the surface should be shaded. The default is `Shading`→`True`.

The next example plots the parabolic cylinder $z = x^2$ several different ways.

EXAMPLE 3

```
Plot3D[x², {x, - 2, 2}, {y, - 2, 2}];
Plot3D[x², {x, - 2, 2}, {y, - 2, 2}, Mesh→False];
Plot3D[x², {x, - 2, 2}, {y, - 2, 2}, Shading→False];
Plot3D[x², {x, - 2, 2}, {y, - 2, 2}, BoxRatios→{1, 1, 1}];
Plot3D[x², {x, - 2, 2}, {y, - 2, 2}, FaceGrids→{{1, 0, 0}, {0, - 1, 0}}];
Plot3D[x², {x, - 2, 2}, {y, - 2, 2}, AxesEdge→{{ - 1, 1}, {1, 1}, {1, - 1}}];
```

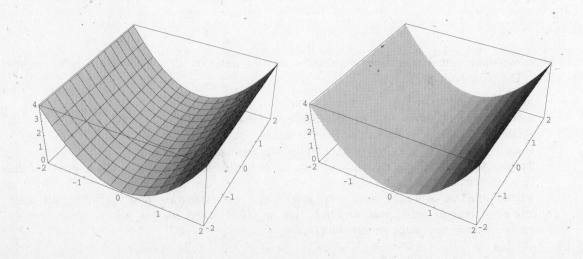

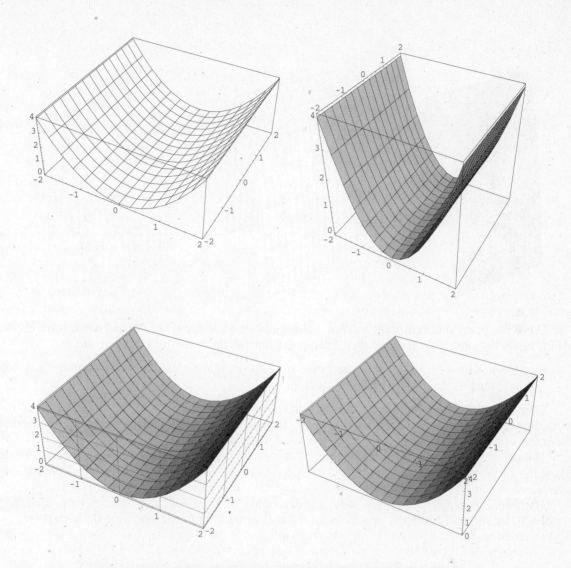

Three-dimensional graphics are generated as a sequence of polygons specially shaded to create a pleasing three-dimensional effect. Although the option Shading→False eliminates the shading, the polygons are drawn opaque so that surfaces behind other surfaces are hidden. The option **HiddenSurface** controls whether the surface is to be drawn opaque or transparent.

- **HiddenSurface→False** draws the polygons transparent (only the connecting lines are drawn) so that all surfaces are visible. The default, which produces opaque polygons, is **HiddenSurface→True**.

EXAMPLE 4

```
Plot3D[1 - y², {x, -5, 5}, {y, -5, 5}, BoxRatios → {1, 1, 2},
      Boxed → False, Axes → False];
Plot3D[1 - y², {x, -5, 5}, {y, -5, 5}, BoxRatios → {1, 1, 2},
      Boxed → False, Axes → False, Shading → False];
Plot3D[1 - y², {x, -5, 5}, {y, -5, 5}, BoxRatios → {1, 1, 2},
      Boxed → False, Axes → False, HiddenSurface → False];
```

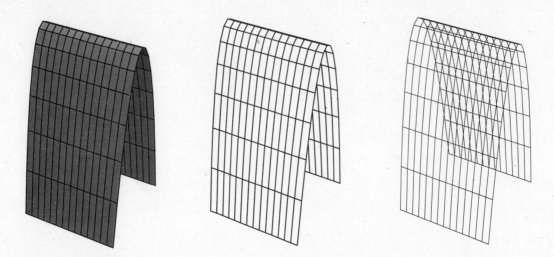

There are many different ways to view a three-dimensional drawing. **ViewPoint** is an option which views the surface from a specified fixed point outside the box which contains it.

- **ViewPoint → {x,y,z}** gives the position of the viewpoint relative to the center of the box that contains the surface being plotted.

The viewpoint coordinates are scaled in such a way that the length of the longest side of the bounding box is 1. The viewpoint must be located outside the bounding box. Generally, the further from the surface the viewpoint is selected, the less the distortion.

The default ViewPoint parameters are {1.3, -2.4, 2.0}. ViewPoint → {0, -2, 0} would show the surface directly from the front. ViewPoint → {0, 0, 2} shows the surface from directly above.

Mathematica supplies an easy way to determine the ViewPoint parameters. Selecting Input⇒3D ViewPoint Selector ... provides an interactive method for determining the viewpoint.

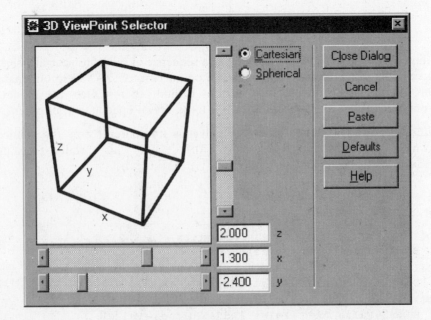

You can adjust the x, y, and z coordinates of the viewpoint directly or you can "drag" the box with the mouse.   When you are finished, click on "Paste" and the text

$$\text{ViewPoint} \rightarrow \{1.300, -2.400, 2.000\}$$

will be inserted at the cursor position.   The next example shows the hyperbolic paraboloid $z = x^2 - y^2$ from several different viewpoints.

EXAMPLE 5

```
Plot3D[x² - y², {x, -5, 5}, {y, -5, 5}, BoxRatios → {1, 1, 1}];
Plot3D[x² - y², {x, -5, 5}, {y, -5, 5}, BoxRatios → {1, 1, 1},
        ViewPoint → {2, 2, 2}];
Plot3D[x² - y², {x, -5, 5}, {y, -5, 5}, BoxRatios → {1, 1, 1},
        ViewPoint → {1.5, -2.6, -1.5}];
```

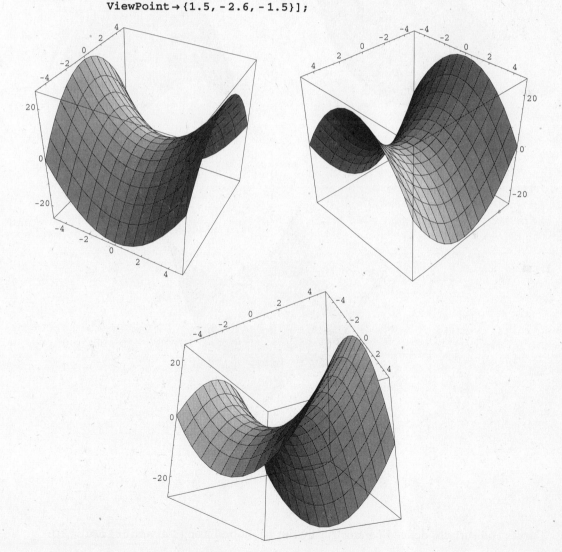

Plot3D can only plot one three-dimensional surface at a time.   However, the Show command, discussed in Chapter 4, can be used to view several surfaces on one set of axes.   The next example shows the intersection of two paraboloids.   The surfaces are shown individually and then together in one bounding box for comparison purposes.   If desired, the option DisplayFunction → Identity can be used to suppress the first two graphs (see p. 93).

EXAMPLE 6

```
paraboloid1 = Plot3D[x² + y², {x, -3, 3}, {y, -3, 3}, BoxRatios → {1, 1, 1}];
paraboloid2 = Plot3D[16 - (x² + y²), {x, -3, 3}, {y, -3, 3}, BoxRatios → {1, 1, 1}];
Show[paraboloid1, paraboloid2];
```

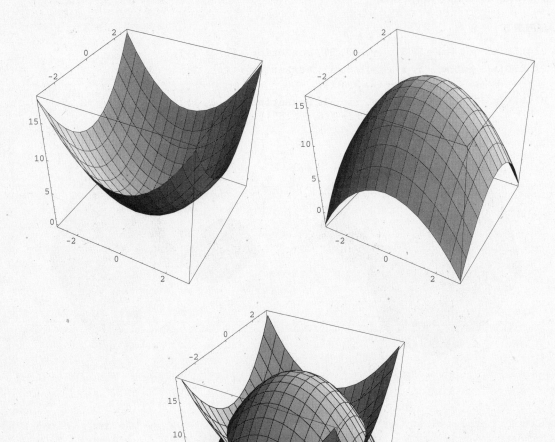

Curves and surfaces defined parametrically can be plotted using `ParametricPlot3D`.

- `ParametricPlot3D[{x[t], y[t], z[t]}, {t, tmin, tmax}]` plots a space curve in three dimensions for $\text{tmin} \le \text{t} \le \text{tmax}$.
- `ParametricPlot3D[{x[s, t], y[s, t], z[s, t]},`
  `                   {s, smin, smax}, {t, tmin, tmax}]`
  plots a surface in three dimensions, $\text{smin} \le \text{s} \le \text{smax}$, $\text{tmin} \le \text{t} \le \text{tmax}$.

EXAMPLE 7

```
ParametricPlot3D[{Cos[t], Sin[t], t/4}, {t, 0, 4π},
            ViewPoint → {1.292, - 2.788, 1.418}];
ParametricPlot3D[{Sin[s + t], Cos[s + t], s}, {s, - 2, 2}, {t, - 2, 2}];
```

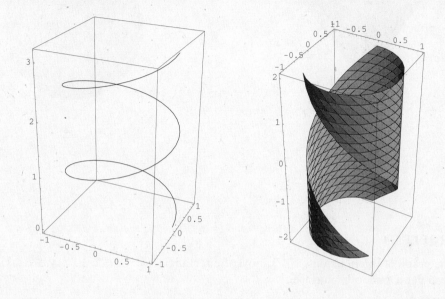

    `Plot3D` allows you to plot surfaces expressed in rectangular coordinates. However, cylindrical and spherical coordinates are useful for problems involving cylinders, spheres, and cones.

    A point $P(x, y, z)$ has cylindrical coordinates (r, θ, z), where $x = r \cos \theta$ and $y = r \sin \theta$, and has spherical coordinates (ρ, θ, ϕ), where $x = \rho \sin \phi \cos \theta$, $y = \rho \sin \phi \sin \theta$, and $z = \rho \cos \phi$. Here r and θ are the polar coordinates of the projection of P in the x-y plane and ϕ is the angle formed by the positive z axis and the line connecting P with the origin. The *Mathematica* commands **CylindricalPlot3D** and **SphericalPlot3D** allow the construction of surfaces given in these coordinate systems. These commands are found in **Graphics`ParametricPlot3D`**. All standard 3D plotting options are available.

Special Note: When dealing with spherical coordinates, *Mathematica*'s convention is to interchange the roles of θ and ϕ from that which is used in many standard calculus textbooks. The description of the command `SphericalPlot3D` described below, although different from the description in *Mathematica*'s help files, agrees with these conventions.

- **CylindricalPlot3D[z[r, θ], {r, rmin, rmax}, {θ, θmin, θmax}]** generates a plot of the surface $z = z(r, \theta)$ described in cylindrical coordinates.
- **SphericalPlot3D[[ρ[φ, θ], {φ, φmin, φmax}, {θ, θmin, θmax}]** generates a plot of the surface $\rho = \rho(\theta, \phi)$ described in spherical coordinates.

EXAMPLE 8

    In cylindrical coordinates, the equation $z = r$ represents a cone and in spherical coordinates $\rho = 1$ represents the unit sphere.

```
<<Graphics`ParametricPlot3D`
CylindricalPlot3D[r, {r, 0, 1}, {θ, 0, 2π}, BoxRatios → {1, 1, 1}];
SphericalPlot3D[1, {φ, 0, π}, {θ, 0, 2π}, BoxRatios → {1, 1, 1}];
```

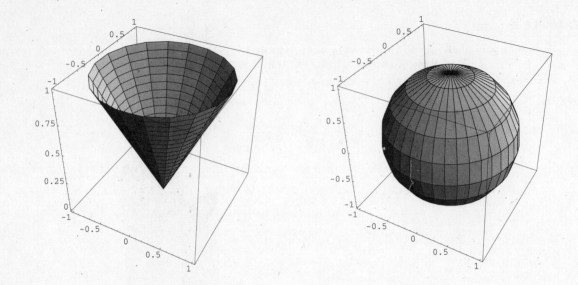

SOLVED PROBLEMS

5.1   Plot the graph of the function $e^{-x^2-y^2}$ above the rectangle $-2 \le x \le 2$, $-2 \le y \le 2$.   Use enough points to get a "smooth" surface.

SOLUTION

```
Plot3D[Exp[-x² - y²], {x, -2, 2}, {y, -2, 2}, PlotPoints → 25];
```

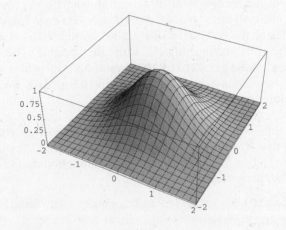

5.2   Obtain a graph of the "saddle"-shaped hyperboloid $z = x^2 - y^2$, $-3 \le x \le 3$, $-3 \le y \le 3$ in a cubic box.   Draw the graph with and without a surface mesh.

SOLUTION

```
Plot3D[x² - y², {x, -5, 5}, {y, -5, 5}, BoxRatios → {1, 1, 1}];
Plot3D[x² - y², {x, -5, 5}, {y, -5, 5}, Mesh → False, BoxRatios → {1, 1, 1}];
```

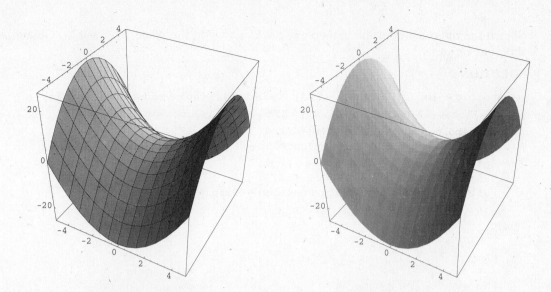

5.3   Draw the graph of the function $f(x, y) = |\sin x \sin y|$ for $-2\pi \le x \le 2\pi$, $-2\pi \le y \le 2\pi$. Do not draw axes.   Use enough points to get a "smooth" surface.

SOLUTION

```
Plot3D[Abs[Sin[x]Sin[y]], {x, -2π, 2π}, {y, -2π, 2π},
    Axes→False, PlotPoints→30];
```

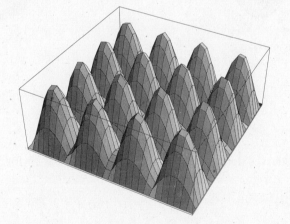

5.4   Draw an *unshaded* graph of the surface $z = |1 - x^2 - y^2|$ for $-1 \le x \le 1$, $-1 \le y \le 1$.   Do not draw axes or a surrounding box.

SOLUTION

```
Plot3D[Abs[1-x²-y²], {x, -1, 1}, {y, -1, 1},
    Shading→False, Axes→False,
    Boxed→False];
```

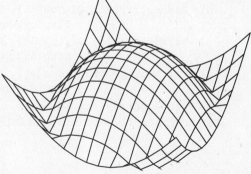

5.5   Sketch the intersection of the paraboloid $z = x^2 + y^2$ with the plane $y + z = 12$.   Obtain two different views.

SOLUTION

```
paraboloid = Plot3D[x² + y², {x, -5, 5}, {y, -5, 5}, DisplayFunction → Identity];
plane = Plot3D[12 - y, {x, -5, 5}, {y, -5, 5}, DisplayFunction → Identity];
Show[paraboloid, plane, BoxRatios → {1, 1, 1},
    PlotRange → {0, 20}, DisplayFunction → $DisplayFunction];
Show[paraboloid, plane, BoxRatios → {1, 1, 1},
    ViewPoint → {1.126, -1.800, 2.634},
    PlotRange → {0, 20}, DisplayFunction → $DisplayFunction];
```

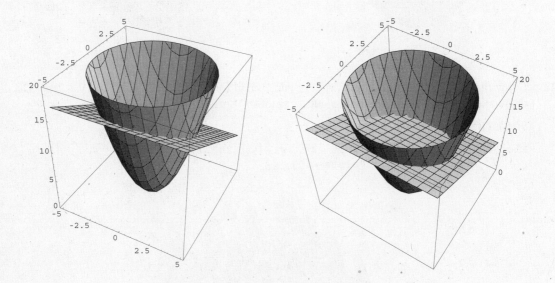

5.6   Sketch the space curves defined by $\begin{cases} x = \cos at \\ y = \sin bt \\ z = \sin ct \end{cases}$ $0 \le t \le 2\pi$ for

(i) $a = 5$, $b = 3$, $c = 1$ (ii) $a = 3$, $b = 3$, $c = 1$ (iii) $a = 2$, $b = 5$, $c = 2$.

(These curves are known as Lissajous curves.)

SOLUTION

```
g1 = ParametricPlot3D[{Cos[5t], Sin[3t], Sin[t]}, {t, 0, 2π},
                    DisplayFunction → Identity];
g2 = ParametricPlot3D[{Cos[3t], Sin[3t], Sin[t]}, {t, 0, 2π},
                    DisplayFunction → Identity];
g3 = ParametricPlot3D[{Cos[2t], Sin[5t], Sin[2t]}, {t, 0, 2π},
                    DisplayFunction → Identity];
g = GraphicsArray[{g1, g2, g3}]
Show[g, DisplayFunction → $DisplayFunction];
```

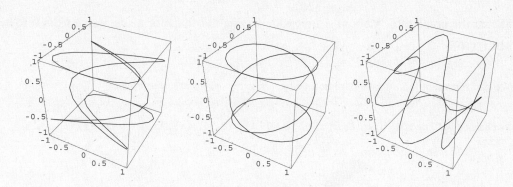

5.7   Sketch the torus defined by $\begin{cases} x = (4 + \sin s) \cos t \\ y = (4 + \sin s) \sin t \\ z = \cos s \end{cases}$ $0 \le s, t \le 2\pi.$

SOLUTION

```
x[s_, t_] = (4 + Sin[s]) Cos[t];
y[s_, t_] = (4 + Sin[s]) Sin[t];
z[s_] = Cos[s];
ParametricPlot3D[{x[s, t], y[s, t], z[s]}, {s, 0, 2π}, {t, 0, 2π}, PlotPoints → 50];
```

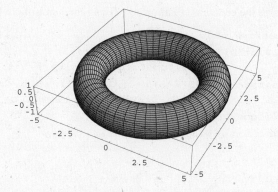

5.8   Sketch the space curve $\begin{cases} x = (4 + \sin 20t) \cos t \\ y = (4 + \sin 20t) \sin t \\ z = \cos 20t \end{cases}$ $0 \le t \le 2\pi.$

This curve is called a toroidal spiral since it lies on the surface of a torus (see previous problem).

SOLUTION

```
x[t_] = (4 + Sin[20t]) Cos[t];
y[t_] = (4 + Sin[20t]) Sin[t];
z[t_] = Cos[20t];
ParametricPlot3D[{x[t], y[t], z[t]},
{t, 0, 2π}, PlotPoints → 250];
```

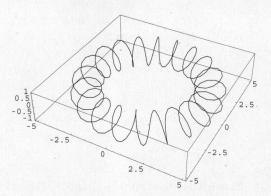

5.9   Sketch the graph of a "ribbon" 1 unit wide having the shape of a sine curve from 0 to 4π.

SOLUTION

We can represent this surface parametrically: $\begin{cases} x = t \\ y = s \\ z = \sin t \end{cases}$ $0 \le s \le 1, 0 \le t \le 4\pi$.

```
ParametricPlot3D[{t, s, Sin[t]}, {s, 0, 1}, {t, 0, 4π}, Axes→False];
```

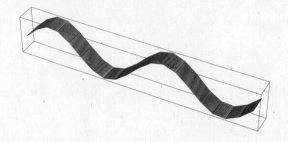

5.10   Draw the "ice cream cone" formed by the cone $z = 3\sqrt{x^2 + y^2}$ and the upper half of the sphere $x^2 + y^2 + (z - 9)^2 = 9$.   Use cylindrical coordinates.

SOLUTION

In cylindrical coordinates the cone has equation $z = 3r$ and the hemisphere has equation $z = 9 + \sqrt{9 - r^2}$.

```
<<Graphics`ParametricPlot3D`
cone = CylindricalPlot3D[3r, {r, 0, 3}, {θ, 0, 2π},
                        DisplayFunction→Identity];
hemisphere = CylindricalPlot3D[9 + √(9 - r²), {r, 0, 3}, {θ, 0, 2π},
                        DisplayFunction→Identity];
g = GraphicsArray[{cone, hemisphere}];
Show[g, DisplayFunction→$DisplayFunction];
```

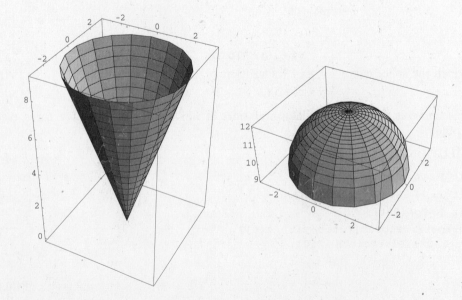

```
Show[cone, hemisphere, DisplayFunction→$DisplayFunction];
```

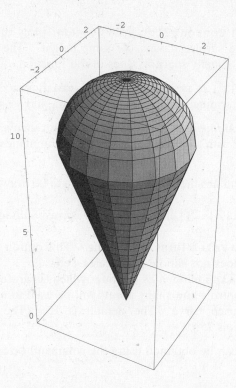

5.11   Sketch the graph of the following surface given in spherical coordinates:

$$\rho = 1 + \sin 4\theta \sin \phi, \qquad 0 \le \theta \le 2\pi, \qquad 0 \le \phi \le \pi$$

SOLUTION

```
<<Graphics`ParametricPlot3D`
SphericalPlot3D[1 + Sin[4θ] Sin[φ], {φ, 0, π}, {θ, 0, 2π}, PlotPoints→50];
```

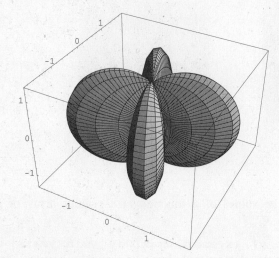

5.2   OTHER GRAPHICS COMMANDS

A level curve of a function of two variables, $f(x, y)$, is a two-dimensional graph of the equation $f(x, y) = k$ for some value of k. A contour plot is a collection of level curves drawn on the same set of axes.

The *Mathematica* command `ContourPlot` draws contour plots of functions of two variables. The contours join points on the surface having the same height. The default is to have contours corresponding to a sequence of equally spaced values of the function.

- `ContourPlot[f[x, y], {x, xmin, xmax}, {y, ymin, ymax}]` draws a contour plot of $f(x, y)$ in a rectangle determined by `xmin`, `xmax`, `ymin`, and `ymax`.

Contour plots produced by *Mathematica* are drawn shaded, in such a way that regions with higher values of $f(x, y)$ are lighter. As with all *Mathematica* graphics commands, options allow control of the appearance of the graph.

- `Contours→n` determines the number of contours to be drawn. The default is 10 equally spaced curves.
- `Contours→{k1, k2, ...}` draws contours corresponding to values of the function $k1, k2, \ldots$.
- `ContourShading→False` turns off shading. This option is particularly useful if your monitor or printer does not handle gray scales well.
- `ContourLines→False` eliminates the lines which separate the shaded contours.
- `PlotPoints→n` controls how many points will be used in each direction in an adaptive algorithm to plot each curve. The default is 15. (The standard default for two-dimensional graphics is 25.)

A complete list of options can be obtained using the command `Options[ContourPlot]`.

EXAMPLE 9

Obtain contour plots of the paraboloid $z = x^2 + y^2$. Note that the level curves are all circles $x^2 + y^2 = k$.

```
ContourPlot[x²+y², {x, -10, 10}, {y, -10, 10}];
ContourPlot[x²+y², {x, -10, 10}, {y, -10, 10}, ContourLines→False];
ContourPlot[x²+y², {x, -10, 10}, {y, -10, 10}, ContourShading→False];
```

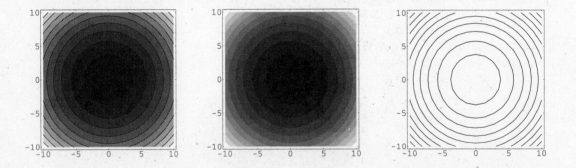

A density plot shows the values of a function at a regular array of points. Lighter regions correspond to higher values.

- `DensityPlot[f[x, y], {x, xmin, xmax}, {y, ymin, ymax}]` draws a density plot of $f(x, y)$ in a rectangle determined by `xmin`, `xmax`, `ymin`, and `ymax`. The option `Mesh→False` eliminates the rectangular mesh which separates the shaded regions.

EXAMPLE 10

```
DensityPlot[x²+y², {x, -10, 10}, {y, -10, 10}];
DensityPlot[x²+y², {x, -10, 10}, {y, -10, 10}, Mesh→False];
```

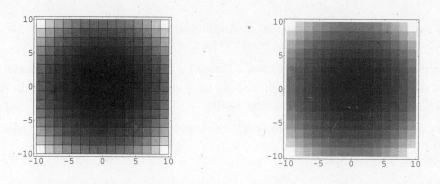

The commands **ListContourPlot** and **ListDensityPlot** are the analogs of **ContourPlot** and **DensityPlot** for discrete lists of numbers. These commands are appropriate for use with functions defined on a lattice of integer coordinates.

- **ListContourPlot[array]** generates a contour plot from a two-dimensional array of numbers.
- **ListDensityPlot[array]** generates a density plot from a two-dimensional array of numbers.

array = {{z_{11}, z_{12}, ...}, {z_{21}, z_{22}, ...}, ...} represents the heights of points in the *x-y* plane and must be a nested array of dimension 2×2 or larger. z_{ij} is the z coordinate of the point (j, i). The options for ListContourPlot and ListDensityPlot are the same as for ContourPlot and DensityPlot, except that the axes are labeled, by default, with positive integers starting with 1. The option **MeshRange** allows you to change the labeling of the axes.

- **MeshRange → {{xmin, xmax}, {ymin, ymax}}** labels the x and y axes from xmin to xmax and from ymin to ymax, respectively.

EXAMPLE 11

```
lst = Table[Random[], {x, 1, 10}, {y, 1, 10}];     ← Generates a 10 × 10 array of random numbers.
ListContourPlot[lst, MeshRange → {{-5, 5}, {3, 7}}];
ListDensityPlot[lst, MeshRange → {{-5, 5}, {3, 7}}];
```

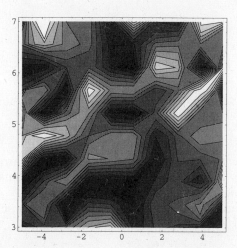
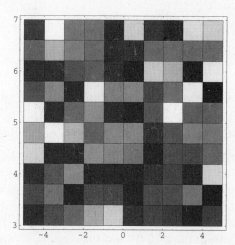

ContourPlot3D is the three-dimensional counterpart to ContourPlot. ContourPlot3D will sketch the level surfaces of f, i.e., the set of points (x, y, z) such that $f(x, y, z) = k$.

- **ContourPlot3D[f[x, y, z], {x, xmin, xmax}, {y, ymin, ymax}, {z, zmin, zmax}]**
  draws a three-dimensional contour plot of the level surface $f(x, y, z) = 0$ in a box determined by xmin, xmax, ymin, ymax, zmin, and zmax.

ContourPlot3D is contained within the package **Graphics`ContourPlot3D`** which must be loaded prior to its use.

The most commonly used options for ContourPlot3D are

- **Contours → {k1, k2, ...}** draws level surfaces corresponding to k1, k2, ... The default is Contours → {0.} so that only one level surface is drawn.
- **PlotPoints → {nx, ny}** determines how many evaluation points will be used in the x and y directions, respectively. **PlotPoints → n** is equivalent to **PlotPoints → {n, n}**. The default is PlotPoints → {3, 5}.

EXAMPLE 12

```
<<Graphics`ContourPlot3D`
ContourPlot3D[z - x² - y², {x, -5, 5}, {y, -5, 5}, {z, 0, 10},
            Contours → {0., 5.},
            BoxRatios → {1, 1, 1},
            PlotPoints → 5];
```

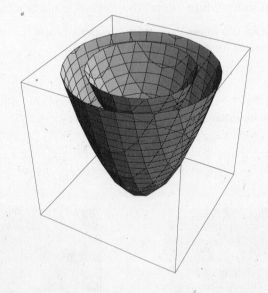

ListPlot3D is the three-dimensional analog of **ListPlot**.

- **ListPlot3D[array]** generates a three-dimensional surface based upon a given array of heights. array = {{z_{11}, z_{12}, ...}, {z_{21}, z_{22}, ...}, ...}, where z_{ij} is the z coordinate of the point (j, i), must be a nested array of dimension at least 2×2.
- **ListPlot3D[array, shades]** generates a surface shaded according to the descriptions in the array shades (GrayLevel, Hue, RGBColor). If array has dimensions m × n, then shades must have dimension $(m-1) \times (n-1)$.

The option **MeshRange** controls the labeling of the *x* and *y* axes.

- **MeshRange → {{xmin, xmax}, {ymin, ymax}}** labels the x and y axes from xmin to xmax and from ymin to ymax, respectively. The default is **Meshrange → Automatic** which assigns integer values starting with 1.

EXAMPLE 13

```
lst = {{1, 5, 2, 2}, {3, 6, 1, 4}, {3, 1, 7, 2}};
shades = {{GrayLevel[.4], GrayLevel[.5], GrayLevel[.6]},
          {GrayLevel[.7], GrayLevel[.8], GrayLevel[.9]}};
ListPlot3D[lst, shades];
```

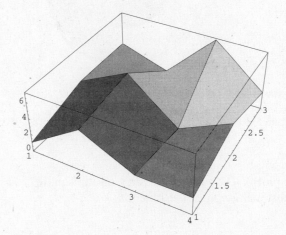

The discrete analog of **ContourPlot3D** is **ListContourPlot3D**.

- **ListContourPlot3D[array]** draws a contour plot of the values in array, a three-dimensional array of numbers representing the values of *f*.

ListContourPlot3D is contained within the package **Graphics`ContourPlot3D`** which must be loaded prior to its use. The most commonly used options for ListContourPlot3D are

- **Contours → {k1, k2, ...}** draws contours corresponding to values of the function k1, k2,.... The default is Contours → {0.} so that only one level surface is drawn.
- **PlotPoints → {nx, ny}** determines how many evaluation points will be used in the *x* and *y* directions, respectively. PlotPoints → n is equivalent to Plot-Points → {n, n}. The default is PlotPoints → {3, 5}.
- **Axes → True** plots axes and corresponding labeling. The default is Axes → False.
- **MeshRange → {{xmin, xmax}, {ymin, ymax}, {zmin, zmax}}** labels the x, y, and z axes from xmin to xmax, from ymin to ymax, and from zmin to zmax, respectively.

EXAMPLE 14

This example generates a *discrete* set of values of the function $f(x, y, z) = x^2 + y^2 + z^2$ and draws two contour plots of $f(x, y, z) = k$ for $k = .5$ and $k = 1.5$. The surfaces generated are spheres, but the larger sphere is drawn in a box which is too small to contain it completely. The result is that both spheres are visible in the picture.

```
<<Graphics`ContourPlot3D`
lst = Table[x² + y² + z², {z, -1, 1, .25}, {y, -1, 1, .25}, {x, -1, 1, .25}];
ListContourPlot3D[lst, MeshRange → {{-1, 1}, {-1, 1}, {-1, 1}},
                  Contours → {.5, 1.5}, Axes → True];
```

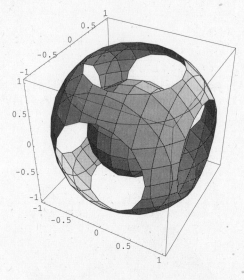

SOLVED PROBLEMS

5.12   Obtain a contour plot of $f(x, y) = \sin x + \sin y$ on the square $-4\pi \leq x, y \leq 4\pi$ and compare it
to the three-dimensional graph of the function.

SOLUTION

```
Plot3D[Sin[x] + Sin[y], {x, -4π, 4π}, {y, -4π, 4π}, PlotPoints → 50];
ContourPlot[Sin[x] + Sin[y], {x, -4π, 4π}, {y, -4π, 4π}, PlotPoints → 50];
```

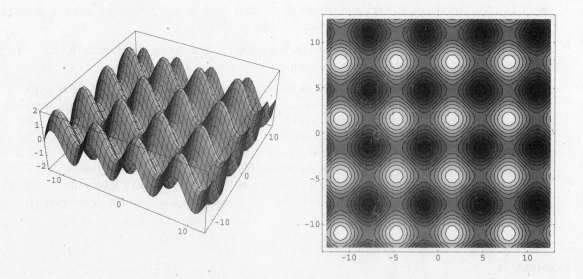

5.13   Compare a contour plot with a density plot for the function $f(x, y) = \sin xy$ over the rectangle
$-\pi \leq x, y \leq \pi$.

SOLUTION

We use `PlotPoints→50` to get a smooth-looking graph.

```
ContourPlot[Sin[xy], {x, -π, π}, {y, -π, π}, PlotPoints→50];
DensityPlot[Sin[xy], {x, -π, π}, {y, -π, π}, PlotPoints→50];
```

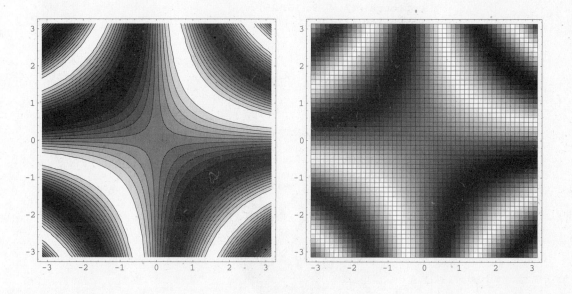

5.14 Obtain a contour plot and a density plot of the discrete function `Quotient[x, y]` as `x` and `y` range from 1 to 10.

SOLUTION

```
lst = Table[Quotient[x, y], {x, 1, 10}, {y, 1, 10}];
ListContourPlot[lst];
ListDensityPlot[lst];
```

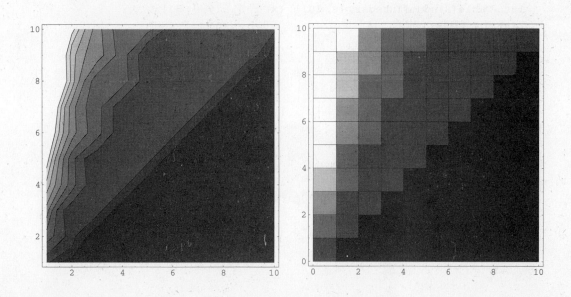

5.15 Let $f(x, y, z) = 5x^2 + 2y^2 + z^2$. Draw the level surfaces $f(x, y, z) = k$ for $k = 1, 4, 9, 16$, and 25. Sketch only for $y \geq 0$ so all the surfaces will be visible.

SOLUTION

```
<<Graphics`ContourPlot3D`
ContourPlot3D[5x² + 2y² + z², {x, - 5, 5}, {y, 0, 5}, {z, - 5, 5},
              Contours → {1, 4, 9, 16, 25},
              BoxRatios → {1, 1, 1},
              PlotPoints → 5];
```

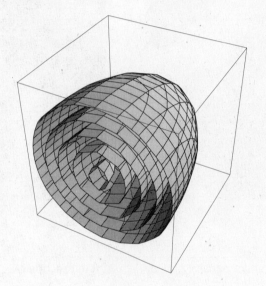

5.16 Construct a 5×5 array of random integers between 0 and 10 and draw a three-dimensional list plot of these values.

SOLUTION

```
lst = Table[Random[Integer, {0, 10}], {x, 1, 5}, {y, 1, 5}];
ListPlot3D[lst];
```

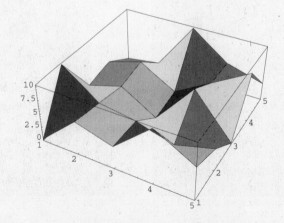

5.17 Draw cylinders $x^2 - y^2 = k$, $k = 0$, 2, and 5, by computing $f(x, y, z) = x^2 - y^2$ at integer values between -5 and 5 for each variable and using `ListContourPlot3D`.

SOLUTION

We use integer values of x, y, and z to construct our list, `lst`.

```
<<Graphics`ContourPlot3D`
lst = Table[x² - y², {z, -5, 5}, {y, -5, 5}, {x, -5, 5}];
ListContourPlot3D[lst, Contours → {0., 2., 5.}, Axes → True,
          MeshRange → {{-5, 5}, {-5, 5}, {-5, 5}}];
```

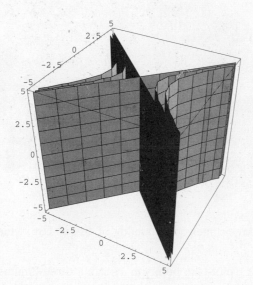

5.3 SPECIAL THREE-DIMENSIONAL PLOTS

The package **Graphics`Graphics3D`** contains a variety of useful graphics commands. Of course, as with all *Mathematica* packages, it must be loaded before the commands can be accessed.

- **BarChart3D[array]** constructs a three-dimensional bar chart using array to determine their heights. $array = \{\{z_{11}, z_{12}, \ldots\}, \{z_{21}, z_{22}, \ldots\}, \ldots\}$ where z_{ij} represents the height of bar (i, j).

Any three-dimensional graphics option may be used with BarChart3D. In addition, the following options are available:

- **XSpacing** and **YSpacing** determine the space between bars in the x and y directions, respectively. Their values must be numbers between 0 and 1. Their default values are **XSpacing → 0** and **YSpacing → 0** (no spacing).
- **SolidBarEdges → False** eliminates the edges between bars.
- **SolidBarEdgeStyle** determines the style of the bar edges. Its default is **SolidBarEdgeStyle → GrayLevel[0]**.
- **SolidBarStyle** determines the style of the bar faces. Its default is **SolidBarStyle → GrayLevel[0.5]**.

EXAMPLE 15

```
<<Graphics`Graphics3D`
1st = {{1, 2, 3, 4}, {5, 6, 7, 8}, {9, 10, 11, 12}};
BarChart3D[1st, AxesLabel → {"x", "y", "z"}];
BarChart3D[1st, XSpacing → .5, YSpacing → .5,
         AxesLabel → {"x", "y", "z"}];
```

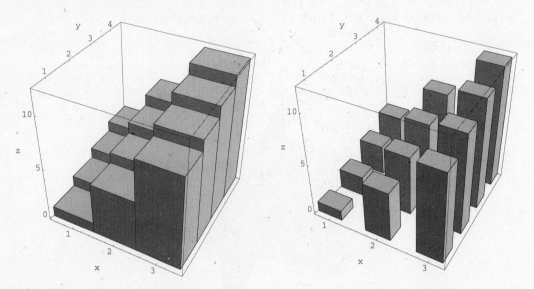

ScatterPlot3D is the three-dimensional analog of **ListPlot**, which plots discrete points in a two-dimensional plane.

- **ScatterPlot3D[1st]** plots the points in 1st in a three-dimensional box. 1st must be a list of sublists, each of which contains three numbers, representing the coordinates of the points.

ScatterPlot3D accepts the PlotStyle option discussed in Chapter 4. In addition **PlotJoined → True** will connect the points. ScatterPlot3D is contained in the **Graphics`Graphics3D`** package.

In the next example we generate 50 random points and plot them three different ways in three-dimensional space.

EXAMPLE 16

```
<<Graphics`Graphics3D`
1st = Table[Random[Integer, {1, 10}], {50}, {3}];      ← Generates 50 random points.
ScatterPlot3D[1st];
ScatterPlot3D[1st, PlotStyle → PointSize[.02]];
ScatterPlot3D[1st, PlotJoined → True];
```

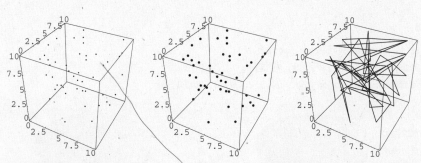

ListSurfacePlot3D creates a mesh of polygons constructed from the vertices specified in a list.

■ **ListSurfacePlot3D[lst]** creates a three-dimensional polygonal mesh from the vertices specified in `lst`, which should be of the form

$$\{\{\{x_{11}, y_{11}, z_{11}\}, \{x_{12}, y_{12}, z_{12}\}, \dots\},$$
$$\{\{x_{21}, y_{21}, z_{21}\}, \{x_{22}, y_{22}, z_{22}\}, \dots\}, \dots\}$$

EXAMPLE 17

```
<<Graphics`Graphics3D`
lst = Table[{x, y, x² - y²}, {x, -3, 3}, {y, -3, 3}];
ListSurfacePlot3D[lst, Axes → True, BoxRatios → {1, 1, 1}];
```

← This generates 49 points on the hyperboloid $z = x^2 - y^2$.

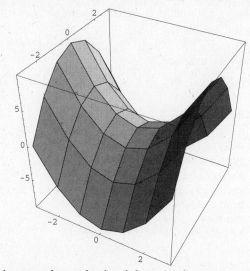

A surface of revolution is a surface obtained by rotating a curve about a given line. The command **SurfaceOfRevolution** constructs such surfaces. There are various forms of the command with several options. **SurfaceOfRevolution** is in the package `<<Graphics`SurfaceOfRevolution`.

■ **SurfaceOfRevolution[f[x], {x, xmin, xmax}]** generates the surface of revolution obtained by rotating the curve $z = f(x)$, xmin ≤ x ≤ xmax, about the z axis.

■ **SurfaceOfRevolution[f[x], {x, xmin, xmax}, {θ, θmin, θmax}]** generates the partial surface of revolution obtained by rotating the curve $z = f(x)$ about the z axis, for θmin ≤ θ ≤ θmax.

■ **SurfaceOfRevolution[{x[t], z[t]}, {t, tmin, tmax}]** generates the surface of revolution obtained by rotating the curve defined parametrically by $x = x(t)$, $z = z(t)$, tmin ≤ t ≤ tmax, about the z axis.

The following example rotates the curve $z = x^2$ about the z axis, completely and partially.

EXAMPLE 18

```
<<Graphics`SurfaceOfRevolution`
SurfaceOfRevolution[x², {x, 0, 3}];
SurfaceOfRevolution[x², {x, 0, 3}, {θ, 0, 3π/2}];
```

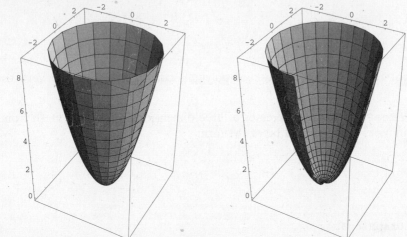

The option **RevolutionAxis** allows rotation about axes other than the z axis.

- **RevolutionAxis → {x, z}** rotates the curve about an axis formed by connecting the origin to the point (x, z) in the x-z plane.
- **RevolutionAxis → {x, y, z}** rotates the curve about an axis formed by connecting the origin to the point (x, y, z) in space.

EXAMPLE 19

```
<<Graphics`SurfaceOfRevolution`
SurfaceOfRevolution[x², {x, 0, 3}, RevolutionAxis→{1, 0},
                BoxRatios→{1, 1, 1},
                Ticks→False,
                AxesLabel→{"x", "y", "z"}];
SurfaceOfRevolution[x², {x, 0, 3}, RevolutionAxis→{1, 1, 1},
                BoxRatios→{1, 1, 1},
                Ticks→False,
                AxesLabel→{"x", "y", "z"}];
```

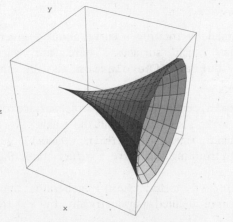 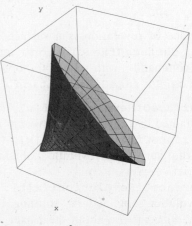

The curve $z = x^2$ is rotated about the line   The curve $z = x^2$ is rotated about the line
connecting the points $(0, 0, 0)$ and $(1, 0, 0)$.   connecting the points $(0, 0, 0)$ and $(1, 1, 1)$.

SOLVED PROBLEMS

5.18   Construct a three-dimensional bar chart representation of Pascal's triangle for $n = 5$.

SOLUTION

Pascal's triangle is a representation of the binomial coefficients $c(n, k) = \dfrac{n!}{k!(n-k)!}$.

		k				
	0	1	2	3	4	5
0	1					
1	1	1				
n \ 2	1	2	1			
3	1	3	3	1		
4	1	4	6	4	1	
5	1	5	10	10	5	1

```
<<Graphics`Graphics3D`
```

$$c[n\_,k\_] = \frac{n!}{k!\,(n-k)!};$$ ← Note: C[n, k] = 0 if k > n.

```
lst = Table[c[n, k], {k, 0, 5}, {n, 0, 5}];
BarChart3D[lst, AxesLabel → {"k", "n", "c[n, k]"},
    Ticks → {Range[0, 5], Range[0, 5], Range[0, 10]}];
```

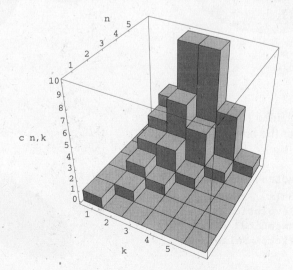

5.19   Construct a scatter plot of the points on the helix $x = \sin 2t$, $y = \cos 2t$, $z = t$, for t between 0 and 10 in increments of .25.

SOLUTION

```
<<Graphics`Graphics3D`
1st = Table[{Sin[2t], Cos[2t], t}, {t, 0, 10, .25}];
ScatterPlot3D[1st, PlotStyle→PointSize[.05], Axes→False];
```

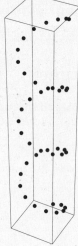

5.20 Construct the surface of revolution obtained by rotating the curve $z = \sin x$, $0 \leq x \leq 2\pi$, about (i) the z axis, and (ii) the x axis.

SOLUTION

```
<<Graphics`SurfaceOfRevolution`
SurfaceOfRevolution[Sin[x], {x, 0, 2π}, Ticks→False,
                AxesLabel→{"x", "y", "z"}];
SurfaceOfRevolution[Sin[x], {x, 0, 2π}, RevolutionAxis→{1, 0}, Ticks→False,
                AxesLabel→{"x", "y", "z"}];
```

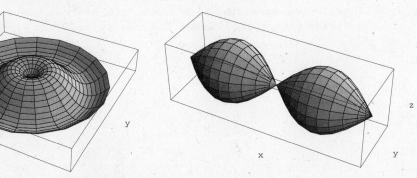

5.21 Sketch the surface obtained by rotating the curve $z = x^2$, $0 \leq x \leq 1$, about the line $z = x$.

SOLUTION

```
<<Graphics`SurfaceOfRevolution`
SurfaceOfRevolution[x², {x, 0, 1},
            RevolutionAxis→{1, 1},
            AxesLabel→{"x", "y", "z"},
            Ticks→False];
```

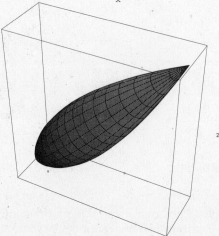

5.4 STANDARD SHAPES—3D GRAPHICS PRIMITIVES

■ **Graphics3D[*primitives*]** or **Graphics3D[*primitives*, *options*]** creates a three-dimensional graphics object which may be viewed with the **Show** command.

The standard primitives are

- **Cuboid[{x, y, z}]** is a three-dimensional graphics primitive that represents a unit cuboid (cube) with corner at (x, y, z), with edges parallel to the axes.
- **Cuboid[{x1, y1, z1}, {x2, y2, z2}]** represents a cuboid (parallelepiped) whose opposite corners are $(x1, y1, z1)$ and $(x2, y2, z2)$.
- **Line[{x1, y1, z1}, {x2, y2, z2}, ...]** draws a sequence of line segments connecting the points $(x1, y1, z1)$, $(x2, y2, z2)$, ...
- **Point[{x, y, z}]** plots a single point at coordinates (x, y, z).
- **Polygon[{x1, y1, z1}, {x2, y2, z2}, ...]** draws a filled polygon with coordinates $(x1, y1, z1)$, $(x2, y2, z2)$, ...
- **Text[*expression*, {x, y, z}]** creates a graphics primitive representing the text *expression*, centered at position (x, y, z).

EXAMPLE 20

```
Show[Graphics3D[{Cuboid[{0, 0, 0}], Cuboid[{1, 1, 1}, {2, 3, 4}]}],
    Axes→True, Ticks→{{0, 1, 2}, {0, 1, 2, 3}, {0, 1, 2, 3, 4}}];
```

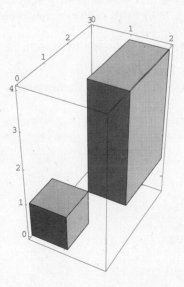

EXAMPLE 21

```
vertices = {{0, 0, 0}, {2, 2, 0}, {0, 2, 1}, {0, 0, 2}};
Show[Graphics3D[Polygon[vertices]],
    Axes→True, Ticks→{{0, 1, 2}, {0, 1, 2}, {0, 1, 2}}];
```

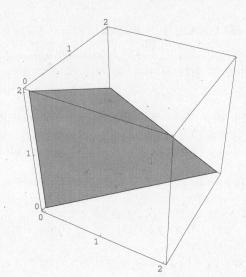

Additional three-dimensional graphics primitives which allow for convenient drawing of standard shapes such as spheres, cones, cylinders, etc., are located within the package **Graphics`Shapes`** which must be loaded prior to use.

- **Cylinder[r, h, n]** draws a cylinder with radius r and half height h using n polygons.
- **Sphere[r, n, m]** draws a sphere of radius r using $n(m-2) + 2$ polygons.
- **Cone[r, h, n]** draws a cone with radius r and half height h using n polygons.
- **Torus[r1, r2, n, m]** draws a torus with radii r1 and r2 using an $n \times m$ mesh.
- **MoebiusStrip[r1, r2, n]** draws a Möbius strip with radii r1 and r2 using 2n polygons.
- **Helix[r, h, m, n]** draws a helix with radius r, half height h, and m turns using an $n \times m$ mesh.
- **DoubleHelix[r, h, m, n]** draws a double helix with radius r, half height h, and m turns using an $n \times m$ mesh.

If the parameters are omitted, e.g., Cone[], *Mathematica*'s defaults are used. The default values are

```
Cylinder[1, 1, 20]
Sphere[1, 20, 15]
Cone[1, 1, 20]
Torus[1, .5, 20, 10]
MoebiusStrip[1, .5, 20]
Helix[1, .5, 2, 20]
DoubleHelix[1, .5, 2, 20]
```

EXAMPLE 22

```
<<Graphics`Shapes`
Show[Graphics3D[Cylinder[]]];
Show[Graphics3D[Sphere[]]];
Show[Graphics3D[Cone[]]];
Show[Graphics3D[Torus[]]];
Show[Graphics3D[MoebiusStrip[]]];
Show[Graphics3D[Helix[]]];
Show[Graphics3D[DoubleHelix[]]];
```

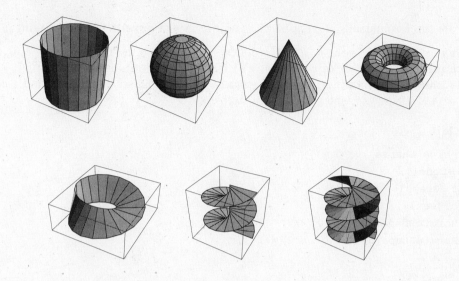

- **WireFrame[object]** shows all polygons used in the construction of object as transparent. It may be used on any Graphics3D object that contains the primitives Polygon, Line, and Point.

EXAMPLE 23

```
<< Graphics`Shapes`
object = Graphics3D[Torus[ ]];
Show[object];
Show[WireFrame[object]];
```

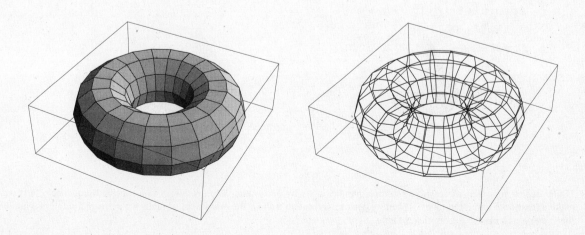

There are three commands in `Graphics`Shapes`` which provide transformations in space:

- `RotateShape[object, `ϕ`, `θ`, `ψ`]` rotates `object` using the Euler angles[1] ϕ, θ, and ψ.
- `TranslateShape[object, {x, y, z}]` translates `object` by the vector `{x, y, z}`.
- `AffineShape[object, {xscale, yscale, zscale}]` scales the x, y, and z coordinates by `xscale`, `yscale`, and `zscale`, respectively.

EXAMPLE 24

```
<<Graphics`Shapes`
shape = Cone[];
object = Graphics3D[shape];
Show[object];
Show[RotateShape[object, 0, π/2, 0]];
Show[RotateShape[object, 0, π/2, π/2]];
```

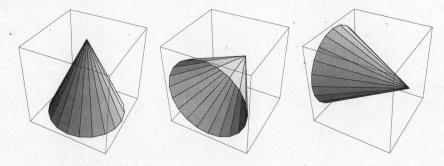

```
Show[object, TranslateShape[object, {1, 2, 3}]];
shrunkenobject = AffineShape[object, {.5, .5, .5}];
Show[object, TranslateShape[shrunkenobject, {1, 2, 3}]];
```

---

[1] Euler angles are a way of describing transformations in $\mathbb{R}^3$ by performing three rotations in a specified sequence. First we make a rotation ϕ about the z axis. Then we perform a rotation θ about the *new y* axis. Finally we perform a rotation ψ about the (new) z axis obtained from this rotation.

SOLVED PROBLEMS

5.22 Draw two cylinders intersecting at right angles.

SOLUTION

```
<<Graphics`Shapes`
cyl1 = Graphics3D[Cylinder[1, 5, 20]];
cyl2 = Graphics3D[RotateShape[Cylinder[1, 5, 20], 0, π/2, 0]];
Show[cyl1, cyl2];
```

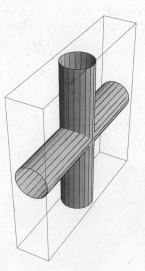

5.23 Construct a cylinder inscribed in a sphere of radius 1.

SOLUTION

Since the sphere has radius 1, we use the default parameters. In order for the cylinder to be inscribed in the sphere, r (radius) and h (half height) must satisfy $r^2 + h^2 = 1$. We choose $r = 1/2$ and $h = \sqrt{3}/2$. In order for the cylinder to be visible, we draw the sphere as a wire frame.

```
<<Graphics`Shapes`
sph = WireFrame[Graphics3D[Sphere[]]];
cyl = Graphics3D[Cylinder[1/2, √3/2, 20]];
Show[sph, cyl, Boxed→False];
```

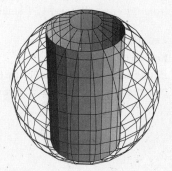

5.24   Draw two interlocking tori.

SOLUTION

We use two tori of default dimensions ($r1 = 1$, $r2 = 0.5$). The second must be rotated 90° and translated 0.5 units so that they interlock without intersecting.

```
<<Graphics`Shapes`
tor1 = Graphics3D[Torus[]];
tor2 = Graphics3D[TranslateShape[
            RotateShape[Torus[], 0, π/2, π/2], {0, .5, 0}]];
Show[tor1, tor2, ViewPoint → {1.75, -2.8, 0.75}, Boxed → False];
```

5.25   Construct an animation showing a helix revolving about the z axis.

SOLUTION

We use a default helix, `Helix[]`. The helix makes one complete revolution as the Euler angle, ϕ, varies from 0 to 2π. We use 12 frames.

```
<<Graphics`Shapes`
helx = Helix[];
rotatedhelx = RotateShape[helx, k, 0, 0];
Do[Show[Graphics3D[rotatedhelx], Boxed → False], {k, 0, 11π/6, π/6}]
```

Select the cell brackets for all the images and go to Cell⇒Animate Selected Graphics.

CHAPTER 6

Equations

6.1 SOLVING ALGEBRAIC EQUATIONS

Solutions of general algebraic equations may be found using the **Solve** command. The command is easy to use, but one must be careful to use a double equal sign, $==$, between the left- and right-hand sides of the equation. (Recall that the double equal sign is a *logical* equality: lhs $==$ rhs has a value of True if and only if lhs and rhs have the same value, False otherwise.)

The roots determined by Solve are expressed as of a list of the form

$$\{\{x \to x1\}, \{x \to x2\}, \dots\}$$

The notation $x \to x1$ indicates that the solution, x, is x1, but x is *not* replaced by the value x1. If the equation has roots of multiplicity $m > 1$, each is repeated m times.

- **Solve[*equations, vars*]** attempts to solve *equations* for *vars*.

For systems of equations, *equations* is a list of the form {*equation1, equation2, ...*} and *vars* represents either a single variable or a list of several. Alternatively, *equations* may be represented by the individual equations separated by $\&\&$ (logical and). If only one variable is present, *vars* may be omitted.

EXAMPLE 1

In this example, there is only one variable so the specification of *vars* is unnecessary.

```
Solve[7 x + 3 == 3 x + 8]
```

$$\left\{\left\{x \to \frac{5}{4}\right\}\right\}$$

If we try to solve an equation which contains two or more variables, we must specify which variable we are looking for.

EXAMPLE 2

```
Solve[a y + b == c x + d]
```
Solve::svars :
  Equations may not give solutions for all "solve" variables.
$\{\{b \to d + c x - a y\}\}$

> Note the space between a and y and between c and x.
> This is important. * may be used instead.

162

We must specify which variable we wish to solve for:

`Solve[a y + b == c x + d, x]`

$$\left\{\left\{x \to -\frac{-b+d-a\,y}{c}\right\}\right\}$$

`Solve[a y + b == c x + d, y]`

$$\left\{\left\{y \to -\frac{b-d-c\,x}{a}\right\}\right\}$$

`Solve[a y + b == c x + d, b]`
`{{b → d + c x - a y}}`

`Solve[a y + b == c x + d, d]`
`{{d → b - c x + a y}}`

EXAMPLE 3

Here is an example which shows how to solve a simple system: $\begin{cases} 2x + 3y = 7 \\ 3x + 4y = 10 \end{cases}$

`Solve[{2 x + 3 y == 7, 3 x + 4 y == 10}, {x, y}]` or
`Solve[2 x + 3 y == 7 && 3 x + 4 y == 10, {x, y}]`
`{{x → 2, y → 1}}`

> The specification of {x, y} is not necessary since we have two equations in two variables.

If you have more unknown variables than equations, you must specify which variable(s) you wish to solve for.   Otherwise you get *Mathematica*'s default.

EXAMPLE 4

`Solve[{x + 2 y + z == 5, 2 x + y + 3 z == 7}, {y, z}]`

$$\left\{\left\{y \to \frac{8-x}{5},\ z \to -\frac{3}{5}(-3+x)\right\}\right\}$$

Of course, `Solve` is not limited to solving only linear equations.

EXAMPLE 5

`Solve[a x² + b x + c == 0, x]`

$$\left\{\left\{x \to \frac{-b-\sqrt{b^2-4\,a\,c}}{2\,a}\right\},\ \left\{x \to \frac{-b+\sqrt{b^2-4\,a\,c}}{2\,a}\right\}\right\}$$

Observe that *Mathematica* gives the general solution in terms of arbitrary *a*, *b*, and *c*.

EXAMPLE 6

`Solve[x³ + y² == 5 && x + y == 3]`
$$\left\{\{y \to 2, x \to 1\},\ \left\{y \to 4-\sqrt{5},\ x \to -1+\sqrt{5}\right\},\ \left\{y \to 4+\sqrt{5},\ x \to -1-\sqrt{5}\right\}\right\}$$

Because *Mathematica* returns the solutions of equations as a nested list, they cannot be used directly as input to other mathematical structures.   There are two techniques, however, that can be used to access their values without unnecessary typing or pasting.

(*a*)   If we wish to compute the value of an expression using the solutions obtained from `Solve`, we can use the `/.` replacement operator and *Mathematica* will substitute the appropriate values.

(*b*)   Since the solutions are nothing more than lists, we can use `Part` or `[[ ]]` to "extract" the solution from the list.

These techniques are illustrated in the next two examples.

EXAMPLE 7

Suppose we wish to solve the equations $\begin{cases} x^2 + y = 5 \\ x + y \ = 3 \end{cases}$ and find the values of the expression $\sqrt{x^2 + y^2}$ at these points. We use the `Solve` command and the object `solutions` for convenience.

```
solutions = Solve[{x²+ y == 5, x + y == 3}, {x, y}]
```
$\{\{y \to 1, x \to 2\}, \{y \to 4, x \to -1\}\}$

```
√x²+y² /. solutions
```
$\left\{\sqrt{5}, \sqrt{17}\right\}$ ← *Mathematica* produces a list containing
                                              both values of the expression.

EXAMPLE 8

Suppose we wish to find the *sum of the squares* of the roots of

$$x^6 - 21x^5 + 175x^4 - 735x^3 + 1624x^2 - 1764x + 720 = 0$$

We use the `Solve` command:

```
solutions = Solve[x⁶ - 21 x⁵ + 175 x⁴ - 735 x³ + 1624 x² - 1764 x + 720 == 0]
```
$\{\{x \to 1\}, \{x \to 2\}, \{x \to 3\}, \{x \to 4\}, \{x \to 5\}, \{x \to 6\}\}$

Look at `solutions` carefully. Basically, it's a list containing sublists. Let's look at its first part.

```
solutions[[1]]
```
$\{x \to 1\}$

Since this list has only one part, we can extract its contents.

```
solutions[[1, 1]]
```
$x \to 1$

To get to the *second* part of this expression (the number after the arrow), we extract further.

```
solutions[[1, 1, 2]]
```
1

To understand why this works, let's look at `solutions[[1, 1]]` closely.

```
FullForm[solutions[[1, 1, 2]]]
```
Rule[x, 1]

$x \to 1$ is equivalent to `Rule[x, 1]` and as such, its second argument is extracted by using `solutions[[1, 1, 2]]`.

The other solutions are obtained similarly as `solutions[[2,1,2]]`, `solutions[[3,1,2]]`, etc. To get the sum of their squares,

```
Sum[solutions[[k, 1, 2]]^2, {k, 1, 6}]
```
or
$$\sum_{k=1}^{6} \text{solutions}[[k,1,2]]^2$$

91

`Solve` is designed to solve algebraic equations, but can sometimes be used to find limited solutions of transcendental equations. A warning message is given to indicate that not all solutions can be found.

EXAMPLE 9

```
Solve[Sin[x] == 1/2, x]
```

```
Solve::ifun :
Inverse functions are being used by Solve, so some solutions may not be found.
```

$$\left\{\left\{x \to \frac{\pi}{6}\right\}\right\}$$

If the equations to be solved are inconsistent, *Mathematica* returns an empty list.

EXAMPLE 10

```
Solve[{2 x + 3 y == 5, 4 x + 6 y == 11}]
```

{ }

If the roots of an equation involves complex numbers, they are represented as rational powers of -1. However, if a more traditional expression is desired, the "phrase"

$$/.(a\_\to b\_) :> (a \to ComplexExpand[b])$$

can be inserted to the right of the Solve command. (:> means RuleDelayed)

If only real roots are desired, The package `<<Miscellaneous`RealOnly`` can be loaded. This effectively labels all complex roots as 'Nonreal'.

EXAMPLE 11

```
Solve[x³==1]
```

$$\{\{x \to 1\}, \{x \to -(-1)^{1/3}\}, \{x \to (-1)^{2/3}\}\}$$

```
Solve[x³ == 1] /. (a_→b_) :> (a → ComplexExpand[b])
```

$$\left\{\{x \to 1\}, \left\{x \to -\frac{1}{2} - \frac{i\sqrt{3}}{2}\right\}, \left\{x \to -\frac{1}{2} - \frac{i\sqrt{3}}{2}\right\}\right\}$$

```
<<Miscellaneous`RealOnly`
Solve[x³ == 1]
```

Nonreal::warning : Nonreal number encountered

$$\{\{x \to 1\}, \{x \to Nonreal\}, \{x \to Nonreal\}$$

Not all algebraic equations are solvable by *Mathematica*, even if theoretical solutions exist. If *Mathematica* is unable to solve an equation, it will represent the solution in a symbolic form. For the most part, such solutions are useless and a numerical approximation is more appropriate. Numerical approximations are obtained with the command **NSolve**.

- **NSolve[*equations, vars*]** solves *equations* numerically for *vars*.
- **NSolve[*equations, vars*, n]** solves *equations* numerically for vars to n digits of precision.

As with Solve, the list of variables may be omitted if there is no ambiguity.

EXAMPLE 12

```
equation = x⁴ - 16 x³ + 61 x² - 22 x - 12 == 0;
Solve[equation]
```

> The equation is represented by **equation** to avoid repetitive typing.

$$\left\{\left\{x \to 3 - \sqrt{5}\right\}, \left\{x \to 3 + \sqrt{5}\right\}, \left\{x \to 5 - 2\sqrt{7}\right\}, \left\{x \to 5 + 2\sqrt{7}\right\},\right.$$

```
NSolve[equation]
```

$$\{\{x \to -0.291503\}, \{x \to 0.763932\}, \{x \to 5.23607\}, \{x \to 10.2915\}\}$$

```
NSolve[equation, x, 24]
```

$$\{\{x \to -0.291502622129181181003232\},$$
$$\{x \to 0.763932022500021030359083\},$$
$$\{x \to 5.236067977499789696964092\},$$
$$\{x \to 10.291502622129181181810032\}\}$$

Occasionally, `NSolve` may fail due to lack of precision, if too many significant digits are called for. This is because *Mathematica* works to a precision of 16 significant digits in its internal calculations. In such cases `N[Solve[`*equations*`]`, *vars*`]`, `n]`, where n is the required number of significant digits, may work successfully.

EXAMPLE 13

> `NSolve[`$\sqrt[3]{x} + \sqrt{x} + x == 10$`, x, 20]`
>
> $\{\{x \to 5.79617\}\}$ ← Only 6 significant digits are returned due to lack of precision.
>
> `N[Solve[`$\sqrt[3]{x} + \sqrt{x} + x == 10$`, x], 20]`
>
> $\{\{x \to 5.7961701257629574569\}\}$

An *extraneous solution* is a number which is technically not a solution of the equation, but evolves from the solution process. When solving radical equations, one typically encounters extraneous solutions. For example, when solving $\sqrt{x} = -3$, which has no real solution, the squaring process yields $x = 9$.

- `VerifySolutions` is an option which determines whether *Mathematica* should verify if solutions obtained are extraneous. The default, `VerifySolutions → True`, eliminates extraneous solutions from the solution list. If such solutions are desired, the option `VerifySolutions → False` should be used.

EXAMPLE 14

> `Solve[`$x + \sqrt{x} == 5$`]`
>
> $\left\{\left\{x \to \frac{1}{2}\left(11 - \sqrt{21}\right)\right\}\right\}$
>
> `Solve[`$x + \sqrt{x} == 5$`, VerifySolutions → False]`
>
> $\left\{\left\{x \to \frac{1}{2}\left(11 - \sqrt{21}\right)\right\}, \left\{x \to \frac{1}{2}\left(11 + \sqrt{21}\right)\right\}\right\}$

If we solve the equation $ax = b$ for x, `Solve` tells us that $x = b/a$. However, if $a = b = 0$, then *every* number x is a solution. The command `Reduce` can be used to describe *all* possible solutions.

- `Reduce[`*equations*`,` *vars*`]` simplifies *equations*, attempting to solve for the variables *vars*. If *equations* is an identity, `Reduce` returns the value `True`. If *equations* is a contradiction, `Reduce` returns the value `False`.

In describing the solutions, `Reduce` uses the symbols `&&` (logical and) and `||` (logical or). `&&` takes precedence over `||`.

EXAMPLE 15

> `Solve[a x == b, x]`
>
> $\left\{\left\{x \to \dfrac{b}{a}\right\}\right\}$
>
> `Reduce[a x == b, x]`
>
> $a == 0 \,\&\&\, b == 0 \,||\, x == \dfrac{b}{a} \,\&\&\, a \neq 0$ ← Either $a = b = 0$ *or* $a \neq 0$ and $x = b/a$.
>
> `Reduce[`$x^2 - 9 == (x + 3)(x - 3)$`, x]`
>
> `True`
>
> `Reduce[`$x^2 - 10 == (x + 3)(x - 3)$`, x]`
>
> `False`

SOLVED PROBLEMS

6.1   Find a line passing through $(2,5)$ and $(7,9)$.

SOLUTION

The general equation of a line is $y = ax + b$. Substituting the coordinates of the given points leads to the equations $2a + b = 5$ and $7a + b = 9$.

```
Solve[2 a + b == 5 && 7 a + b == 9]
```
$$\left\{\left\{a \to \frac{4}{5}, b \to \frac{17}{5}\right\}\right\}$$

The line has equation $y = \frac{4}{5}x + \frac{17}{5}$.

6.2   Find an equation of the circle passing through $(1,4)$, $(2,7)$, and $(4,11)$.

SOLUTION

The general equation of a circle is $x^2 + y^2 + ax + by + c = 0$. We substitute the coordinates of the given points into the equation to obtain $17 + a + 4b + c = 0$, $53 + 2a + 7b + c = 0$, and $137 + 4a + 11b + c = 0$.

```
Solve[{17 + a + 4 b + c == 0, 53 + 2 a + 7 b + c == 0, 137 + 4 a + 11 b + c == 0}]
```
$$\{\{a \to -54, b \to 6, c \to 13\}\}$$

The equation of the circle is $x^2 + y^2 - 54x + 6y + 13 = 0$.

6.3   Solve the equation $x^5 + x^4 + x^3 + x^2 + x + 2 = 0$.

SOLUTION

```
Solve[x⁵ + x⁴ + x³ + x² + x + 2 == 0]
```
$$\{\{x \to Root[2 + \#1 + \#1^2 + \#1^3 + \#1^4 + \#1^5\&, 1]\},$$
$$\{x \to Root[2 + \#1 + \#1^2 + \#1^3 + \#1^4 + \#1^5\&, 2]\},$$
$$\{x \to Root[2 + \#1 + \#1^2 + \#1^3 + \#1^4 + \#1^5\&, 3]\},$$
$$\{x \to Root[2 + \#1 + \#1^2 + \#1^3 + \#1^4 + \#1^5\&, 4]\},$$
$$\{x \to Root[2 + \#1 + \#1^2 + \#1^3 + \#1^4 + \#1^5\&, 5]\}\}$$

Mathematica cannot solve this equation exactly so it returns a symbolic solution. However, we can obtain a numerical approximation.

```
NSolve[x⁵ + x⁴ + x³ + x² + x + 2 == 0]
```
$$\{\{x \to -1.21486\}, \{x \to -0.522092 - 1.06118\,i\}, \{x \to -0.522092 + 1.06118\,i\},$$
$$\{x \to 0.629523 - 0.883585\,i\}, \{x \to 0.629523 + 0.883585\,i\}\}$$

6.4   Solve the following system for w, x, and y and then determine the solution when $z = 1$, $z = 2$, and $z = 3$.

$$w + x + y + z = 3$$
$$2w + 3x + 4y + 5z = 10$$
$$w - x + y - z = 4$$

SOLUTION

```
equations = {w + x + y + z == 3, 2 w + 3 x + 4 y + 5 z == 10, w - x + y - z == 4};
```

```
solution = Solve[equations, {w, x, y}]
```

$$\left\{\left\{w \to \frac{1}{4}(5 + 4 z), x \to \frac{1}{2}(-1 - 2 z), y \to \frac{1}{4}(9 - 4 z)\right\}\right\}$$

```
solution /. z → 1
```

$$\left\{\left\{w \to \frac{9}{4}, x \to -\frac{3}{2}, y \to \frac{5}{4}\right\}\right\}$$

```
solution /. z → 2
```

$$\left\{\left\{w \to \frac{13}{4}, x \to -\frac{5}{2}, y \to \frac{1}{4}\right\}\right\}$$

```
solution /. z → 3
```

$$\left\{\left\{w \to \frac{17}{4}, x \to -\frac{7}{2}, y \to -\frac{3}{4}\right\}\right\}$$

6.5 Find, to 20 significant digits, a number such that the sum of itself, its square, and its cube is 30.

SOLUTION

```
NSolve[x + x² + x³ == 30, x, 20]
```

```
{{x → -1.8557621138713175532 - 2.7604410593413850003 i),
 {x → -1.8557621138713175532 + 2.7604410593413850003 i),
 {x → 2.7115242277426351064}}
```

6.6 Solve the trigonometric equation $2 \sin^2 x + 1 = 3 \sin x$ for $\sin x$ and then for x.

SOLUTION

To solve for $\sin x$, we can use .

```
Solve[2 Sin[x]² + 1 == 3 Sin[x]]
```

$$\left\{\left\{\mathrm{Sin}[x] \to \frac{1}{2}\right\}, \{\mathrm{Sin}[x] \to 1\}\right\}$$

If we solve for x, only the principal solutions (using inverse functions) are obtained.

```
Solve[2 Sin[x]² + 1 == 3 Sin[x], x]
```

```
Solve::ifun :
  Inverse functions are being used by Solve, so some solutions may not be found.
```

$$\left\{\left\{x \to \frac{\pi}{6}\right\}, \left\{x \to \frac{\pi}{2}\right\}\right\}$$

6.7 Solve for x: $e^{2x} + e^x = 3$.

SOLUTION

```
Solve[Exp[2 x] + Exp[x] == 3, x]
```

```
Solve::ifun :
  Inverse functions are being used by Solve, so some solutions may not be found.
```

$$\left\{\left\{x \to \mathrm{Log}\left[\frac{1}{2}\left(-1 + \sqrt{13}\right)\right]\right\}, \left\{x \to i\,\pi + \mathrm{Log}\left[\frac{1}{2}\left(1 + \sqrt{13}\right)\right]\right\}\right\}$$

6.8   Sketch the graphs of $f(x) = x^3 - 7x^2 + 2x + 20$ and $g(x) = x^2$ on the same set of axes and find their points of intersection exactly and approximately.

SOLUTION

```
f[x_] = x^3 - 7x^2 + 2x + 20;
g[x_] = x^2;
Plot[{f[x],g[x]},{x,-10,10}];
```

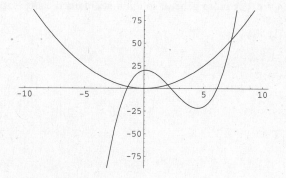

```
xvalues = Solve[f[x] == g[x],x];
{x, f[x]} /. xvalues //Simplify
```

$$\left\{ \{2, 4\}, \; \left\{3 - \sqrt{19}, \; 28 - 6\sqrt{19}\right\}, \; \left\{3 + \sqrt{19}, \; 28 + 6\sqrt{19}\right\} \right\}$$

```
% //N
{{2., 4.}, {-1.3589, 1.84661}, {7.3589, 54.1534}}
```

6.9   A theorem from algebra says that if $p(x) = a_n x^n + a_{n-1} x^{n-1} + a_{n-2} x^{n-2} + \ldots + a_1 x + a_0$, the sum of the roots of the equation $p(x) = 0$ is $-\dfrac{a_{n-1}}{a_n}$ and their product is $(-1)^n \dfrac{a_0}{a_n}$.   Verify this for the equation

$$20x^7 + 32x^6 - 221x^5 - 118x^4 + 725x^3 - 18x^2 - 726x + 252 = 0$$

SOLUTION

```
solution = Solve[20x^7 + 32x^6 - 221x^5 - 118x^4 + 725x^3 - 18x^2 - 726x + 252 == 0]
```

$$\left\{ \left\{x \to -\frac{7}{2}\right\}, \; \left\{x \to \frac{2}{5}\right\}, \; \left\{x \to \frac{3}{2}\right\}, \; \left\{x \to -\sqrt{2}\right\}, \; \left\{x \to \sqrt{2}\right\}, \; \left\{x \to -\sqrt{3}\right\}, \; \left\{x \to \sqrt{3}\right\} \right\}$$

```
Sum[solution[[k,1,2]],{k,1,7}]   or   ∑_{k=1}^{7} solution[[k,1,2]]
```

$$-\frac{8}{5} \qquad\qquad \leftarrow n = 7; \quad -\frac{a_{n-1}}{a_n} = -\frac{32}{20} = -\frac{8}{5}$$

```
Product[solution[[k,1,2]],{k,1,7}]   or   ∏_{k=1}^{7} solution[[k,1,2]]
```

$$-\frac{63}{5} \qquad\qquad \leftarrow n = 7; \quad -(1)^n \frac{a_0}{a_n} = -(1)^7 \frac{252}{20} = -\frac{63}{5}$$

6.10   Find all possible solutions, x, for the equation $ax + b = cx + d$.

SOLUTION

```
Solve[a x + b == c x + d, x]
```

$$\left\{\left\{x \rightarrow -\frac{b-d}{a-c}\right\}\right\}$$

This solution presumes $a \neq c$.   A more general solution is obtained using `Reduce`.

```
Reduce[a x + b == c x + d, x]
```

$$a == c \,\&\&\, b == d \,||\, x == \frac{-b+d}{a-c} \,\&\&\, a - c \neq 0$$

6.2   SOLVING TRANSCENDENTAL EQUATIONS

A transcendental equation is one which is nonalgebraic.   Although `Solve` and `NSolve` can be used in a limited way to handle simple trigonometric or exponential equations, they were not designed to handle equations involving more complicated transcendental functions.   The *Mathematica* command **`FindRoot`** is better equipped to handle these.

`FindRoot` uses iterative methods to find solutions.   A starting value, sometimes called the *initial guess*, must be specified.   For best results the initial guess should be as close to the desired root as possible.

- **`FindRoot[`***lhs* == *rhs*, `{x, x0}]` solves the equation *lhs* = *rhs* using Newton's method with starting value x0.
- **`FindRoot[`***lhs* == *rhs*, `{x, {x0, x1}}]` solves the equation *lhs* == *rhs* using (a variation of) the secant method with starting values x0 and x1.
- **`FindRoot[`***lhs* == *rhs*, `{x, x0, xmin, xmax}]` attempts to solve the equation, but stops if the iteration goes outside the interval [xmin, xmax].

If a *function* is specified in place of the equation *lhs* == *rhs*, `FindRoot` will determine a zero of the function.   A zero of f is a number, x, such that $f(x) = 0$.

EXAMPLE 16

The   equation $\sin x = x^2 - 1$ has   two   solutions.   The graph of the two functions shows that they intersect near $x = -1$ and $x = 1$.

```
Plot[{Sin[x], x² - 1}, {x, -π, π}];
FindRoot[Sin[x] == x² - 1, {x, -1}]
{x → -0.636733}
FindRoot[Sin[x] == x² - 1, {x, 1}]
{x → 1.40962}
```

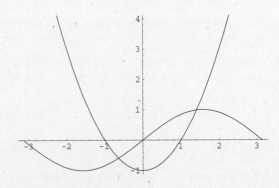

EXAMPLE 17

The equation $x + \sin x = 0$ has a unique solution $x = 0$. Observe what happens if we take our initial guess too far from the solution.

```
FindRoot[x + Sin[x] == 0, {x, 100}]
FindRoot::cvnwt :
Newton's method failed to converge to the prescribed accuracy after 15 iterations.
{x → -0.000774542}
```

By default, 15 iterations are performed before FindRoot is aborted. The number of iterations performed before quitting is controlled by the option **MaxIterations**.

- **MaxIterations → n** instructs *Mathematica* to use a maximum of n iterations in the iterative process before aborting.

EXAMPLE 18

Using 25 iterations in the previous example leads to a more successful conclusion.

```
FindRoot[x + Sin[x] == 0, {x, 100}, MaxIterations → 25]
{x → 7.74432 ×10⁻¹¹}
```

FindRoot attempts to find real solutions. However, if a complex initial value is specified, or if the equation contains complex numbers, complex solutions will be sought. The equation in the next example has no real solutions.

EXAMPLE 19

```
FindRoot[x² + x + 1 == 0, {x, 1}]
FindRoot::cvnwt :
Newton's method failed to converge to the prescribed accuracy after 15 iterations.
FindRoot[x² + x + 1 == 0, {x, I}]
{x → -0.5 + 0.866025 i}
```

If Newton's method fails to solve a given problem, the *secant method* is a good alternative.[1]

EXAMPLE 20

The equation $\sqrt{|x|} + x - 1 = 0$ has a solution between 0 and 1. However, Newton's method fails because *Mathematica* cannot compute the derivative. Specifying two starting values causes the secant method to be used instead.

```
FindRoot[√Abs[x] + x - 1, {x, 1}]
FindRoot::frjc : Could not symbolically find the Jacobian of {√Abs[x] + x - 1}.  Try
    giving two starting values for each variable.

FindRoot[√Abs[x] + x - 1, {x, {0, 1}}]
{x → 0.381966}
```

As an alternative, if the derivative of the function can be determined, the **Jacobian** option will allow successful computation of the root by Newton's method. For functions of a single variable, the Jacobian is simply the derivative.

---

[1]Newton's method uses the x intercept of the tangent line to improve the accuracy of the initial guess. Thus, Newton's method fails if the derivative of the function cannot be computed. The secant method, although a bit slower, uses the values of the function at two distinct points, computing the x intercept of the *secant* line.

- Jacobian → *derivative* "injects" the derivative, *derivative*, of the function into the algorithm. The default is Jacobian → Automatic which specifies that the derivative be computed symbolically, if possible.

EXAMPLE 21

The derivitive of $\sqrt{|x|}$ is $\dfrac{sign(x)}{2\sqrt{|x|}}$ where $sign(x) = \begin{cases} 1 & \text{if } x > 0 \\ -1 & \text{if } x < 0 \end{cases}$ (verify!).

```
FindRoot[√Abs[x] + x - 1, (x, 1), Jacobian → Sign[x]/(2√Abs[x]) + 1]
{x → 0.381966}
```

There are two options which control the calculation in FindRoot and other numerical algorithms.

- WorkingPrecision is an option which specifies how many digits of precision should be maintained internally in computation. The default (on most computers) is WorkingPrecision → 16.
- AccuracyGoal is an option which specifies how many significant digits of accuracy are to be obtained. If |lhs - rhs| < AccuracyGoal, iteration stops. The default is AccuracyGoal → Automatic, which is 10 less than the value of WorkingPrecision.

EXAMPLE 22

We wish to obtain an approximation to the solution of the equation $\cos\left(\dfrac{100}{x}\right) = \dfrac{x}{x+1}$, nearest to 5000, accurate to 10 decimal places.

```
FindRoot[Cos[100/x] == x/(x+1), {x, 5000}]
{x → 5000.}
```

Mathematica's defaults are insufficient to give the required accuracy. By increasing WorkingPrecision, we can obtain the desired accuracy.

```
FindRoot[Cos[100/x] == x/(x+1), {x, 5000}, WorkingPrecision → 24]

{x → 5000.833191159556095809817}
```

Since AccuracyGoal is, by default, 10 less than WorkingPrecision, only the first 14 significant digits can be trusted. Thus $x \approx 5000.8331911596$.

A slight variation of FindRoot allows intermediate calculations to be performed and displayed. Simply place the extra instructions to the left of the equation to be solved, separated by semicolons. This technique is illustrated in the next two examples in solving the equation $e^{-x} = x$.

EXAMPLE 23

To see how quickly the sequence of approximations converges when we solve the equation $e^{-x} = x$, we can have FindRoot print the results of intermediate calculations.

```
n = -1;
FindRoot[If[n ≥ 0, Print[n, " ", x]]; n++; Exp[-x] == x, {x, 2}]
0    2.
1    0.357609
2    0.558708
3    0.56713
4    0.567143
{x → 0.567143}
```

EXAMPLE 24

To obtain a comparison between Newton's method and the secant method, we can ask `FindRoot` to print the number of iterations needed to converge, using `WorkingPrecision → 100` and `Timing` to determine the speed of the procedure (timings will vary depending upon your computer's speed).

```
n = 0;
FindRoot[n++; Exp[-x] == x, {x, 1}, WorkingPrecision → 100] //Timing
n
{0.11 Second,
   {x → 0.5671432904097838729999686622103555497538157871865125081
        3513107922304579308668456669321944469617522946}}.
9
```

> Newton's method, 9 iterations, 0.11 sec.

```
n = 0;
FindRoot[n++; Exp[-x] == x, {x, {1, 2}}, WorkingPrecision → 100] //Timing
n
{0.33 Second,
   {x → 0.5671432904097838729999686622103555497538157871865125081
        3513107922304579308668456669321944469617522946}}
28
```

> Secant method, 28 iterations, 0.33 sec.

If the equation to be solved has a root of multiplicity 2 or greater, Newton's method may converge slowly or not at all. In this situation, convergence can sometimes be improved by a judicious choice of **DampingFactor**.

- **DampingFactor → *factor*** is an option which controls the behavior of convergence in Newton's method. The size of each step taken in Newton's method is multiplied by the value of *factor*. The default is `DampingFactor → 1`.

EXAMPLE 25

```
FindRoot[(Exp[x] - 1)², {x, 2}]        ← 0 is a zero of multiplicity 2.
FindRoot::cvnwt :
   Newton's method failed to converge to the prescribed accuracy after 15 iterations.
{x → 0.000911355}
FindRoot[(Exp[x] - 1)², {x, 2}, DampingFactor → 2]
{x → 7.73452 × 10⁻⁶}
```

Although not shown, convergence is achieved in 8 iterations.

FindRoot can also be used to determine the solution of simultaneous equations.

- **FindRoot[*equations*, {*var1*, a1}, {*var2*, a2}, ...]** attempts to solve *equations* using initial values a1, a2, ... for *var1*, *var2*, ..., respectively. The equations are enclosed in a list: {*equation1*, *equation2*, ...}. Alternatively, the equations may be separated by && (logical and).

Newton's method for functions of several variables is much more sensitive to choice of starting values than its counterpart for single variables. Therefore, a graph of the functions involved is quite helpful.

EXAMPLE 26

Solve the system of equations $\begin{cases} e^x + \ln y = 2 \\ \sin x + \cos y = 1 \end{cases}$

First we graph the equations.

```
<<Graphics`ImplicitPlot`
ImplicitPlot[{Exp[x] + Log[y] == 2, Sin[x] + Cos[y] == 1}, {x, 0, 2},
            PlotRange → {0, 3}]
```

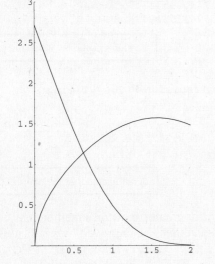

Warning messages may be generated when executing this command. These may be safely ignored.

It appears that there is only one solution. We use $x = 1$, $y = 1$ for our initial guess.

```
FindRoot[{Exp[x] + Log[y] == 2, Sin[x] + Cos[y] == 1}, {x, 1}, {y, 1}]
{x → 0.624295, y → 1.14233}
```

If the function in an equation is such that its evaluation is costly, particularly if high precision is desired, there is another procedure which may be beneficial.

- **InterpolateRoot[**_lhs_ == _rhs_, **{x, a, b}]** solves the equation _lhs_ == _rhs_ using initial values a and b.

As with FindRoot, the equation may be replaced by a function, in which case its zero is computed.

Whereas FindRoot uses linear functions (straight lines) to approximate the root of the equation, InterpolateRoot uses polynomials of degree 3 or less. The result is that higher precision can be achieved with less function evaluations. InterpolateRoot is contained within the package **NumericalMath`InterpolateRoot`** which must be loaded prior to use.

EXAMPLE 27

This example computes the zero (between 2 and 3) of the Bessel function[2] $J_0(x)$, using a working precision of 1000 significant digits. For comparison, the *Mathematica* function Timing is used. The actual numerical approximation is suppressed to save space. (Delete the semicolon and run the instruction to see the result of the calculation.)

---

[2] $J_0(x)$ is a solution of the differential equation $x^2 y'' + xy' + x^2 y = 0$.

```
FindRoot[BesselJ[0,x],{x,2},WorkingPrecision→1000]; //Timing
{7.69 Second, Null}
<<NumericalMath`InterpolateRoot`
InterpolateRoot[BesselJ[0,x],{x,2,3},WorkingPrecision→1000]; //Timing
{4.95 Second,Null}
```

SOLVED PROBLEMS

6.11   Solve the equation $5\cos x = 4 - x^3$.   Make sure you find all solutions.

SOLUTION

Since $5\cos x = 4 - x^3$ if and only if $5\cos x - 4 + x^3 = 0$, we introduce the function $f(x) = 5\cos x - 4 + x^3$ and look for x intercepts.   (Although we could look for the intersection of two curves, it is easier to determine where points intercept an axis.)

```
f[x_] = 5 Cos [x] - 4 + x³;
Plot[f[x], {x, -3, 3}];
```

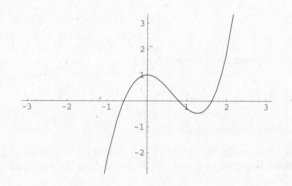

It appears that there are three solutions, near -0.5, 0.8, and 1.6.

```
FindRoot[f[x], {x,-.5}]
{x → -0.576574}
FindRoot[f[x], {x, 0.8}]
{x → 0.797323}
FindRoot[f[x], {x, 1.6}]
{x → 1.61805}
```

6.12   Find a solution[3] of the equation $\sin x = 2$.

SOLUTION

Since $-1 \le \sin x \le 1$ for all real x, this problem has no real solutions.   We can force FindRoot to search for a complex solution by using a complex initial guess.

```
FindRoot[Sin[x] == 2, {x, I}]
{x → 1.5708 + 1.31696 i}
```

---

[3]This problem may be omitted by those unfamiliar with functions of a complex variable.

6.13   Find a 20-significant-digit approximation to the solution of $x + |\sin(x-1)| = 5$.

SOLUTION

First we plot the function $f(x) = x + |\sin(x-1)| - 5$.

```
f[x_] = x + Abs[Sin[x - 1]] - 5
Plot[f[x], {x, -10, 10}];
```

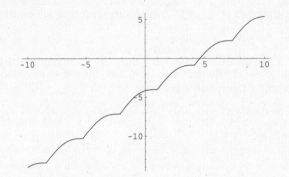

It appears that the only solution lies between 4 and 5.

```
FindRoot[f[x], {x, 5}]
```
FindRoot::frjc : Could not symbolically find the Jacobian of {f[x]}.   Try giving two
     starting values for each variable.

Newton's method fails because the derivative of $f(x)$ cannot be determined by *Mathematica*.   We try the secant method with **WorkingPrecision → 30**, 10 more than the required number of significant digits. (by default AccuracyGoal is 10 less than WorkingPrecision)

```
FindRoot[f[x], {x, 4, 5}, WorkingPrecision → 30]
```
{x → 4.57764001198757729525937359564}

To 20 significant digits, the solution is 4.5776400119875772953.

6.14   Find the points of intersection of the parabola $y = x^2 + x - 10$ with the circle $x^2 + y^2 = 25$.

SOLUTION

First, plot the two graphs.

```
g1 = Graphics[Circle[{0, 0}, 5]];
g2 = Plot[x² + x - 10, {x, -5, 5}, DisplayFunction → Identity];
Show[g1, g2,
     AspectRatio → Automatic, PlotRange → {-10, 10},
     Axes → True, DisplayFunction → $DisplayFunction];
```

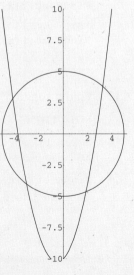

The parabola $y = x^2 + x - 10$ intersects the circle $x^2 + y^2 = 25$ at four points.   Now solve for the intersection points.   Because of the complicated structure of the exact solution, we obtain a numerical approximation.

```
NSolve[y == x² + x - 10 && x² + y² == 25]
```
{{y → 2.83654, x → -4.11753}, {y → 3.80098, x → 3.24846},
 {y → -4., x → -3.}, {y → -4.63752, x → 1.86907}}

6.15 Find the points of intersection of the limacon $r = 5 - 4\cos\theta$ and the parabola $y = x^2$.

SOLUTION

First we plot both curves on the same set of axes.

```
<<Graphics`Graphics`
limacon = PolarPlot[5 - 4Cos[t], {t, 0, 2π}, DisplayFunction → Identity];
parabola = Plot[x², {x, -3, 3}, DisplayFunction → Identity];
Show[limacon, parabola, DisplayFunction → $DisplayFunction];
```

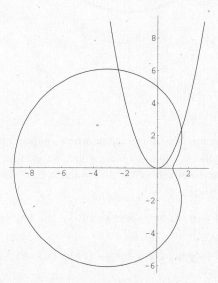

We convert the equation of the limacon to rectangular coordinates:

$$r = 5 - 4\cos\theta$$
$$r^2 = 5r - 4r\cos\theta$$
$$x^2 + y^2 = 5\sqrt{x^2 + y^2} - 4x$$

$$\boxed{\begin{array}{l} r = \sqrt{x^2 + y^2} \\ x = r\cos\theta \end{array}}$$

The first intersection point appears to be near $(2, 2)$:

```
FindRoot[{y == x², x² + y² == 5√(x² + y²) - 4 x}, {x, 2}, {y, 2}]
{x → 1.53711, y → 2.3627}
```

The second point lies near $(-3, 6)$:

```
FindRoot[{y == x², x² + y² == 5√(x² + y²) - 4 x}, {x, -3}, {y, 6}]
{x → -2.4552, y → 6.02802}
```

6.16 Where does the spiral of Archimedes, $r = \theta$, intersect the ellipse $4x^2 + 9y^2 = 400$?

SOLUTION

```
<<Graphics`Graphics`
<<Graphics`ImplicitPlot`
spiral = PolarPlot[θ, {θ, 0, 6π}, DisplayFunction → Identity];
ellipse = ImplicitPlot[4 x² + 9 y² == 400, {x, -10, 10},
                    PlotStyle → Thickness[.01],
                    DisplayFunction → Identity];
Show[spiral, ellipse, DisplayFunction → $DisplayFunction];
```

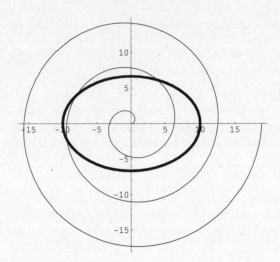

The graph shows three points of intersection which appear to be near $(4, 6)$, $(-8, 4)$, and $(-9, -2)$. To convert the polar equation to rectangular, we use the transformations $r = \sqrt{x^2 + y^2}$ and $\theta = \tan^{-1}(y/x)$. This leads to the equation $\sqrt{x^2 + y^2} = \tan^{-1}(y/x)$. However, Newton's method is more stable if we use $\tan(\sqrt{x^2 + y^2}) = y/x$.

```
FindRoot[{Tan[√(x²+y²)] == y/x, 4 x² + 9 y² == 400}, {x, 4}, {y, 6}]
{x → 3.93476, y → 6.1289}
```

```
FindRoot[{Tan[√(x²+y²)] == y/x, 4 x² + 9 y² == 400}, {x, -8}, {y, 4}]
{x → -8.04703, y → 3.95785}
```

```
FindRoot[{Tan[√(x²+y²)] == y/x, 4 x² + 9 y² == 400}, {x, -9}, {y, -2}]
{x → -9.38786, y → -2.29668}
```

6.17   Find a solution of the system of equations

$$x + y + z = 6$$
$$\sin x + \cos y + \tan z = 1$$
$$e^x + \sqrt{y} + \frac{1}{z} = 5$$

near the point $(1, 2, 3)$.

SOLUTION

```
FindRoot[{x + y + z == 6, Sin[x] + Cos[y] + Tan[z] == 1,
   Exp[x] + √y + 1/z == 5}, {x, 1}, {y, 2}, {z, 3}]
{x → 1.23382, y → 1.5696, z → 3.19658}
```

CHAPTER 7

Algebra and Trigonometry

7.1 POLYNOMIALS

Because they are so prevalent in algebra, *Mathematica* offers commands which are devoted exclusively to polynomials.

- **PolynomialQ[*expression, variable*]** yields `True` if *expression* is a polynomial in *variable*, and `False` otherwise. If only one variable is involved, *variable* may be omitted.
- **Variables[*polynomial*]** gives a list of all independent variables in *polynomial*.
- **Coefficient[*polynomial, form*]** gives the coefficient of *form* in *polynomial*. **Coefficient[*polynomial, form*, n]** gives the coefficient of *form* to the nth power in *polynomial*.
- **CoefficientList[*polynomial, variable*]** gives a list of the coefficients of powers of *variable* in *polynomial*, starting with the 0th power.

EXAMPLE 1

```
PolynomialQ[x² + 3x + 2]
```
True
```
PolynomialQ[x² + 3x + 2/x]
```
False
```
PolynomialQ[x² + 3x + 2/y, x]          ← 2/y is treated as a constant with respect to x.
```
True

EXAMPLE 2

```
poly1 = (x + 1)¹⁰;
poly2 = x³ - 5 x²y + 3 x y² - 7 y³;
Variables[poly2]
```
{x, y}
```
Coefficient[poly1, x, 5]
```
252
```
Coefficient[poly2, x]
```
3y²

```
Coefficient[poly2, y, 2]
```
3x
```
Coefficient[poly2, x y²]
```
3
```
CoefficientList[poly1, x]
```
{1, 10, 45, 120, 210, 252, 210, 120, 45, 10, 1}
```
CoefficientList[poly2, x]
```
{-7y³, 3y², -5y, 1}
```
CoefficientList[poly2, y]
```
{x³, -5x², 3x, -7}

Often it is convenient to write the solution of a polynomial equation as a *logical* expression. For example, if $x^2 - 4 = 0$, then $x = -2$ *or* $x = 2$. Roots of polynomial equations can be expressed in this form using two specialized commands, **Roots** and **NRoots**. The solutions are given in disjunctive form separated by the symbol $||$ (logical *or*).

- **Roots[*lhs* == *rhs*, *variable*]** produces the solutions of a polynomial equation.
- **NRoots[*lhs* == *rhs*, *variable*]** produces numerical approximations of the solutions.

EXAMPLE 3

Find all the solutions of $x^4 + x^3 - 8x^2 - 5x + 15 = 0$ which are greater than 2.

```
solutions = Roots[x⁴ + x³ - 8x² - 5x + 15 == 0, x]
```

$$x == \frac{1}{2}\left(-1 - \sqrt{13}\right) \; || \; x == \frac{1}{2}\left(-1 + \sqrt{13}\right) \; || \; x == -\sqrt{5} \; || \; x == \sqrt{5}$$

```
solutions && x > 2 //Simplify
```
$x == \sqrt{5}$

> **&&** is *Mathematica*'s logical *and*.
> See Section 7.4 for a discussion of **Simplify**.

```
numericalsolutions = NRoots[x⁴ + x³ - 8x² - 5x + 15 == 0, x]
```
$x == -2.30278 \; || \; x == -2.23607 \; || \; x == 1.30278 \; || \; x == 2.23607$

```
numericalsolutions && x > 2 //Simplify
```
$x == 2.23607$

The division algorithm for polynomials guarantees that given two polynomials, p and s, where $\deg(p) \geq \deg(s)$, there exist uniquely determined polynomials, q and r, such that

$$p(x) = q(x)s(x) + r(x) \qquad \text{where} \qquad \deg(r) < \deg(s)$$

The *Mathematica* commands which produce the quotient and remainder are

- **PolynomialQuotient[p, s, x]** gives the quotient upon division of p by s expressed as a function of x. Any remainder is ignored.
- **PolynomialRemainder[p, s, x]** returns the remainder when p is divided by s. The degree of the remainder is less than the degree of s.

EXAMPLE 4

```
p = x⁵ - 7x⁴ + 3x² - 5x + 9;
s = x² + 1;
q = PolynomialQuotient[p, s, x]
```
$10 - x - 7x^2 + x^3$
```
r = PolynomialRemainder[p, s, x]
```
$-1 - 4x$

- **Expand[*polynomial*]** expands products and powers, writing *polynomial* as a sum of individual terms.
- **Factor[*polynomial*]** attempts to factor *polynomial* over the integers. If factoring is unsuccessful, ***polynomial*** is unchanged.
- **FactorTerms[*polynomial*]** factors out common constants which appear in the terms of *polynomial*. FactorTerms[*polynomial*, *var*] factors out any common monomials containing variables other than *var*.
- **Collect[*polynomial*, *var*]** takes a polynomial having two or more variables and expresses it as a polynomial in *var*.

EXAMPLE 5

```
poly = 6x²y³z⁴ + 8x³y²z⁵ + 10x²y⁴z³;
Factor[poly]
```
$2x^2y^2z^3(5y^2 + 3yz + 4xz^2)$ ← poly is factored completely
```
FactorTerms[poly, x]
```
$2y^2z^3(5x^2y^2 + 3x^2yz + 4x^3z^2)$ ← only the terms *not* involving x are factored
```
FactorTerms[poly, y]
```
$2x^2z^3(5y^4 + 3y^3z + 4xy^2z^2)$ ← only the terms *not* involving y are factored
```
FactorTerms[poly, z]
```
$2x^2y^2(5y^2z^3 + 3yz^4 + 4xz^5)$ ← only the terms *not* involving z are factored

EXAMPLE 6

```
poly = 1 + 2 x + 3 y + 4 x y + 5 x²y + 6x y² + 7 x²y²;
Collect[poly, x]
```
$1 + 3y + x(2 + 4y + 6y^2) + x^2(5y + 7y^2)$ ← powers of x are factored out
```
Collect[poly, y]
```
$1 + 2x + (3 + 4x + 5x^2)y + (6x + 7x^2)y^2$ ← powers of y are factored out

By default, Factor allows factorization only over the integers. There are options which allow this default to be overridden.

- **Extension → {*extension1*, *extension2*, ...}** can be used to specify a list of algebraic numbers which may be included as well. (The braces, **{}**, are not necessary if only one extension is needed.) **Extension → Automatic** extends the field to include any algebraic numbers which appear in the polynomial.
- **GaussianIntegers → True** allows the factorization to take place over the set of integers with i adjoined. Alternatively, i or I may be included in the list of extensions.

EXAMPLE 7

```
Factor[x⁸ - 41x⁴ + 400]
```
$(-2 + x)(2 + x)(-5 + x^2)(4 + x^2)(5 + x^2)$
```
Factor[x⁸ - 41x⁴ + 400, GaussianIntegers → True]
```
$(-2 + x)(-2i + x)(2i + x)(2 + x)(-5 + x^2)(5 + x^2)$
```
Factor[x⁸ - 41x⁴ + 400, Extension → √5]
```
$-\left(\sqrt{5} - x\right)(-2 + x)(2 + x)\left(\sqrt{5} + x\right)(4 + x^2)(5 + x^2)$
```
Factor[x⁸ - 41x⁴ + 400, Extension → {I, √5}]
```
$-\left(\sqrt{5} - x\right)\left(\sqrt{5} - ix\right)\left(\sqrt{5} + ix\right)(-2 + x)(-2i + x)(2i + x)(2 + x)\left(\sqrt{5} + x\right)$

The greatest common divisor (GCD) of polynomials, $p_1, p_2, \ldots$ is the polynomial of largest degree which can be divided evenly (remainder 0) into $p_1, p_2, \ldots$. The least common multiple (LCM) of polynomials $p_1, p_2, \ldots$ is the polynomial of smallest degree which can be divided evenly by $p_1, p_2, \ldots$.

- **PolynomialGCD[p1, p2, ...]** computes the greatest common divisor of the polynomials **p1, p2, ...**.
- **PolynomialLCM[p1, p2, ...]** computes the least common multiple of the polynomials **p1, p2, ...**.

EXAMPLE 8

```
p = (x - 1)(x - 2)²(x - 3)³;
q = (x - 1)²(x - 2)(x - 3)⁴;
PolynomialGCD[p, q]
```
$(-3 + x)^3 (-2 + x) (-1 + x)$
```
PolynomialLCM[p, q]
```
$(-3 + x)^4 (-2 + x)^2 (-1 + x)^2$

By default, both `PolynomialGCD` and `PolynomialLCM` assume the coefficients of the polynomials to be rational numbers. As with `Factor`, the option `Extension` can be used to specify a list of algebraic numbers (and/or `I`) which may be allowed.

EXAMPLE 9

```
p = x² - 5;
q = x + √5
PolynomialGCD[p, q]
```
1
```
PolynomialLCM[p, q]
```
$(\sqrt{5} + x)(-5 + x^2)$ ← Without **Extension → Automatic**, $\sqrt{5}$ is
                                        treated as a separate algebraic number.
```
PolynomialGCD[p, q, Extension → Automatic]
```
$\sqrt{5} + x$
```
PolynomialLCM[p, q, Extension → Automatic]
```
$(-5 + x^2)$

Although *Mathematica* will automatically expand integer exponents of products and quotients, if the exponent is noninteger, the expression will be left unexpanded. To force the "distribution" of the exponent, the command **PowerExpand** is available.

- **PowerExpand[*expression*]** expands nested powers, powers of products and quotients, roots of products and quotients, and their logarithms.

EXAMPLE 10

```
(a b)⁵
```
$a^5 b^5$ ← *Mathematica* distributes the exponent because it is an integer.
```
(a b)ˣ
```
$(a\ b)^x$ ← *Mathematica* does nothing because the exponent is undefined.
```
PowerExpand[(a b)ˣ]
```
$a^x b^x$ ← We force the expansion with **PowerExpand**.

One must be very careful with `PowerExpand` when multivalued functions are involved.

EXAMPLE 11

```
√a b /.{a→ -1,b→ -1}
1
PowerExpand[√a b]
√a√b
PowerExpand[√a b] /.{a→ -1,b→ -1}
-1                          ←PowerExpand expands and then replaces the values of a and b by −1.
```

Here are a few additional examples illustrating PowerExpand:

EXAMPLE 12

```
(aˣ)ʸ // PowerExpand
aˣʸ
(a/b)ˣ // PowerExpand
aˣb⁻ˣ
Log[xy] // PowerExpand
Log[x] + Log[y]
Log[x/y] // PowerExpand
Log[x] - Log[y]
Log[xʸ] // PowerExpand
y Log[x]
```

SOLVED PROBLEMS

7.1　Test to see if $1 + x \sin y + x^2 \cos y + x^5 e^y$ is a polynomial in x. Is it a polynomial in y?

SOLUTION

```
PolynomialQ[1 + x Sin[y] + x² Cos[y] + x⁵ Exp[y], x]      ←y is treated as a constant in this expression.
True
PolynomialQ[1 + x Sin[y] + x² Cos[y] + x⁵ Exp[y], y]
False
```

7.2　What are the coefficients of the polynomial expansion of $(2x + 3)^5$?

SOLUTION

```
poly = (2x + 3y)⁵;
CoefficientList[poly, x]
{243, 810, 1080, 720, 240, 32}
```

7.3　What is the coefficient of $x y^2 z^3$ in the expansion of $(x + y + z)^6$?

SOLUTION

```
poly = (x + y + z)⁶;
Coefficient[poly, x y² z³]
60
```

7.4   Expand $(x + a + 1)^4$ completely.

SOLUTION

```
Expand[(x + a + 1)⁴]
```
$1 + 4a + 6a^2 + 4a^3 + a^4 + 4x + 12ax + 12a^2x + 4a^3x + 6x^2 + 12ax^2 + 6a^2x^2 + 4x^3 + 4ax^3 + x^4$

7.5   Express $(x + a + 1)^4$ as a polynomial in x.

SOLUTION

```
Collect[(x + a + 1)⁴, x]
```
$1 + 4a + 6a^2 + 4a^3 + a^4 + (4 + 12a + 12a^2 + 4a^3)x + (6 + 12a + 6a^2)x^2 + (4 + 4a)x^3 + x^4$

7.6   Factor the polynomial

$$poly = 6x^3 + x^2y - 11xy^2 - 6y^3 - 5x^2z + 11xyz + 11y^2z - 2xz^2 - 6yz^2 + z^3$$

and solve for z so that $poly = 0$.

SOLUTION

```
poly = 6x³ + x²y - 11xy² - 6y³ - 5x²z + 11xyz + 11y²z - 2xz² - 6yz² + z³;
Factor[poly]
```
$(x + y - z)(3x + 2y - z)(2x - 3y + z)$
```
Roots[poly == 0, z]
```
$z == x + y \,||\, z == 3x + 2y \,||\, z == -2x + 3y$

7.7   Find the quotient and remainder when $x^5 + 2x^4 - 3x^3 + 7x^2 - 10x + 5$ is divided by $x^2 - 4$ and verify that this answer is correct.

SOLUTION

```
p = x⁵ + 2x⁴ - 3x³ + 7x² - 10x + 5
s = x² - 4
q = PolynomialQuotient[p, s, x]
```
$15 + x + 2x^2 + x^3$
```
r = PolynomialRemainder[p, s, x]
```
$65 - 6x$
```
checkpoly = q * s + r // Expand
```
$5 - 10x + 7x^2 - 3x^3 + 2x^4 + x^5$
```
checkpoly == p
```
True

7.8   Express $(x + y + z)^3$ as a polynomial in z.

SOLUTION

```
Collect[(x + y + z)³, z]
```
$x^3 + 3x^2y + 3xy^2 + y^3 + (3x^2 + 6xy + 3y^2)z + (3x + 3y)z^2 + z^3$

7.9   Let $p = 2x^4 - 15x^3 + 39x^2 - 40x + 12$ and $q = 4x^4 - 24x^3 + 45x^2 - 29x + 6$. Compute their GCD and LCM and show that their product is equal to $p\,q$.

SOLUTION

```
p = 2 x⁴ - 15 x³ + 39 x² - 40 x + 12;
q = 4 x⁴ - 24 x³ + 45 x² - 29 x + 6;
a = PolynomialGCD[p, q]
```
$-6 + 17 x - 11 x^2 + 2 x^3$
```
b = PolynomialLCM[p, q]
```
$(-2 + x) (6 - 29 x + 45 x^2 - 24 x^3 + 4 x^4)$
```
Expand[a * b] == Expand[p * q]
```
True

7.10 Factor $x^4 - 25$ over the integers and then over the field containing $\sqrt{5}$ and I.

SOLUTION

```
Factor[x⁴ - 25]
```
$(-5 + x^2) (5 + x^2)$
```
Factor[x⁴ - 25, Extension → { √5, I }]
```
$- \left(\sqrt{5} - x\right) \left(\sqrt{5} - i\, x\right) \left(\sqrt{5} + i\, x\right) \left(\sqrt{5} + x\right)$

7.11 Expand $\ln\left[\sqrt{\dfrac{x^a y^b}{z^c}}\right]$.

SOLUTION

```
Log [ √(xᵃyᵇ/zᶜ) ] // PowerExpand
```
$\dfrac{1}{2} (a \operatorname{Log}[x] + b \operatorname{Log}[y] - c \operatorname{Log}[z])$

7.2 RATIONAL AND ALGEBRAIC FUNCTIONS

There are a few commands appropriate for use with rational functions (fractions).

- **Numerator[***fraction***]** returns the numerator of *fraction*.
- **Denominator[***fraction***]** returns the denominator of *fraction*.
- **Cancel[***fraction***]** cancels out common factors in the numerator and denominator of *fraction*. The option **Extension → Automatic** allows operations to be performed on algebraic numbers which appear in *fraction*.
- **Together[***expression***]** combines the terms of *expression* using a common denominator. Any common factors in numerator and denominator are canceled.
- **Apart[***fraction***]** writes *fraction* as a sum of partial fractions.

EXAMPLE 13

```
Cancel [ (x² + 5 x + 6) / (x² + 3 x + 2) ]
```
$\dfrac{3 + x}{1 + x}$

EXAMPLE 14

$$\text{Together}[\frac{1}{x+1}+\frac{2}{x^2-1}]$$

$$\frac{1}{-1+x}$$

EXAMPLE 15

$$\text{Apart}\left[\frac{x^2+5x}{x^4+x^3-x-1}\right]$$

$$\frac{1}{-1+x}+\frac{2}{1+x}+\frac{-1-3x}{1+x+x^2}$$

Since *Mathematica*, by default, converts factors with negative exponents to their positive exponent equivalents, the result of `Numerator` or `Denominator` may be different than expected.

EXAMPLE 16

$$\text{frac}=\frac{x^{-1}y^{-2}}{z^{-3}};$$

`Numerator[frac]`

z^3

`Denominator[frac]`

$x\,y^2$

- `ExpandNumerator[`*expression*`]` expands the numerator of *expression* but leaves the denominator alone.
- `ExpandDenominator[`*expression*`]` expands the denominator of *expression* but leaves the numerator alone.
- `ExpandAll[`*expression*`]` expands both numerator and denominator of *expression*, writing the result as a sum of fractions with a common denominator.

EXAMPLE 17

$$\text{expr}=\frac{(x+1)(x+2)}{(x+3)(x+4)};$$

`ExpandNumerator[expr]`

$$\frac{2+3x+x^2}{(3+x)(4+x)}$$

`ExpandDenominator[expr]`

$$\frac{(1+x)(2+x)}{12+7x+x^2}$$

`ExpandAll[expr]`

$$\frac{2}{12+7x+x^2}+\frac{3x}{12+7x+x^2}+\frac{x^2}{12+7x+x^2}$$

`ExpandNumerator[ExpandDenominator[expr]]`

$$\frac{2+3x+x^2}{12+7x+x^2}$$

The commands described in this section are not limited to rational functions (quotients of polynomials) but will work for algebraic expressions involving radicals, and nonalgebraic expressions involving functions or undefined objects. In addition, if the option **Trig → True** is set within the command, *Mathematica* will use standard trigonometric identities to simplify the expression. This will be discussed further in Section 7.3.

EXAMPLE 18

$$\texttt{Expand}\left[\left(1+\sqrt{x}\right)^6\right]$$

$1 + 6\sqrt{x} + 15\,x + 20\,x^{3/2} + 15\,x^2 + 6\,x^{5/2} + x^3$

EXAMPLE 19

$$\texttt{Apart}\left[\frac{1}{\left(\sqrt{x}+1\right)\left(\sqrt{x}+2\right)}\right]$$

$\dfrac{1}{1+\sqrt{x}} - \dfrac{1}{2+\sqrt{x}}$

SOLVED PROBLEMS

7.12 The expression $\dfrac{f(x)-f(a)}{x-a}$ appears in calculus in connection with the derivative. Simplify this expression for $f(x) = x^9$, $a = -3$.

SOLUTION

```
f[x_] = x⁹;
a = -3;
Cancel[ f[x] - f[a]
        ─────────── ]
          x - a
```

$6561 - 2187\,x + 729\,x^2 - 243\,x^3 + 81\,x^4 - 27\,x^5 + 9\,x^6 - 3\,x^7 + x^8$

7.13 Express the sum of $\dfrac{a}{b}, \dfrac{c}{d}$, and $\dfrac{e}{f}$ as a single fraction.

SOLUTION

```
Together[a/b + c/d + e/f]
```

$\dfrac{b\,d\,e + b\,c\,f + a\,d\,f}{b\,d\,f}$

7.14 Write $\dfrac{(x+2)(x^2+3)(2x-7)}{(x^2+5x+2)(x-5)(x+6)}$ with expanded numerator and denominator.

SOLUTION I

```
ExpandNumerator[ExpandDenominator[ (x+2) (x²+3) (2x-7)
                                    ───────────────────── ]]
                                   (x²+5x+2) (x-5) (x+6)
```

$\dfrac{-42 - 9\,x - 8\,x^2 - 3\,x^2 + 2\,x^4}{-60 - 148\,x - 23\,x^2 + 6\,x^3 + x^4}$

SOLUTION II

```
ExpandAll[ (x+2) (x²+3) (2x-7) ]//Together
           ───────────────────────
           (x²+5x+2) (x-5) (x+6)
```

$$\frac{-42-9x-8x^2-3x^2+2x^4}{-60-148x-23x^2+6x^3+x^4}$$

7.15 Add $\dfrac{2x+3}{5x-7}, \dfrac{7x-2}{3x+1}$, and $\dfrac{x^2}{x^2+1}$ and express as a single fraction with expanded numerator and denominator.

SOLUTION

```
   2x+3
p = ──── ;
   5x-7

   7x-2
q = ──── ;
   3x+1

    x²
r = ──── ;
   x²+1

Together[p+q+r]//ExpandDenominator
```

$$\frac{17-48x+51x^2-64x^3+56x^4}{-7-16x+8x^2-16x^3+15x^4}$$

> Without //ExpandDenominator, the denominator would be expressed in factored form.

7.16 What is the partial fraction expansion of $\dfrac{(x-1)^6}{(x^2+1)(x+1)^2(x-4)}$?

SOLUTION

```
Apart[    (x-1)⁶        ]
      ─────────────────
      (x²+1) (x+1)²(x-4)
```

$$-4+\frac{729}{425\,(-4+x)}+x-\frac{32}{5\,(1+x)^2}+\frac{288}{25\,(1+x)}-\frac{4\,(4+x)}{17\,(1+x^2)}$$

7.17 Find the partial fraction expansion of the function in the previous problem with linear complex denominators.

SOLUTION

```
Apart[      (x-1)⁶          ]
      ─────────────────────
      (x+I) (x-I) (x+1)²(x-4)
```

> To force *Mathematica* to express the result using linear complex denominators, we factor x^2+1 as $(x+I)(x-I)$.

$$-4+\frac{729}{425\,(-4+x)}+x-\frac{\frac{2}{17}-\frac{8i}{17}}{-i+x}-\frac{\frac{2}{17}+\frac{8i}{17}}{-i+x}-\frac{32}{5\,(1+x)^2}+\frac{288}{25\,(1+x)}$$

7.18 Express $(e^x+e^{2x})^4$ as a sum of exponentials.

SOLUTION

```
Expand[(Eˣ+E²ˣ)⁴]
```

$$e^{4x}+4\,e^{5x}+6\,e^{6x}+4\,e^{7x}+e^{8x}$$

7.3   TRIGONOMETRIC FUNCTIONS

Although the commands discussed in the previous section may be applied to trigonometric functions, doing so does not take advantage of the simplification offered by trigonometric identities. To incorporate these into the calculation, the option `Trig → True` must be set. (The default is `Trig → False` for all but the `Simplify` command.) The following examples show the difference.

EXAMPLE 20

$$\text{Cancel}\left[\frac{\text{Sin}[x]}{1 - \text{Cos}[x]^2}\right]$$

$$-\frac{\text{Sin}[x]}{-1 + \text{Cos}[x]^2}$$

$$\text{Cancel}\left[\frac{\text{Sin}[x]}{1 - \text{Cos}[x]^2}, \text{Trig} \to \text{True}\right]$$

$$\text{Csc}[x]$$

EXAMPLE 21

$$\text{Together}\left[\frac{\text{Cos}[x]^2}{1 - \text{Sin}[x]^2} + \frac{\text{Sin}[x]^2}{1 - \text{Cos}[x]^2}\right]$$

$$\frac{\text{Cos}[x]^2 - \text{Cos}[x]^4 + \text{Sin}[x]^2 - \text{Sin}[x]^4}{(-1 + \text{Cos}[x]^2) \ (-1 + \text{Sin}[x]^2)}$$

$$\text{Together}\left[\frac{\text{Cos}[x]^2}{1 - \text{Sin}[x]^2} + \frac{\text{Sin}[x]^2}{1 - \text{Cos}[x]^2}, \text{Trig} \to \text{True}\right]$$

$$2$$

`Trig → True` applies to hyperbolic as well as circular functions.

EXAMPLE 22

`Expand[(Cosh[x]² + Sinh[x]²)(Cosh[x]² - Sinh[x]²)]`

$\text{Cosh}[x]^4 - \text{Sinh}[x]^4$

`Expand[(Cosh[x]² + Sinh[x]²)(Cosh[x]² - Sinh[x]²), Trig → True]`

$\text{Cosh}[x]^2 + \text{Sinh}[x]^2$

To allow additional manipulation of trigonometric expressions, *Mathematica* offers the following specialized commands, which apply to both circular and hyperbolic functions.

- `TrigExpand[expression]` expands *expression*, taking advantage of certain trigonometric identities.
- `TrigReduce[expression]` rewrites products and powers of trig functions in *expression* as trigonometric expressions with combined arguments, reducing *expression* to a linear trig function (i.e., without powers or products).
- `TrigFactor[expression]` converts *expression* into a factored expression of trigonometric functions of a single argument.

The next example shows the difference between `Expand` and `TrigExpand`.

EXAMPLE 23

`Expand[(Sin[x] + Cos[x])²]`

$\text{Cos}[x]^2 + 2\,\text{Cos}[x]\,\text{Sin}[x] + \text{Sin}[x]^2$

`TrigExpand[(Sin[x] + Cos[x])²]`

$1 + 2\,\text{Cos}[x]\,\text{Sin}[x]$

EXAMPLE 24

```
TrigExpand[Sin[x+y]]
```
Cos[y] Sin[x] + Cos[x] Sin[y]
```
TrigExpand[Sin[2 x]]
```
2 Cos[x] Sin[x]
```
TrigExpand[Sin[2 x+y]]
```
2 Cos[x] Cos[y] Sin[x] + Cos[x]² Sin[y] - Sin[x]² Sin[y]

TrigExpand can also be applied to hyperbolic functions.

EXAMPLE 25

```
TrigExpand[Cosh[x+y]]
```
Cosh[x] Cosh[y] + Sinh[x] Sinh[y]

EXAMPLE 26

> **TrigReduce** rewrites the original expression as a linear trig expression.

```
TrigReduce[Sin[2 x]² + Sin[x] Cos[3 x]³]
```

$\frac{1}{8}$ (4 - 4 Cos[4x] - 3Sin[2x] + 3Sin[4x] - Sin[8x] + Sin[10x])

```
TrigReduce[Sinh[2 x]² + Sinh[x]Cosh[3 x]³]
```

$\frac{1}{8}$ (-4 + 4 Cosh[4x] - 3Sinh[2x] + 3Sinh[4x] - Sinh[8x] + Sinh[10x])

The next example shows the difference between TrigFactor and TrigReduce. Notice that TrigFactor writes the expression as a product while TrigReduce writes the expression as a sum.

EXAMPLE 27

```
expr = TrigExpand[4 Sin[x]² Cos [2 x]³]
```
3 Cos[x]² Sin[x]² + Cos[x]⁶ Sin[x]² - 3 Sin[x]⁴ -
  15 Cos[x]⁴ Sin[x]⁴ + 15 Cos[x]² Sin[x]⁶ - Sin[x]⁸
```
TrigFactor[expr]
```
4 (Cos[x] - Sin[x])³ Sin[x]² (Cos[x] + Sin[x])³
```
TrigReduce[expr]
```
$\frac{1}{4}$ (-3 + 6 Cos[2x] - 4 Cos[4x] + 2 Cos[6x] - Cos[8x])

The Solve command can be used to solve trigonometric equations. However, because only principal values of inverse trigonometric functions are returned, not all solutions will be obtained.

EXAMPLE 28

Consider the equation $1 - 2\cos x - \sin x + \sin 2x = 0$.

```
equation = 1 - 2 Cos[x] - Sin[x] + Sin[2x] == 0
```
```
Solve[equation, x]
```
Solve::ifun:
  Inverse functions are being used by Solve, so some solutions may not be found.

$\left\{ \left\{ x \to -\frac{\pi}{3} \right\}, \left\{ x \to \frac{\pi}{3} \right\}, \left\{ x \to \frac{\pi}{2} \right\} \right\}$

Since trigonometric and hyperbolic functions can be represented in terms of exponential functions (complex exponentials in the case of circular trig functions), *Mathematica* offers two conversion functions:

- **TrigToExp[*expression*]** converts trigonometric and hyperbolic functions to exponential form.
- **ExpToTrig[*expression*]** converts exponential functions to trigonometric and/or hyperbolic functions.

TrigToExp and ExpToTrig may also be used to convert *inverse* trigonometric and hyperbolic functions.

EXAMPLE 29

```
TrigToExp[Cos[x]]
```
$$\frac{e^{-ix}}{2} + \frac{e^{ix}}{2}$$
```
TrigToExp[Sinh[x]]
```
$$-\frac{e^{-x}}{2} + \frac{e^{x}}{2}$$
```
ExpToTrig[Exp[x]]
Cosh[x] + Sinh[x]
```

SOLVED PROBLEMS

7.19   Simplify the trigonometric function $\dfrac{1}{\cos^2 x - \sin^2 x}$.

SOLUTION

```
TrigReduce[ 1 / (Cos[x]² - Sin[x]²) ]
Sec[2 x]
```

7.20   Factor and simplify: $\sin^2 x \cos^2 x + \cos^4 x$.

SOLUTION

```
TrigFactor[Sin[x]² Cos[x]² + Cos[x]⁴]
Cos[x]²
```

7.21   Solve the trigonometric equation $1 - 2\cos x - 2\sin x + 4\sin 2x = 0$.

SOLUTION

```
equation = 1 - 2 Cos[x] - 2 Sin[x] + 4 Sin[2 x] == 0;
Solve[equation, x]
Solve::ifun :
  Inverse functions are being used by Solve, so some solutions may not be found.
```

$$\left\{\left\{x \to \text{ArcCos}\left[\frac{1}{8}+\frac{\sqrt{13}}{8}-\frac{1}{2}\sqrt{\frac{9}{8}-\frac{\sqrt{13}}{8}}\right]\right\},\right.$$

$$\left\{x \to \text{ArcCos}\left[\frac{1}{8}+\frac{\sqrt{13}}{8}+\frac{1}{2}\sqrt{\frac{9}{8}-\frac{\sqrt{13}}{8}}\right]\right\},$$

$$\left\{x \to \text{ArcCos}\left[\frac{1}{8}-\frac{\sqrt{13}}{8}-\frac{1}{2}\sqrt{\frac{9}{8}+\frac{\sqrt{13}}{8}}\right]\right\},$$

$$\left.\left\{x \to -\text{ArcCos}\left[\frac{1}{8}-\frac{\sqrt{13}}{8}+\frac{1}{2}\sqrt{\frac{9}{8}+\frac{\sqrt{13}}{8}}\right]\right\}\right\}$$

`% //N`   ←A numerical solution would probably be more useful.

`{{x → 1.40492}, {x → 0.165873}, {x → 2.83487}, {x → -1.26407}}`

7.22 Add and simplify: $\dfrac{\cos x}{1+\sin x}+\tan x$.

SOLUTION

`Together[`$\dfrac{\text{Cos[x]}}{\text{1 + Sin[x]}}$`+ Tan[x], Trig → True]`

$$\frac{1}{\left(\text{Cos}\left[\frac{x}{2}\right]-\text{Sin}\left[\frac{x}{2}\right]\right)\left(\text{Cos}\left[\frac{x}{2}\right]+\text{Sin}\left[\frac{x}{2}\right]\right)}$$

`TrigReduce[%]`

`Sec[x]`

> Sometimes you have to apply two or more trig commands to simplify completely.

7.23 Combine and simplify: $\dfrac{\sinh x}{\cosh x-\sinh x}+\dfrac{\cosh x}{\cosh x+\sinh x}$.

SOLUTION

`Together[`$\dfrac{\text{Sinh[x]}}{\text{Cosh[x] - Sinh[x]}}$`+`$\dfrac{\text{Cosh[x]}}{\text{Cosh[x] + Sinh[x]}}$`, Trig → True]`

`Cosh[2x]`

7.24 Construct a table of multiple angle formulas for $\sin nx$ and $\cos nx$, $n = 2, 3, 4$, and 5.

SOLUTION

`trigtable = Table[{n, TrigExpand[Sin[n x]], TrigExpand[Cos[n x]]}, {n, 2, 5}];`
`TableForm[trigtable, TableHeadings → {None, {"n", "    sin nx", "    cos nx"}}]`

n	sin nx	cos nx
2	$2\,\text{Cos}[x]\,\text{Sin}[x]$	$\text{Cos}[x]^2 - \text{Sin}[x]^2$
3	$3\,\text{Cos}[x]^2\,\text{Sin}[x] - \text{Sin}[x]^3$	$\text{Cos}[x]^3 - 3\,\text{Cos}[x]\,\text{Sin}[x]^2$
4	$4\,\text{Cos}[x]^3\,\text{Sin}[x] - 4\,\text{Cos}[x]\,\text{Sin}[x]^3$	$\text{Cos}[x]^4 - 6\,\text{Cos}[x]^2\,\text{Sin}[x]^2 + \text{Sin}[x]^4$
5	$5.\text{Cos}[x]^4\,\text{Sin}[x] - 10\,\text{Cos}[x]^2\,\text{Sin}[x]^3 + \text{Sin}[x]^5$	$\text{Cos}[x]^5 - 10\,\text{Cos}[x]^3\,\text{Sin}[x]^2 + 5\,\text{Cos}[x]\,\text{Sin}[x]^4$

7.25   Construct a table of linear trig formulas for $\sin^n x$ and $\cos^n x$, $n = 2, 3, 4$, and 5.

SOLUTION

```
trigtable = Table[{n, TrigReduce[Sin[x]ⁿ], TrigReduce[Cos[x]ⁿ]}, {n, 2, 5}];
TableForm[trigtable, TableHeadings → {None, {"n", "    sinⁿx", "    cosⁿx"}}]
```

n	$\sin^n x$	$\cos^n x$
2	$\frac{1}{2}(1 - \cos[2x])$	$\frac{1}{2}(1 + \cos[2x])$
3	$\frac{1}{4}(3\sin[x] - \sin[3x])$	$\frac{1}{4}(3\cos[x] + \cos[3x])$
4	$\frac{1}{8}(3 - 4\cos[2x] + \cos[4x])$	$\frac{1}{8}(3 + 4\cos[2x] + \cos[4x])$
5	$\frac{1}{16}(10\sin[x] - 5\sin[3x] + \sin[5x])$	$\frac{1}{16}(10\cos[x] + 5\cos[3x] + \cos[5x])$

7.26   Express e^{x+y} in terms of hyperbolic functions and expand.

SOLUTION

```
ExpToTrig[Eˣ⁺ʸ]
```
$$\cosh[x+y] + \sinh[x+y]$$
```
TrigExpand[%]
```
$$\cosh[x]\cosh[y] + \cosh[y]\sinh[x] + \cosh[x]\sinh[y] + \sinh[x]\sinh[y]$$

7.27   Express $\sinh^{-1} x$ and $\tanh^{-1} x$ in exponential form.

SOLUTION

```
TrigToExp[ArcSinh[x]]
```
$$\text{Log}\left[x + \sqrt{1 + x^2}\right]$$
```
TrigToExp[ArcTanh[x]]
```
$$-\frac{1}{2}\text{Log}[1-x] + \frac{1}{2}\text{Log}[1+x]$$

7.4   THE ART OF SIMPLIFICATION

There are many different ways to write any particular algebraic or trigonometric expression. Obviously one person's interpretation of "simple" may not agree with another's. For example, in dealing with rational functions, $(x + 3)^2$ may be preferable to $x^2 + 6x + 9$, but when manipulating polynomials, the latter is clearly more desirable.

As you have seen from reading this chapter, *Mathematica* offers a variety of commands which allow full control of how an expression will appear. With practice, you will learn to use these commands to reshape its appearance to suit your needs.

As a step in the direction towards simplification, *Mathematica* offers two commands which can be used to simplify complex structures.

- **Simplify[*expression*]** performs a sequence of transformations on *expression*, and returns the simplest form it finds.
- **FullSimplify[*expression*]** tries a wider range of transformations on *expression* involving elementary and special functions, and returns the simplest form it finds.

Simplify tries expanding, factoring, and other standard mathematical transformations to reduce the complexity of *expression*. Because of its general nature, Simplify tends to be quite slow in comparison to more direct instructions. FullSimplify always produces an expression *at least* as simple as Simplify, but may take somewhat longer.

You can specify a time limitation (in seconds) with the option **TimeConstraint**. The default for Simplify is **TimeConstraint → 300** and for FullSimplify, **TimeConstraint → Infinity**. For both commands, **Trig → True** is the default for trigonometric evaluation.

EXAMPLE 30

First let's generate a messy algebraic expression.

$$\texttt{messyexpression} = \texttt{Expand}\left[\left(\frac{1}{x+1}+\frac{1}{x+2}+\frac{1}{x+3}\right)^5\right]$$

$$\frac{1}{(1+x)^5}+\frac{1}{(2+x)^5}+\frac{5}{(1+x)(2+x)^4}+\frac{10}{(1+x)^2(2+x)^3}+\frac{10}{(1+x)^3(2+x)^2}+\frac{5}{(1+x)^4(2+x)}+$$

$$\frac{1}{(3+x)^5}+\frac{5}{(1+x)(3+x)^4}+\frac{5}{(2+x)(3+x)^4}+\frac{10}{(1+x)^2(3+x)^3}+\frac{10}{(2+x)^2(3+x)^3}+$$

$$\frac{20}{(1+x)(2+x)(3+x)^3}+\frac{10}{(1+x)^3(3+x)^2}+\frac{10}{(2+x)^3(3+x)^2}+\frac{30}{(1+x)(2+x)^2(3+x)^2}+$$

$$\frac{30}{(1+x)^2(2+x)(3+x)^2}+\frac{5}{(1+x)^4(3+x)}+\frac{5}{(2+x)^4(3+x)}+\frac{20}{(1+x)(2+x)^3(3+x)}+$$

$$\frac{30}{(1+x)^2(2+x)^2(3+x)}+\frac{20}{(1+x)^3(2+x)(3+x)}$$

Now we will simplify. Of course, *Mathematica* does not "remember" how messyexpression was generated.

```
Simplify[messyexpression]
```

$$\frac{(11+12x+3x^2)^5}{(1+x)^5(2+x)^5(3+x)^5}$$

EXAMPLE 31

```
messytrigexpression = Expand[(Tan[x]² + Sin[x]² + Cos[x]²)⁵]
```

Cos[x]¹⁰ + 5 Cos[x]⁶ Sin[x]² + 5 Cos[x]⁸ Sin[x]² + 10 Cos[x]² Sin[x]⁴ + 20 Cos[x]⁴ Sin[x]⁴ +
10 Cos[x]⁶ Sin[x]⁴ + 30 Sin[x]⁶ + 30 Cos[x]² Sin[x]⁶ + 10 Cos[x]⁴ Sin[x]⁶ +
20 Sin[x]⁸ + 5 Cos[x]² Sin[x]⁸ + Sin[x]¹⁰ + 10 Sin[x]⁴ Tan[x]² + 30 Sin[x]⁶ Tan[x]² +
5 Sin[x]⁸ Tan[x]² + 20 Sin[x]⁴ Tan[x]⁴ + 10 Sin[x]⁶ Tan[x]⁴ + 5 Sin[x]² Tan[x]⁶ +
10 Sin[x]⁴ Tan[x]⁶ + 5 Sin[x]² Tan[x]⁸ + Tan[x]¹⁰

```
Simplify[messytrigexpression]
```

Sec[x]¹⁰

CHAPTER 8

Differential Calculus

8.1 LIMITS

The limit of a function is the foundation stone of differential calculus. For a complicated function, the calculation of a limit can be quite difficult and can require specialized techniques for its evaluation. *Mathematica* has built-in procedures for accomplishing this task and always attempts to determine the *exact* value of the limit.

- **Limit[f[x], x→a]** computes the value of $\lim_{x \to a} f(x)$.

EXAMPLE 1

We wish to compute $\lim_{x \to 2} \dfrac{x^5 - 32}{x^3 - 8}$. Because both numerator and denominator approach zero as $x \to 2$, the limit is not immediately obvious.

Limit$\left[\dfrac{\text{x}^5 - 32}{\text{x}^3 - 8}, \text{x} \to 2\right]$

$\dfrac{20}{3}$

Left- and right-hand limits can be computed with the **Direction** option.

- **Direction→1** causes the limit to be computed as a *left-hand limit* with values of x approaching a from below.
- **Direction→-1** causes the limit to be computed as a *right-hand limit* with values of x approaching a from above.

The default for the Limit command is **Direction→Automatic**, which provides Direction→-1 except for limits at ∞. Thus *Mathematica may give a misleading representation of the limit of a discontinuous function* if the Direction option is omitted.

195

EXAMPLE 2

Consider $\lim\limits_{x \to 0} \dfrac{|x|}{x}$.

```
Limit[Abs[x]/x, x → 0]
```
1

By default, only the right-hand limit has been computed, since no direction was specified. To fully analyze the limit we must compute the left-hand limit as well.

```
Limit[Abs[x]/x, x → 0, Direction → 1]
```
−1

The limit does not exist since the left- and right-hand limits are different numbers.

Mathematica can compute infinite limits and limits at ∞.

EXAMPLE 3

```
Limit[1/x, x → 0, Direction → -1]
```
∞
```
Limit[1/x, x → 0, Direction → 1]
```
−∞
```
Limit[(2 x² + 3 x + 4)/(x² + 1), x → ∞]
```
2

The functions in the next example exhibit a different behavior. As $x \to 0$, the function oscillates an infinite number of times. *Mathematica* returns the limit as an `Interval` object. `Interval[{min, max}]` represents the range of values between *min* and *max*.

EXAMPLE 4

```
Limit[Sin[1/x], x → 0]
Interval[{-1, 1}]
Limit[Tan[1/x], x → 0]
Interval[{-∞, ∞}]
```

SOLVED PROBLEMS

8.1   Compute $\lim\limits_{x \to 0} \dfrac{2^x + x - 1}{3x}$.

      SOLUTION

```
Limit[(2^x + x - 1)/(3 x), x → 0]
```
$\dfrac{1}{3}$ (1 + Log[2])

8.2   Compute $\lim\limits_{x \to 0} \dfrac{\tan x - x}{x^3}$.

SOLUTION

$$\text{Limit}\left[\frac{\text{Tan}[x]-x}{x^3}, x \to 0\right]$$

$$\frac{1}{3}$$

8.3   Compute $\lim_{x\to 0}(1+\sin x)^{\cot 2x}$.

SOLUTION

$$\text{Limit}[(1+\text{Sin}[x])^{\cot[2x]}, x \to 0]$$

$$\sqrt{e}$$

8.4   Compute $\lim_{x\to\infty}(e^x+x)^{1/x}$ and $\lim_{x\to-\infty}(e^x+x)^{1/x}$.

SOLUTION

$$\text{Limit}[(\text{Exp}[x]+x)^{1/x}, x \to \infty]$$

e

$$\text{Limit}[(\text{Exp}[x]+x)^{1/x}, x \to -\infty]$$

1

8.5   Compute $\lim_{x\to 1}(2-x)^{\tan(\frac{\pi}{2}x)}$.

SOLUTION

$$\text{Limit}\left[(2-x)^{\text{Tan}[\frac{\pi}{2}x]}, x \to 1\right]$$

$e^{2/\pi}$

8.6   If p dollars is compounded n times per year at an annual interest rate of r, the money will be worth $p\left(1+\dfrac{r}{n}\right)^{nt}$ dollars after t years. How much will the money be worth after t years if it is compounded continuously ($n\to\infty$)?

SOLUTION

$$\text{Limit}[p(1+r/n)^{nt}, n \to \infty]$$

$e^{rt}\,p$

8.7   The derivative of a function is defined to be $\lim_{h\to 0}\dfrac{f(x+h)-f(x)}{h}$.   Use this definition to compute the derivative of $f(x) = \ln x + x^5 + \sin x$.

SOLUTION

$$f[x\_] = \text{Log}[x] + x^5 + \text{Sin}[x];$$

$$\text{Limit}\left[\frac{f[x+h]-f[x]}{h}, h \to 0\right]$$

$$\frac{1}{x} + 5\,x^4 + \text{Cos}[x]$$

8.8   The second derivative of a function can be computed as the limit

$$\lim_{h \to 0} \frac{f(x+h) - 2f(x) + f(x-h)}{h^2}$$

Use this limit to compute the second derivative of $f(x) = \ln x + x^5 + \sin x$.

SOLUTION

```
f[x_] = Log[x] + x^5 + Sin[x];
Limit[ (f[x+h] - 2f[x] + f[x-h])/h^2 , h -> 0]
```

$-\dfrac{1}{x^2} + 20\,x^3 - \text{Sin}[x]$

8.2 DERIVATIVES

There are several ways derivatives can be computed in *Mathematica*.   Each has its advantages and disadvantages, so the proper choice for a particular situation must be determined.

- If `f[x]` represents a function, its derivative is represented by `f'[x]`. Higher-order derivatives are represented by `f''[x]`, `f'''[x]`,....

EXAMPLE 5

```
f[x_] = x^5 + x^4 + x^3 + x^2 + x + 1;
f'[x]
```
$1 + 2x + 3x^2 + 4x^3 + 5x^4$
```
f''[x]
```
$2 + 6x + 12x^2 + 20x^3$
```
f'''[x]
```
$6 + 24x + 60x^2$

The prime notation can also be used for "built-in" functions, as illustrated in the next example. If the argument is omitted, *Mathematica* returns a pure function representing the required derivative. (Pure functions are discussed in the Appendix.)

EXAMPLE 6

```
Sqrt'
```
$\dfrac{1}{2\sqrt{\#1}}\,\&$
```
Sqrt'[x]
```
$\dfrac{1}{2\sqrt{x}}$
```
Sqrt''
```
$-\dfrac{1}{2\,(2\,\#1^{3/2})}\,\&$
```
Sqrt''[x]
```
$-\dfrac{1}{4\,x^{3/2}}$

- `D[f[x], x]` returns the derivative of f with respect to x.
- `D[f[x], {x, n}]` returns the nth derivative of f with respect to x.

EXAMPLE 7

```
D[x⁵ + x⁴ + x³ + x² + x + 1, x]
```
$1 + 2x + 3x^2 + 4x^3 + 5x^4$
```
D[x⁵ + x⁴ + x³ + x² + x + 1, {x, 2}]
```
$2 + 6x + 12x^2 + 20x^3$
```
D[x⁵ + x⁴ + x³ + x² + x + 1, {x, 3}]
```
$6 + 24x + 60x^2$

- $\partial_\square$, which can be found in the BasicInput palette, is equivalent to D. ∂_x will return the derivative with respect to x. The nth derivative is represented $\partial_{\{x, n\}}$.

EXAMPLE 8

```
∂ₓ(x⁵ + x⁴ + x³ + x² + x + 1)
```
$1 + 2x + 3x^2 + 4x^3 + 5x^4$
```
∂₍ₓ, ₂₎(x⁵ + x⁴ + x³ + x² + x + 1)
```
$2 + 6x + 12x^2 + 20x^3$
```
∂₍ₓ, ₃₎(x⁵ + x⁴ + x³ + x² + x + 1)
```
$6 + 24x + 60x^2$

- **Derivative[n]** is a *functional operator* which acts on a function to produce a new function, namely, its nth derivative. **Derivative[n][f]** gives the nth derivative of f as a pure function and **Derivative[n][f][x]** will evaluate the nth derivative of f at x.

It is useful to remember that f' is converted to Derivative[1][f]. Thus f'[x] becomes Derivative[1][f][x]. f'', f''', etc., are handled similarly.

EXAMPLE 9

```
f[x_] = x⁵ + x⁴ + x³ + x² + x + 1;
Derivative[1][f]
```
$1 + 2\#1 + 3\#1^2 + 4\#1^3 + 5\#1^4 \&$

← *Mathematica* returns a pure function representing the derivative of f. Pure functions are discussed in the Appendix.

```
Derivative[1][f][x]
```
$1 + 2x + 3x^2 + 4x^3 + 5x^4$

The *value* of a derivative at a specific point can be computed several different ways, depending upon how the derivative is computed. The next example illustrates the most common techniques.

EXAMPLE 10

```
f[x_] = (x² - x + 1)⁵;
f''[1]
```
30
```
D[f[x], {x, 2}] /. x → 1
```
30
```
∂₍ₓ, ₂₎f[x] /. x → 1
```
30

← In each of the first three parts of this example, the second derivative is computed first, and then x is replaced by 1.

```
g := Derivative[2][f]
g[1]
```
30
```
f[x_] = x³;
g[1]
```
6

← Here we have defined a new function, g, as the second derivative of f. If f is changed, g will be the second derivative of the new function. Note the use of := here. This is crucial if g is to reflect the change in f.

Mathematica computes derivatives of combinations of functions, sums, differences, products, quotients, and composites, by "memorizing" the various rules.   If we do not define the functions, we can see what the rules are.

EXAMPLE 11

```
Clear[f, g]
D[f[x] +g[x], x]
  f'[x] +g'[x]
```
← The derivative of a sum is the sum of the derivatives of its terms.

```
D[f[x]g[x], x]
  g[x]f'[x] + f[x]g'[x]
```
← This is the familar product rule.

```
D[f[x]/g[x], x]//Together
```
$$\frac{g[x]f'[x] - f[x]g'[x]}{g[x]^2}$$
← quotient rule

```
D[f[g[x]], x]
  f'[g[x]]g'[x]
```
← chain rule

We can use *Mathematica* to investigate some basic theory from a graphical perspective.   <u>Rolle's Theorem</u> guarantees, under certain conditions, the existence of a point where the derivative of a function is 0 :

Let f be continuous on the closed interval $[a, b]$ and differentiable on the open interval (a, b) and suppose $f(a) = f(b) = 0$.   Then there exists a number, c, between a and b, such that $f'(c) = 0$.

In other words, if a smooth (differentiable) function vanishes (has a value of 0) at two distinct locations, a and b, its derivative must vanish somewhere in between.

EXAMPLE 12

Show that the function $f(x) = (x^3 + 2x^2 + 15x + 2) \sin \pi x$ satisfies Rolle's Theorem on the interval $[0, 1]$ and find the value of c referred to in the theorem.

Since f is the product of a polynomial and a trigonometric sine function, f is continuous and differentiable everywhere.

```
f[x_] = (x³ + 2x² + 15x + 2) Sin[πx];
f[0]
  0
f[1]
  0
FindRoot[f'[c] == 0, {c, 0.5}]
  {c → 0.640241}
```
← We used 0.5 as our initial guess since it is halfway between 0 and 1.

```
Plot[{f[x], f[.640241]}, {x, 0, 1}];
```

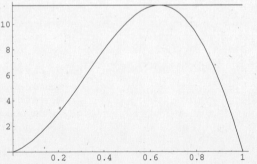

<u>The Mean Value Theorem</u> is similar to Rolle's theorem and does not require f to be 0 at each endpoint of the interval:

Let f be continuous on the closed interval $[a, b]$ and differentiable on the open interval (a, b). Then there exists a number, c, between a and b such that $f(b) - f(a) = f'(c)(b - a)$.

If we write the conclusion of the theorem in the form $\dfrac{f(b) - f(a)}{b - a} = f'(c)$, we see that the Mean Value Theorem guarantees the existence of a number, c, between a and b, such that the tangent line at $(c, f(c))$ is parallel to the line segment connecting the endpoints of the curve.

Note: Rolle's theorem and the Mean Value Theorem guarantee the existence of *at least* one number c. In actuality there may be several.

EXAMPLE 13

Find the value(s), c, guaranteed by the Mean Value Theorem for the function $f(x) = \sqrt{x} + \sin 2\pi x$ on the interval $[0, 2]$.

```
f[x_] = √x + Sin[2πx];
a = 0; b = 2;
    f[b] - f[a]
m = ──────────;
      b - a
Plot[f'[x] - m, {x, 0, 2}, PlotRange → {-8, 8}]
```

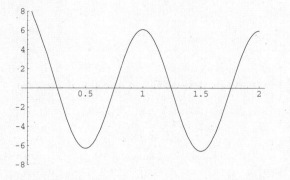

We estimate the zeros of the function $f'(x) - m$ to determine the approximate locations of c. There appear to be four values: 0.3, 0.7, 1.3, and 1.7 (approximately).

```
FindRoot[f'[c] == m, {c, .3}]
FindRoot[f'[c] == m, {c, .7}]
FindRoot[f'[c] == m, {c, 1.3}]
FindRoot[f'[c] == m, {c, 1.7}]
{c → 0.257071}
{c → 0.753319}
{c → 1.24344}
{c → 1.75836}
```

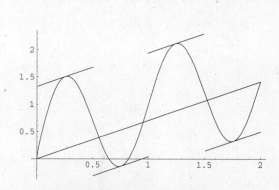

The tangent lines are parallel to the secant connecting the endpoints of the curve.

SOLVED PROBLEMS

8.9   Compute the 3rd derivative of $\tan x$.

SOLUTION

```
Tan'''[x] //Expand
```
$2\,\text{Sec}[x]^4 + 4\,\text{Sec}[x]^2\,\text{Tan}[x]^2$

$\leftarrow$ // Expand is needed for *Mathematica* version 3; versions 4 and 5 expand automatically.

8.10   Compute the values of the first 10 derivatives of $f(x) = e^{x^2}$ at $x = 0$.   Put the results in a tabular form.

SOLUTION

```
f[x_] = Exp[x²];
derivtable = Table[{n, D[f[x], {x, n}] /. x→0}, {n, 1, 10}];
TableForm[derivtable,
        TableAlignments→Center,
        TableDirections→Row,
        TableSpacing→2,
        TableHeadings→{None, {"n", "f⁽ⁿ⁾(0) "}}]
```

n	1	2	3	4	5	6	7	8	9	10
$f^{(n)}(0)$	0	2	0	12	0	120	0	1680	0	30240

8.11   Sketch the graph of $f(x) = x^4 - 50x^2 + 300$ and its derivative, on one set of axes, for $-10 \leq x \leq 10$.

SOLUTION

```
<<Graphics`Legend`
f[x_] = x⁴ - 50x² + 300;
Plot[{f[x], f'[x]}, {x, -10, 10},
     PlotStyle→{GrayLevel[0], Dashing[{.015}]},
     PlotLegend→{"f(x)", "f'{x}"}];
```

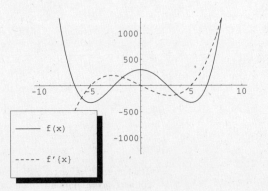

Observe that $f'(x) = 0$ precisely where $f(x)$ has a relative (local) maximum or minimum.

8.12   Given $f(x)$ whose graph is C, the slope of the line tangent to C at a is $f'(a)$. Let $f(x) = \sin x$.   Sketch the graph and its tangent line at $a = \pi/3$.

SOLUTION

```
f[x_] = Sin[x];
a = π/3;
l[x_] = f[a] + f'[a] (x-a);
Plot[{f[x], l[x]}, {x, 0, 2π}];
```

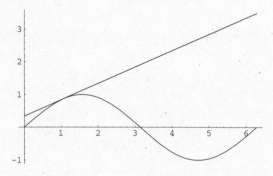

> Recall that the equation of a line having slope m passing through (x_1, y_1) is
> $$y - y_1 = m(x - x_1)$$
> or $\qquad y = y_1 + m(x - x_1)$
> Here, $x_1 = a, y_1 = f(a)$, and $m = f'(a)$ so
> $$y = f(a) + f'(a)(x - a)$$

8.13   Construct an animation showing the tangent line moving along the curve $y = \sin x$, $0 \le x \le 2\pi$.

SOLUTION

We would like to design the graphic so that the tangent line has a constant length, say 2, centered at the point of tangency.   To this end we construct a function $l[x]$ which effectively defines the tangent line for values of x between $a - 1/\sqrt{1 + m^2}$ and $a + 1/\sqrt{1 + m^2}$ (m represents the slope of the tangent line).

There is one small problem, however.   *Mathematica* returns an error message when we try to plot $l[x]$ on the interval $0 \le x \le 2\pi$ because $l[x]$ is only defined on a subset of this interval.   We use the command `Off[Plot::plnr]` to suppress the error message.

```
f[x_] = Sin[x];
m := f'[a]
l[x_] := f[a] + m(x-a) /; a - 1/√(1+m²) ≤ x ≤ a + 1/√(1+m²)
Do[Plot[{f[x], l[x]}, {x, 0, 2π},
    PlotRange → {-1.5, 1.5}], {a, 0, 2π, π/6}]
```
```
Plot::plnr :                    ← This message is repeated many times.
   l[x] is not a machine-size real number at x = 1.04500937601917009`.
```
```
Off[Plot::plnr]                 ← This eliminates the error message.
Do[Plot[{f[x], l[x]}, {x, 0, 2π},
    PlotRange → {-1.5, 1.5}], {a, 0, 2π, π/6}]
On[Plot::plnr]                  ← This restores the error messaging capability.
```

The generated images are shown on the next page.   Select the graphics cells, go to the Cell menu, and click on Cell ⇒ Animate Selected Graphics to start the animation.

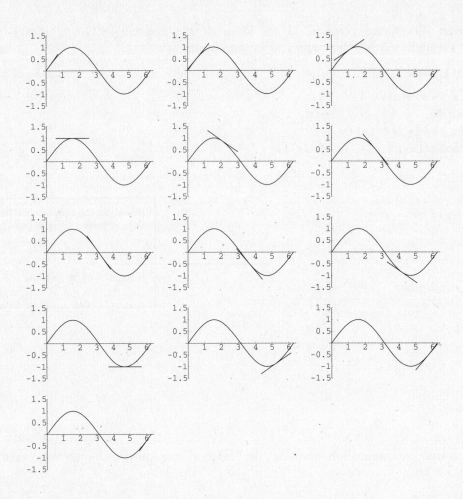

8.14   Find the value(s) of c guaranteed by Rolle's Theorem for the function
$f(x) = 4x + 39x^2 - 46x^3 + 17x^4 - 2x^5$ on the interval $[0, 4]$.

SOLUTION

Since $f(x)$ is a polynomial, it is continuous and differentiable everywhere. First we verify that
$f(0) = f(4) = 0$.

f[x_] = 4x + 39x² - 46x³ + 17x⁴ - 2x⁵;

f[0]

0

f[4]

0

Now we look to see where $f'(c) = 0$. Since
f' is a polynomial we can use **NSolve**.

NSolve [f'[c] == 0]

{{c→-0.0472411}, {c→1.05962},
  {c→2.27466}, {c→3.51296}}

There are three values of c between 0 and 4.
(Rolle's Theorem guarantees *at least* one.)
A plot of the graph confirms our result.

Plot[f[x], {x, 0, 4}];

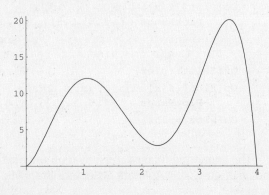

8.15   Verify the Mean Value Theorem for the function $f(x) = x + \sin 2x$ on the interval $[0, \pi]$.

SOLUTION

$f(x)$ is continuous and differentiable everywhere. Define $a = 0$, $b = 2\pi$ and solve the equation $f(b) - f(a) = f'(c)(b - a)$ for c. To approximate their values we look at the graph with the endpoints connected by a line segment.

```
f[x_] = x + Sin[2x];
a = 0; b = π;
    f[b] - f[a]
m = ─────────── ;                    ←slope of the secant connecting the endpoints
       b - a
l[x_] = f[a] + m(x - a);             ←function representing the secant line
Plot[{f[x], l[x]}, {x, a, b}];
```

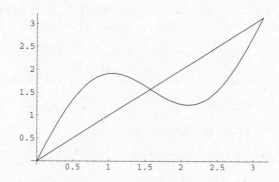

It looks like the tangent to the curve will be parallel to the secant near $x = 1$ and $x = 2.5$.

```
FindRoot[f[b] - f[a] == f'[c](b - a), {c, 1}]
FindRoot[f[b] - f[a] == f'[c](b - a), {c, 2.5}]
{c → 0.785398}
{c → 2.35619}                        Both values lie between 0 and π.
```

8.3 MAXIMUM AND MINIMUM VALUES

A function, f, has an *absolute (global) maximum* over an interval, I, at a point c if $f(x) \leq f(c)$ for all x in I. In other words, $f(c)$ is the largest value of $f(x)$ in I. A similar definition (with the inequality reversed) holds for an *absolute minimum*. One of the most important applications in differential calculus is optimization, i.e., finding the maximum and minimum values of a function, subject to certain constraints.

Not all functions have absolute maxima and minima. However the <u>Extreme Value Theorem</u> gives conditions sufficient to guarantee their existence:

If f is continuous on a closed bounded interval then f has both an absolute maximum and an absolute minimum in that interval.

A critical number of a function f is a number, c, for which $f'(c) = 0$ or $f'(c)$ fails to exist.

It can be shown that if a function is continuous on the closed interval $[a, b]$, then the absolute maximum and minimum will be found either at a critical number or at the endpoint of the interval. We can use *Mathematica* to help us find the maximum and/or minimum values.

EXAMPLE 14

We wish to find the absolute maximum and minimum values of the function $f(x) = x^4 - 4x^3 + 2x^2 + 4x + 2$ on the interval $[0, 4]$. First we find the critical numbers.

```
f[x_] = x⁴ - 4x³ + 2x² + 4x + 2;
Solve[f'[x] == 0]
```
$$\left\{\{x \to 1\}, \{x \to 1 - \sqrt{2}\}, \{x \to 1 + \sqrt{2}\}\right\}$$

Of these three numbers only two are in the interval $[0, 4]$. We compute the value of the function at these numbers as well as the endpoints of the interval.

```
c1 = 0; c2 = 1; c3 = 1 + √2; c4 = 4;
pointstocheck = {{c1, f[c1]}, {c2, f[c2]}, {c3, f[c3]}, {c4, f[c4]}}//Expand;
TableForm[pointstocheck, TableHeadings → {None, {"x", "f[x]"}}]
```

x	f[x]
0	2
1	5
$1 + \sqrt{2}$	1
4	50

The absolute maximum of f is 50 and the absolute minimum is 1.

EXAMPLE 15

A wire, 100 inches long, is to be used to form a square and a circle. Determine how the wire should be distributed in order for the combined area of the two figures to be (*a*) as large as possible and (*b*) as small as possible.

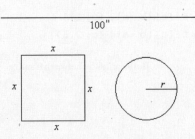

The combined area of the two figures is $A(x) = x^2 + \pi r^2$. The circle has a circumference of $2\pi r$, so it follows that $4x + 2\pi r = 100$. Since the wire is 100 inches long, $0 \le x \le 25$.

```
Solve[4 x + 2π r == 100, r]
```
$$\left\{\left\{r \to -\frac{2(-25 + x)}{\pi}\right\}\right\}$$

```
a[x_] = x² + π r² /. r → -  2(-25 + x)                    ←replace r in terms of x
                          ―――――――
                             π
```

$$\frac{4(-25 + x)^2}{\pi} + x^2$$

```
Solve[a'[x] == 0]                            ← find critical value(s)
```

$$\left\{\left\{x \to \frac{100}{4+\pi}\right\}\right\}$$

```
x1 = 0;
```

$$x2 = \frac{100}{4+\pi};$$

```
x3 = 25;
```

Compute the values of $a(x)$ at these three values. The values are placed in a table with numerical approximations for comparison.

```
pointstocheck = {{x1, a[x1], N[a[x1]]}, {x2, a[x2], N[a[x2]]},
        {x3, a[x3], N[a[x3]]}} //Together;
TableForm[pointstocheck, TableAlignments → Center,
        TableHeadings → {None, {"x", "a[x]", "N[a[x]]"}}]
```

x	a[x]	N[a[x]]
0	$\frac{2500}{\pi}$	795.775
$\frac{100}{4+\pi}$	$\frac{2500}{4+\pi}$	350.062
25	625	625.

The largest combined area occurs when $x = 0$ (all the wire is used to form the circle). The smallest area occurs when one side of the square is $\frac{100}{4+\pi}$ (cut the wire $\frac{400}{4+\pi}$ from one end).

To further confirm that $x = \frac{100}{4+\pi}$ gives a minimum area we can apply the second-derivative test.

$$\mathtt{Sign}\left[a''\left[\frac{100}{4+\pi}\right]\right]$$

```
1
```

> Since the sign of the second derivative at the critical point is positive, we have a relative minimum at $100/(4 + \pi)$. Since this is the only relative extremum, it must be the absolute minimum.

A function has a *relative (local) maximum* at c if there exists an open interval I containing c such that $f(x) \leq f(c)$ for all x in I. In other words, there exists an open interval containing c such that $f(c)$ is the largest value of f for all x in this interval. A similar definition holds for a relative minimum.

Unlike absolute maximum (minimum), a function may have several relative maxima (minima). If a numerical approximation of their location is all that is required, the *Mathematica* command **FindMinimum** offers an efficient (and convenient) procedure.

- **FindMinimum[f[x], {x, x0}]** finds the relative minimum of $f(x)$ near $x0$.

FindMinimum uses variations of the *Method of Steepest Descent* to find the relative minimum of a function. The option **Method → Newton** or **Method → QuasiNewton** can be used to change the method used.

As with FindRoot, the options AccuracyGoal and WorkingPrecision can be set if greater accuracy is desired. In addition, PrecisionGoal can be set to determine the precision in the value of the function at the minimum point. (Precision is the number of significant digits in the answer; accuracy is the number of significant digits to the right of the decimal point.)

Although *Mathematica* does not provide a command for computing relative maxima, **-FindMinimum[-f[x], {x, x0}]** will effectively do that job (an extra minus sign will appear in the answer, but can be ignored).

EXAMPLE 16

The function $f(x) = x + \sin(5x)$ has three relative maxima and two relative minima in the interval $[0, \pi]$. A quick look at the graph of the function gives good approximations to their locations.

```
f[x_] = x + Sin[5x];
Plot[f[x], {x, 0, π}];
```

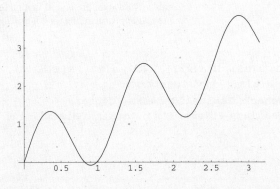

```
FindMinimum[f[x], {x, 1}]
{-0.0775897, {x→0.902206}}          ←The value of the function comes first, followed by the value of x.
FindMinimum[f[x], {x, 2}]
{1.17905, {x→2.15884}}
-FindMinimum[-f[x], {x, 0.4}]
{1.33423, {-(x→0.354431)}}          ←Ignore this minus (−) sign.
-FindMinimum[-f[x], {x, 1.6}]
{2.59086, {-(x→1.61107)}}
-FindMinimum[-f[x], {x, 2.8}]
{3.8475, {-(x→2.8677)}}
```

Thus the relative maxima are at $(0.354431, 1.33423)$, $(1.61107, 2.59086)$, and $(2.8677, 3.8475)$. The minima are at $(0.902206, -0.0775897)$ and $(2.15884, 1.17905)$.

SOLVED PROBLEMS

8.16 Find two positive numbers whose sum is 50, such that the square root of the first added to the cube root of the second is as large as possible.

SOLUTION

```
y = 50 - x;                            Plot[f[x], {x, 0, 50}];
f[x_] = √x + ∛y;
NSolve[f'[x] == 0]
{{x→41.1553}}
-FindMinimum[-f[x], {x, 40}]   ←alternate method
{8.48329, {-(x→41.1553)}}
y /. x→41.1553
8.8447
```

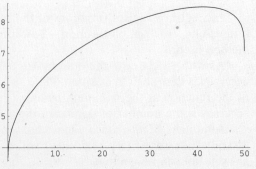

The two numbers are $x = 41.1553$ and $y = 8.8447$.
The maximum sum is 8.48329.

8.17 A right circular cylinder is inscribed in a unit sphere.

(a) Find the largest possible volume.
(b) Find the largest possible surface area.

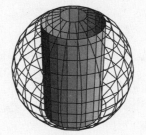

SOLUTION

(a) We consider a two-dimensional perspective of the problem. Label the radius and height of the inscribed cylinder r and h, respectively. The volume of the inscribed cylinder is $V = \pi r^2 h$ and, by the Theorem of Pythagoras, $r^2 + \left(\dfrac{h}{2}\right)^2 = 1$. It is easily seen (even without *Mathematica*) that $r^2 = 1 - \left(\dfrac{h}{2}\right)^2$. Thus the volume, as a function of h, becomes $V(h) = \pi\left[1 - \left(\dfrac{h}{2}\right)^2\right]h$.

```
v[h_] = π(1 - (h/2)²) h;

Solve[v'[h] == 0, h]
```
$$\left\{\left\{h \to -\frac{2}{\sqrt{3}}\right\}, \left\{h \to \frac{2}{\sqrt{3}}\right\}\right\}$$

```
vmax = v[2/√3]
```
Obviously only the positive value of h is appropriate.

$$\frac{4\pi}{3\sqrt{3}}$$

```
Sign[v''[2/√3]]
```
−1

Since the sign of the second derivative at the critical point is negative, we have a relative maximum at $2/\sqrt{3}$. Since this is the only relative extremum, it must be the absolute maximum.

(b) The surface area of the cylinder (including top and bottom) is $S = 2\pi rh + 2\pi r^2$. As in part (a), $r^2 + \left(\dfrac{h}{2}\right)^2 = 1$ but because r and r^2 both appear in the equation for S, it is easier to solve for h in terms of r.

```
Solve[r² + (h/2)² == 1, h]
```
$$\left\{\left\{h \to -2\sqrt{1-r^2}\right\}, \left\{h \to 2\sqrt{1-r^2}\right\}\right\}$$

Now substitute the (positive) value of h into the formula for S:

```
s[r_] = 2 π r h + 2 π r² /. h → 2√(1-r²)
```
$2\pi r^2 + 4\pi r\sqrt{1-r^2}$

```
s'[r]
```

$$4 \pi r - \frac{4 \pi r^2}{\sqrt{1-r^2}} + 4 \pi \sqrt{1-r^2}$$

Solve for the critical value of *r*:

```
Solve[s'[r] == 0, r]//Simplify
```

$$\left\{\left\{r \to - \sqrt{\frac{1}{10}\left(5-\sqrt{5}\right)}\right\}, \left\{r \to \sqrt{\frac{1}{10}\left(5+\sqrt{5}\right)}\right\}\right\}$$

Only the positive value of *r* is acceptable.   We use it to compute the maximum surface area.

```
s[√(1/10 (5 + √5))]//Simplify
```

$$\left(1+\sqrt{5}\right)\pi$$

```
Sign[s''[√(1/10 (5 + √5))]]
```
   ←confirmation of an absolute maximum

$$-1$$

8.18   Find the points on the circle $x^2 + y^2 - 2x - 4y = 0$ closest to and furthest from P(4,4).

SOLUTION

First we draw a diagram.

```
<<Graphics`ImplicitPlot`
circ = ImplicitPlot[x² + y² - 2x - 4y == 0, {x, -5, 5}, DisplayFunction → Identity];
p = Graphics[{PointSize[.02], Point[{4, 4}]}];
Show[circ, p, DisplayFunction → $DisplayFunction];
```

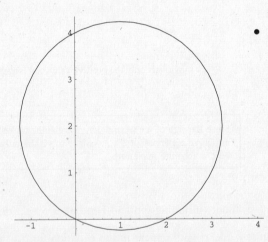

Let (x, y) represent a point on the circle.   First, we need to solve for *y* in terms of *x*.

```
Solve[x² + y² - 2x - 4y == 0, y] //Simplify
```
$$\left\{\left\{y \to 2 - \sqrt{4 + 2x - x^2}\right\}, \left\{y \to 2 + \sqrt{4 + 2x - x^2}\right\}\right\}$$

We shall minimize the *square* of the distance from (x, y) to $(4, 4)$.   We call this d2.   It is clear from the picture that the point closest to P lies on the upper semicircle.

```
y = 2 + √(4 + 2x - x²);
d2[x_] = (x - 4)² + (y - 4)²;
```

```
Solve [d2' [x] == 0]
```

$$\left\{\left\{x \to \frac{1}{13}\left(13 + 3\sqrt{65}\right)\right\}\right\}$$

```
{x, y} /. x → 1/13 (13 + 3 √65) //Simplify
```

$$\left\{1 + 3\sqrt{\frac{5}{13}}, 2 + 2\sqrt{\frac{5}{13}}\right\}$$

```
%//N
{2.86052, 3.24035}
```

The point furthest from P lies on the lower semicircle.

```
y = 2 - √(4 + 2 x - x²);
d2[x_] = (x - 4)² + (y - 4)²;
Solve [d2' [x] == 0]
```

$$\left\{\left\{x \to \frac{1}{13}\left(13 - 3\sqrt{65}\right)\right\}\right\}$$

```
{x, y} /. x → 1/13 (13 - 3 √65) //Simplify
```

$$\left\{1 - 3\sqrt{\frac{5}{13}}, 2 - 2\sqrt{\frac{5}{13}}\right\}$$

```
%//N
{-0.860521, 0.759653}
```

8.19 A river is 100 feet wide. A local telephone company wants to run a cable from point A on one side of the river to a point B on the opposite side, 500 feet along the shore from point C which is opposite A. However, it costs 3 times as much money to run the cable underwater as on land. How should the company run the cable in order to minimize cost?

SOLUTION

If we let a represent the cost per foot to run the cable on land, $3a$ is the cost to run a foot of cable underwater. The total cost is then $c(x) = 3a\sqrt{x^2 + 100^2} + a(500 - x)$. Of course, $0 \le x \le 500$.

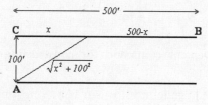

```
c[x_] = 3 a √(x² + 100²) + a (500 - x);
Solve[c' [x] == 0, x]
```

$$\left\{\left\{x \to 25\sqrt{2}\right\}\right\}$$

Now we must compare the cost corresponding to this solution with the cost at the endpoints of the interval.

```
c[0] //N
800.a
c[25 √2] //N
782.843 a
c[500] //N
1529.71 a
```

For minimum cost run the cable underwater to the point $25\sqrt{2}$ feet from C, then on land to B.

8.4 POWER SERIES

The nicest functions to work with are polynomials. They are continuous and can easily be differentiated and integrated. If a difficult function is encountered in a problem, one approach is to approximate it by a polynomial.

If the value of the function and its derivatives are known at a single point, a, the function can often be represented by a power series. This, however, is usually an infinite series which must be truncated for practical application. The trick is to truncate it in such a way that it accurately approximates the given function, at least in some neighborhood of a.

The following series, known as a Taylor series, gives a representation of an analytic[1] function, $f(x)$. If $a = 0$, the series is known as a Maclaurin series.

$$f(x) = \sum_{k=0}^{\infty} \frac{f^{(k)}(a)}{k!} (x - a)^k$$

$f^{(k)}(a)$ represents the kth derivative of f evaluated at a. If $k = 0$, it represents $f(a)$.

If we truncate this infinite series by omitting all terms of degree greater than n, we obtain the *Taylor polynomial* of f about a of degree n. We shall represent this polynomial as $p_n(x)$. If $a = 0$, the polynomial is called a Maclaurin polynomial.

EXAMPLE 17

To obtain the Maclaurin polynomial of degree 5 for the function $f(x) = e^x$, we can use the Sum command or the Σ symbol from the BasicInput palette. Here are three different ways the polynomial can be generated.

```
f[x_] = Exp[x];
```
(a) `Sum[(D[f[x], {x, k}]/.x→0)/k! * x^k, {k, 0, 5}]`

$$1 + x + \frac{x^2}{2} + \frac{x^3}{6} + \frac{x^4}{24} + \frac{x^5}{120}$$

(b) $\sum_{k=0}^{5} \dfrac{\partial_{\{x,k\}} \texttt{f[x]} /. x \to 0}{k!} x^k$

$$1 + x + \frac{x^2}{2} + \frac{x^3}{6} + \frac{x^4}{24} + \frac{x^5}{120}$$

(c) $\sum_{k=0}^{5} \dfrac{\texttt{Derivative[k][f][0]}}{k!} x^k$

$$1 + x + \frac{x^2}{2} + \frac{x^3}{6} + \frac{x^4}{24} + \frac{x^5}{120}$$

Mathematica includes a convenient command for constructing a series approximation to a function.

- `Series[f[x], {x, a, n}]` generates a `SeriesData` object representing the Taylor polynomial of `f[x]` of degree n about a.

EXAMPLE 18

```
f[x_] = Exp[x];
Series[f[x], {x, 0, 5}]
```

$$1 + x + \frac{x^2}{2} + \frac{x^3}{6} + \frac{x^4}{24} + \frac{x^5}{120} + O[x]^6$$

---

[1] An analytic function of a real variable is one which has a Taylor series expansion. Most functions encountered in applications are of this type; however, even if a function has derivatives of all orders, it may not be analytic.

The symbol $O[x]^6$ in the above expansion represents the "order" of omitted terms in the (infinite) expansion. $O[x]^6$ means that the omitted terms have powers of x of degree ≥ 6.

A SeriesData object is a *representation* of a power series but does not have a numerical value. If we try to compute the value of a series obtained from the Series command, we run into trouble.

EXAMPLE 19

```
f[x_] = Exp[x];
p[x_] = Series[f[x], {x, 0, 5}]
```

$$1 + x + \frac{x^2}{2} + \frac{x^3}{6} + \frac{x^4}{24} + \frac{x^5}{120} + O[x]^6$$

```
p[1]
```
```
SeriesData::ssdn: Attempt to evaluate a series at the number
    1; returning Indeterminate.
Indeterminate
```

To see what a SeriesData object looks like, the command **InputForm** can be used.

- **InputForm[*expression*]** prints *expression* in a form suitable for input to *Mathematica*.

EXAMPLE 20

```
f[x_] = Exp[x];
s = Series[f[x], {x, 0, 5}];
InputForm[s]
SeriesData[x, 0, {1, 1, 1/2, 1/6, 1/24, 1/120}, 0, 6, 1]
```

In order to convert the series into one which can be evaluated, the function **Normal** can be used to transform it to an ordinary polynomial.

- **Normal[*series*]** returns a polynomial representation of *series* which can be evaluated. The $O[x]^n$ term is omitted.

EXAMPLE 21

```
f[x_] = Exp[x];
s = Series[f[x], {x, 0, 5}]
```

$$1 + x + \frac{x^2}{2} + \frac{x^3}{6} + \frac{x^4}{24} + \frac{x^5}{120} + O[x]^6$$

```
p[x_] = Normal[s]
```

$$1 + x + \frac{x^2}{2} + \frac{x^3}{6} + \frac{x^4}{24} + \frac{x^5}{120}$$

> Normal has converted the SeriesData object into an ordinary polynomial, whose value can be computed.

```
p[1]
```

$$\frac{163}{60}$$

The number obtained in the previous example, 163/60, is approximately 2.71667.   If we compare this to the (known) value of $e \approx 2.71828$ we see a small error in our approximation.   We would expect the error to diminish as the degree of the polynomial increases. This is shown to be the case in the next example.

EXAMPLE 22

```
f[x_] = Exp[x];
exactvalue = f[1];
p[n_] := Normal[Series[f[x], {x, 0, n}]] /. x → 1
data = Table[{n, N[p[n]], N[Abs[p[n] - exactvalue]]}, {n, 1, 10}];
TableForm[data, TableHeadings → {None, {"n", "p(1)", "Error"}}]
```

n	p(1)	Error
1	2.	0.718282
2	2.5	0.218282
3	2.66667	0.0516152
4	2.70833	0.0099485
5	2.71667	0.00161516
6	2.71806	0.000226273
7	2.71825	0.0000278602
8	2.71828	3.05862×10^{-6}
9	2.71828	3.02886×10^{-7}
10	2.71828	2.73127×10^{-8}

EXAMPLE 23

To see the convergence of a power series even more dramatically, we can construct an animation showing the sequence of Maclaurin polynomials converging to e^x. We consider the interval $[0, 5]$. Once the images are generated, select the graphics cells and go to Cell⇒Animate Selected Graphics to start the animation. The individual frames are shown below.

```
f[x_] = Exp[x];
p[n_, x_] := Normal[Series[f[t], {t, 0, n}]] /. t → x
Do[Plot[{p[n, x], f[x]}, {x, 0, 5}, PlotRange → {0, Exp[5]}], {n, 1, 10}]
```

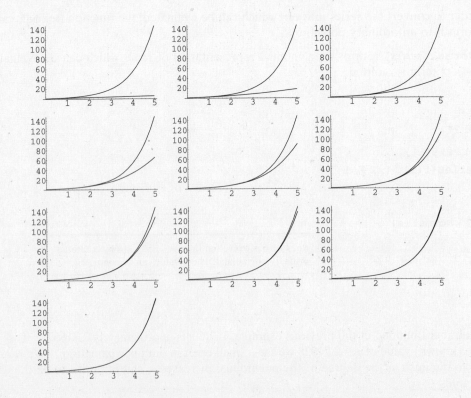

If only the *coefficient* of a particular term of a series is needed, the command **SeriesCoefficient** may be used. The actual series, which may be quite long, need not be printed. SeriesCoefficient is the SeriesData equivalent of the command Coefficient for polynomials.

- **SeriesCoefficient[*series*, n]** returns the coefficient of the nth degree term of a SeriesData object.

EXAMPLE 24

Find the coefficient of the 10th term of the Maclaurin series of e^x.

```
f[x_] = Exp[x];
s = Series[f[x], {x, 0, 10}];
SeriesCoefficient[s, 10]
```
$$\frac{1}{3628800}$$

SOLVED PROBLEMS

8.20 Obtain the Maclaurin polynomial of degree 10 for the function $f(x) = \tan^{-1} x$ using a direct summation and then by using the Series command.

SOLUTION

```
f[x_] = ArcTan[x];
```
$$\sum_{k=0}^{10} \frac{Derivative[k][f][0]}{k!} x^k$$
$$x - \frac{x^3}{3} + \frac{x^5}{5} - \frac{x^7}{7} + \frac{x^9}{9}$$
```
Series[f[x], {x, 0, 10}]
```
$$x - \frac{x^3}{3} + \frac{x^5}{5} - \frac{x^7}{7} + \frac{x^9}{9} + O[x]^{11}$$

8.21 Obtain a representation of x^5 in powers of $x - 2$.

SOLUTION

```
Series[x^5, {x, 2, 5}] //Normal
```
$$32 + 80(-2 + x) + 80(-2 + x)^2 + 40(-2 + x)^3 + 10(-2 + x)^4 + (-2 + x)^5$$

8.22 Construct a Taylor polynomial of degree 5 about $a = 1$ for the function $\sqrt{x}$ and use it to approximate $\sqrt{3/2}$.

SOLUTION

```
p[x_] = Series[√x, {x, 1, 5}] //Normal
```
$$1 + \frac{1}{2}(-1 + x) - \frac{1}{8}(-1 + x)^2 + \frac{1}{16}(-1 + x)^3 - \frac{5}{128}(-1 + x)^4 + \frac{7}{256}(-1 + x)^5$$
```
approx = p[3/2] //N
```
1.22498
```
exact = √(3/2) //N
```
1.22474
```
Abs[approx - exact]
```
0.000230715   ← This is the absolute error of the approximation.

8.23 Let $f(x) = \sin x$ and construct the Maclaurin polynomials of degrees 7, 9, and 11.   Then plot $f(x)$ and the three polynomials on one set of axes, $0 \le x \le 2\pi$, and observe their behavior.

SOLUTION

```
f[x_] = Sin[x];
p7[x_] = Series[f[x], {x, 0, 7}]//Normal;
p9[x_] = Series[f[x], {x, 0, 9}]//Normal;
p11[x_] = Series[f[x], {x, 0, 11}]//Normal;
Plot[{f[x], p7[x], p9[x], p11[x]}, {x, 0, 2π},
     PlotStyle → {Thickness[.01], Thickness[.001],
                  Thickness[.001], Thickness[.001]}];
```

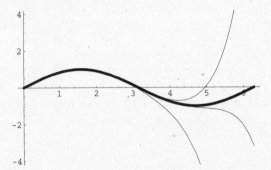

As the degree of the polynomial increases, the better the polynomial approximates $f(x) = \sin x$.

8.24 Let $f(x) = \sin x$ and construct the Maclaurin polynomial of degree 11.   Construct an error function and compute its value from $x = 0$ to $x = 6$ in unit increments.   Place the results in the form of a table and comment on the values of the error as x gets further from 0.

SOLUTION

```
f[x_] = Sin[x];
p11[x_] = Normal[Series[f[x], {x, 0, 11}]];
error[x_] = Abs[f[x] - p11[x]];
errorvalues = Table[{x, error[x]}, {x, 0, 6, 1.}];
TableForm[errorvalues, TableHeadings → {None, {"x", " error[x]"}}]
```

x	error[x]
0	0
1.	1.59828×10^{-10}
2.	1.29086×10^{-6}
3.	0.000245414
4.	0.0100021
5.	0.174693
6.	1.78084

As x gets further from 0, the error gets larger.

8.25 Let $f(x) = \sin x$.   Construct the Maclaurin polynomials of degrees 1, 3, 5, 7, and 9 and compute their value at $x = 1$.   Determine the error in the approximations and express in a tabular form.

SOLUTION

```
f[x_] = Sin[x];
exactvalue = f[1];
value[n_] : = Normal[Series[f[x], {x, 0, n}]] /. x → 1
data = Table[{n, N[value[n]], N[Abs[value[n] - exactvalue]]}, {n, 1, 9, 2}];
TableForm[data, TableHeadings → {None, {"n", " p(1)", " Error"}}]
```

n	p(1)	Error
1	1.	0.158529
3	0.833333	0.00813765
5	0.841667	0.000195682
7	0.841468	2.73084×10^{-6}
9	0.841471	2.48923×10^{-8}

> As n gets larger, the error gets smaller.

8.26 Let $f(x) = \ln x$ and construct the Taylor polynomials about $a = 1$ for f of degrees 5, 10, and 15. Then plot $f(x)$ and the three polynomials on one set of axes, $1 \le x \le 2$.

SOLUTION

```
f[x_] = Log[x];
p5[x_] = Series[f[x], {x, 1, 5}]//Normal;
p10[x_] = Series[f[x], {x, 1, 10}]//Normal;
p15[x_] = Series[f[x], {x, 1, 15}]//Normal;
Plot[{f[x], p5[x], p10[x], p15[x]}, {x, 1, 2},
     PlotStyle → {Thickness[.01], Thickness[.001],
                  Thickness[.001], Thickness[.001]}];
```

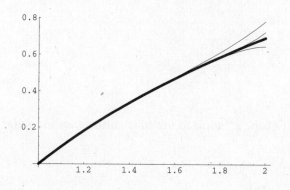

8.27 Let $f(x) = \ln x$ and construct the Taylor polynomial of degree 5 about $a = 1$. Construct an error function and compute its value from $x = 1$ to $x = 2$ in increments of 0.1. Place the results in the form of a table and comment on the values of the error as x gets further from 1.

SOLUTION

```
f[x_] = Log[x];
p5[x_] = Normal[Series[f[x], {x, 1, 5}]];
error[x_] = Abs[f[x]-p5[x]];
errorvalues = Table[{x, error[x]}, {x, 1, 2, .1}];
TableForm[errorvalues, TableHeadings → {None, {"x", " error[x]"}}]
```

```
x               error[x]
1               0
1.1             1.53529 × 10⁻⁷
1.2             9.10987 × 10⁻⁶
1.3             0.0000967355
1.4             0.000509097
1.5             0.00182656
1.6             0.00514837
1.7             0.0122941
1.8             0.026016
1.9             0.0502191
2.              0.0901862
```

> As x gets further from 1, the error gets larger.

8.28 Let $f(x) = \ln x$. Construct the Taylor polynomials of degrees $1, 2, 3, \ldots, 10$ about $a = 1$ and compute their value at $x = 1.5$. Determine the error in the approximations and express in a tabular form.

SOLUTION

```
f[x_] = Log[x];
exactvalue = f[1.5];
value[n_] := Normal[Series[f[x], {x, 1, n}]] /. x → 1.5
data = Table[{n, N[value[n]], N[Abs[value[n] - exactvalue]]}, {n, 1, 10}];
TableForm[data, TableHeadings → {None, {"n", "p(1.5)", " Error"}}]
```

n	p(1.5)	Error
1	0.5	0.0945349
2	0.375	0.0304651
3	0.416667	0.0112016
4	0.401042	0.00442344
5	0.407292	0.00182656
6	0.404688	0.000777608
7	0.405804	0.000338463
8	0.405315	0.000149818
9	0.405532	0.000067196
10	0.405435	0.0000304603

8.29 What is the coefficient of the x^{20} term of the Maclaurin series for $\sin(x^2 + 1)$?

SOLUTION

```
s = Series[Sin[x² + 1], {x, 0, 20}];
SeriesCoefficient[s, 20]
```

$$-\frac{\mathrm{Sin}[1]}{3628800}$$

Integral Calculus

9.1 ANTIDERIVATIVES

An antiderivative of a function f is a function F, such that $F'(x) = f(x)$. The **Integrate** command produces antiderivatives. You will notice, however, that the constant of integration, C, is omitted from the answer.

- **Integrate[f[x], x]** computes the antiderivative (indefinite integral) $\int f(x)\,dx$. The symbol $\int \Box\, d\Box$ from the BasicInput palette may also be used.

Mathematica can compute antiderivatives of elementary integrals found in standard tables, but if unable to evaluate an antiderivative in terms of elementary functions, the software will try to express the antiderivative in terms of special functions. If this is not possible, *Mathematica* returns the antiderivative unevaluated.

EXAMPLE 1

$\int x^2 \, \text{Exp}[x] \, \text{Sin}[x] \, dx$ or   **Integrate[x^2 Exp[x] Sin[x], x]**

$-\dfrac{1}{2}\, e^x \, (-1+x)^2 \, \text{Cos}[x] + \dfrac{1}{2}\, e^x \, (-1+x^2) \, \text{Sin}[x]$

EXAMPLE 2

$\int \text{Sin}[x^2] \, dx$ or   **Integrate[Sin[x^2], x]**

$\sqrt{\dfrac{\pi}{2}}\ \text{FresnelS}\left[\sqrt{\dfrac{2}{\pi}}\,x\right]$

> This integral has no simple antiderivative so *Mathematica* expresses it as a Fresnel sine integral: $\text{Fresnel}S[x] = \displaystyle\int_0^x \sin\!\left(\pi \tfrac{t^2}{2}\right) dt$

EXAMPLE 3

$\int \text{Sin}[\text{Sin}[x]] \, dx$ or   **Integrate[Sin[Sin[x]], x]**

$\int \text{Sin}[\text{Sin}[x]] \, dx$ ← *Mathematica* cannot evaluate this antiderivative.

Care must be taken when *general* antiderivatives involving parameters are requested.

219

EXAMPLE 4

$$\int x^n \, dx$$

$$\frac{x^{1+n}}{1+n}$$

Of course, this result is valid only if $n \neq -1$, but if the value of n is specified, *Mathematica* knows what to do.

$$n = -1;$$
$$\int x^n \, dx$$

$$\text{Log}[x]$$

SOLVED PROBLEMS

9.1   Compute $\int \sqrt{x} \, dx$.

SOLUTION

$$\int \sqrt{x} \, dx \quad \text{or} \quad \texttt{Integrate}\left[\sqrt{x}, \, x\right]$$

$$\frac{2 x^{3/2}}{3}$$

9.2   Compute $\int \sqrt{a^2 + x^2} \, dx$.

SOLUTION

$$\int \sqrt{a^2 + x^2} \, dx \quad \text{or} \quad \texttt{Integrate}\left[\sqrt{a^2 + x^2}, \, x\right]$$

$$\frac{1}{2} x \sqrt{a^2 + x^2} + \frac{1}{2} a^2 \text{Log}\left[x + \sqrt{a^2 + x^2}\right]$$

9.3   Compute $\int \dfrac{1}{\sqrt{u^2 - a^2}} \, du$.

SOLUTION

$$\int \frac{1}{\sqrt{u^2 - a^2}} \, du \quad \text{or} \quad \texttt{Integrate[1/Sqrt[u}^2 - \texttt{a}^2\texttt{], u]}$$

$$\text{Log}\left[u + \sqrt{-a^2 + u^2}\right]$$

9.4   Compute $\int \tanh x \, dx$.

SOLUTION

$$\int \texttt{Tanh[x]} \, dx \quad \text{or} \quad \texttt{Integrate[Tanh[x], x]}$$

$$\text{Log}[\text{Cosh}[x]]$$

9.5   Evaluate (a) $\int f'(x)\,dx$ and (b) $\int g'(f(x))\,f'(x)\,dx$.

SOLUTION

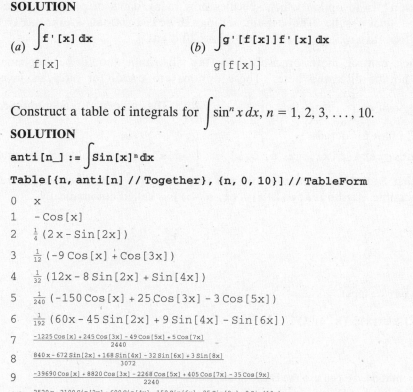

(a) $\displaystyle\int \mathtt{f'[x]\,dx}$

    `f[x]`

(b) $\displaystyle\int \mathtt{g'[f[x]]f'[x]\,dx}$

    `g[f[x]]`

9.6 Construct a table of integrals for $\displaystyle\int \sin^n x\, dx$, $n = 1, 2, 3, \ldots, 10$.

SOLUTION

```
anti[n_] := ∫Sin[x]ⁿ dx
```

```
Table[{n, anti[n] // Together}, {n, 0, 10}] // TableForm
```

```
0   x
1   - Cos[x]
2   1/4 (2 x - Sin[2x])
3   1/12 (-9 Cos[x] + Cos[3x])
4   1/32 (12x - 8 Sin[2x] + Sin[4x])
5   1/240 (-150 Cos[x] + 25 Cos[3x] - 3 Cos[5x])
6   1/192 (60x - 45 Sin[2x] + 9 Sin[4x] - Sin[6x])
```

7    $\dfrac{-1225\,\mathtt{Cos[x]} + 245\,\mathtt{Cos[3x]} - 49\,\mathtt{Cos[5x]} + 5\,\mathtt{Cos[7x]}}{2440}$

8    $\dfrac{840\,x - 672\,\mathtt{Sin[2x]} + 168\,\mathtt{Sin[4x]} - 32\,\mathtt{Sin[6x]} + 3\,\mathtt{Sin[8x]}}{3072}$

9    $\dfrac{-39690\,\mathtt{Cos[x]} + 8820\,\mathtt{Cos[3x]} - 2268\,\mathtt{Cos[5x]} + 405\,\mathtt{Cos[7x]} - 35\,\mathtt{Cos[9x]}}{2240}$

10   $\dfrac{2520\,x - 2100\,\mathtt{Sin[2x]} + 600\,\mathtt{Sin[4x]} - 150\,\mathtt{Sin[6x]} + 25\,\mathtt{Sin[8x]} - 2\,\mathtt{Sin[10x]}}{10240}$

9.2 DEFINITE INTEGRALS

A definite integral is computed one of two ways: exactly, using the Fundamental Theorem of Calculus or approximately, using numerical methods. You can instruct *Mathematica* which method you wish to use by choosing from two commands.

- `Integrate[f[x], {x, a, b}]` computes, whenever possible, the exact value of $\displaystyle\int_a^b f(x)\,dx$.

  The symbol $\displaystyle\int_\square^\square \square\, d\square$ on the BasicInput palette may be used as well.

- `NIntegrate[f[x], {x, a, b}]` computes an approximation to the value of $\displaystyle\int_a^b f(x)\,dx$ using strictly numerical methods.

`NIntegrate` evaluates the integral using an adaptive algorithm, subdividing the interval of integration until a desired degree of accuracy is achieved. The interval is divided recursively until the value of `AccuracyGoal` or `PrecisionGoal` is achieved.

- `AccuracyGoal` is an option which specifies how many digits to the right of the decimal point should be sought in the final result. The default is `AccuracyGoal → Infinity`, which specifies that accuracy should *not* be used as the criterion for terminating the numerical procedure.
- `WorkingPrecision` is an option which specifies how many digits of precision should be maintained in internal computations. The default value is `WorkingPrecision → $WorkingPrecision`, which is usually set to 16.

- **PrecisionGoal** is an option which specifies how many digits of precision should be sought in the final result. The default setting, **PrecisionGoal→Automatic**, sets PrecisionGoal to WorkingPrecision minus 10 digits.

Other options, which control more precisely how the algorithm should be implemented, are available, but will not be discussed here. These options are useful for integrals involving "pathological" functions such as $\int_{.0001}^{1} \sin\left(\frac{1}{x}\right) dx$ or $\int_{-1000}^{1000} e^{-x^2} dx$. The interested reader should consult the *Mathematica* help files for details.

The sequence **N[Integrate[f[x], {x, a, b}]]** or $\int_{a}^{b}$ **f[x] dx //N** evaluates the integral, whenever possible, by first finding the antiderivative and then using the Fundamental Theorem of Calculus. If this is impossible, **NIntegrate[f[x], {x, a, b}]** is called automatically.

EXAMPLE 5

To evaluate $\int_{0}^{1} x\, e^x \sin x \, dx$ we input

Integrate[x Exp[x] Sin[x], {x, 0, 1}]

$-\dfrac{1}{2}+\dfrac{1}{2}\, e\, \text{Sin}[1]$

As an alternate representation, we can use the BasicInput palette.

$\int_{0}^{1}$ **x Exp[x] Sin[x] dx**

$-\dfrac{1}{2}+\dfrac{1}{2}\, e\, \text{Sin}[1]$

If a numerical approximation is desired, we can type

$\int_{0}^{1}$ **x Exp[x] Sin[x] dx //N** or **N[Integrate[x Exp[x] Sin[x], {x, 0, 1}]]**

0.643678

Here, the antiderivative of the function $x\, e^x \sin x$ was computed and then evaluated from 0 to 1. If a strictly numerical procedure is preferred, we use **NIntegrate**.

NIntegrate[x Exp[x] Sin[x], {x, 0, 1}]
0.643678

EXAMPLE 6

Suppose we want to obtain a 20-significant-digit approximation to $\int_{0}^{1} \sin(\sin x)\, dx$.

$\text{N}\!\left[\int_{0}^{1} \text{Sin[Sin[x]] dx}, 20\right]$

0.43060610312069060491

Mathematica automatically adjusts WorkingPrecision and PrecisionGoal to achieve the desired result.

Mathematica can handle certain *improper* integrals. An improper integral of type I is an integral with one or two infinite limits of integration. We define $\int_a^\infty f(x)dx = \lim_{t\to\infty} \int_a^t f(x)dx$ and $\int_{-\infty}^b f(x)dx = \lim_{t\to\infty} \int_{-t}^b f(x)dx$ provided the limits exist. Such an integral is said to be *convergent*. If both $\int_{-\infty}^a f(x)dx$ and $\int_a^\infty f(x)dx$ exist, we define $\int_{-\infty}^\infty f(x)dx = \int_{-\infty}^a f(x)dx + \int_a^\infty f(x)dx$.

EXAMPLE 7

$$\int_0^\infty \text{E}^{-\text{x}}\,\text{dx}$$

1

EXAMPLE 8

$$\int_0^\infty \text{x}\,\text{dx} \qquad \leftarrow \text{This integral is divergent.}$$

∞

EXAMPLE 9

$$\int_{-\infty}^\infty \frac{1}{1+\text{x}^2}\,\text{dx}$$

π

The value of a type I improper integral may depend upon the values of parameters within the integrand. The option **Assumptions** allows the specification of conditions to be imposed upon these parameters.

- **Assumptions → *conditions*** specifies *conditions* to be applied to parameters within the integral.

EXAMPLE 10

$\int_1^\infty x^n\,dx$ converges if $n < -1$ and diverges otherwise.

```
Integrate[x^n, {x, 1, ∞}, Assumptions→n < -1]
```

$$\frac{1}{-1-\text{n}}$$

An improper integral of type II is an integral whose function is discontinuous on the interval of integration. If f is continuous on $[a, b)$ but not at b, we define $\int_a^b f(x)dx = \lim_{t\to b^-} \int_a^t f(x)dx$ and if f is continuous on $(a, b]$ but not at a, we define $\int_a^b f(x)dx = \lim_{t\to a^+} \int_t^b f(x)dx$. If the limit exists, we say the integral is *convergent*. If f has a discontinuity at $c\,\varepsilon\,(a, b)$ and both $\int_a^c f(x)dx$ and $\int_c^b f(x)dx$ are convergent, then $\int_a^b f(x)dx = \int_a^c f(x)dx + \int_c^b f(x)dx$.

EXAMPLE 11

$$\int_0^1 \text{Log[x] dx}$$

-1

EXAMPLE 12

$$\int_{-2}^3 \frac{1}{\text{x}} \text{dx} \quad \text{or} \quad \text{Integrate[1/x, \{x, -2, 3\}]}$$

Integrate::idiv :

   Integral of $\frac{1}{\text{x}}$ does not converge on $\{-2, 3\}$.

$$\int_{-2}^3 \frac{1}{\text{x}} \text{dx}$$

Sometimes an integral diverges but its *Cauchy Principal Value* exists.

- **PrincipalValue→True** specifies that the Cauchy Principal Value of an integral is to be determined.

The next example describes the concept behind Cauchy Principal Value.

EXAMPLE 13

Consider the integral $\int_{-2}^3 \frac{dx}{x}$ of Example 12. Because of the discontinuity at 0, the integral is improper. If we break up the integral into the sum of two integrals, $\int_{-2}^0 \frac{dx}{x} + \int_0^3 \frac{dx}{x}$, each integral, evaluated separately, diverges. However,

$$\lim_{t \to 0^+}\left[\int_{-2}^{-t} \frac{dx}{x} + \int_t^3 \frac{dx}{x}\right] = \lim_{t \to 0^+}\left[\ln|x|\Big|_{-2}^{-t} + \ln x\Big|_t^3\right]$$

$$= \lim_{t \to 0^+}\left[\ln t - \ln 2 + \ln 3 - \ln t\right]$$

$$= \ln 3 - \ln 2$$

$$= \ln \frac{3}{2}$$

The idea behind the Cauchy Principal Value is that the limits are taken *simultaneously*, not separately.

Integrate[1/x, {x, -2, 3}, PrincipalValue→True]

$$\text{Log}\left[\frac{3}{2}\right]$$

SOLVED PROBLEMS

9.7   Compute the area bounded by the curves $f(x) = 1 - x^2$ and $g(x) = x^4 - 3x^2$.

    SOLUTION

```
f[x_] = 1 - x²;
g[x_] = x⁴ - 3x²;
<<Graphics`FilledPlot`
FilledPlot[{f[x], g[x]}, {x, -2, 2}];
```

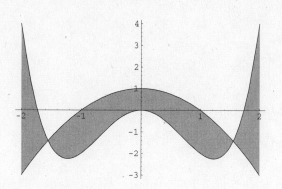

```
intersectionpoints = Solve[f[x] == g[x]]
```

$$\left\{\left\{x \to -i\sqrt{-1+\sqrt{2}}\right\}, \left\{x \to i\sqrt{-1+\sqrt{2}}\right\}, \left\{x \to -\sqrt{1+\sqrt{2}}\right\}, \left\{x \to \sqrt{1+\sqrt{2}}\right\}\right\}$$

The points of intersection correspond to the real solutions of this equation.

```
a = intersectionpoints[[3, 1, 2]];
b = intersectionpoints[[4, 1, 2]];
```

$$\int_a^b (f[x] - g[x]) \, dx \, //Simplify$$

$$\frac{8}{15}\sqrt{1+\sqrt{2}}\,(4+\sqrt{2})$$

```
%//N
4.48665
```

9.8 The volume of the solid of revolution obtained in rotating about the x axis the area bounded by the curve $y = f(x)$, the x-axis, and the lines $x = a$ and $x = b$ is $\pi \displaystyle\int_a^b [f(x)]^2 \, dx$. Compute the volume of the sphere obtained if the semicircle $y = \sqrt{r^2 - x^2}$, $-r \le x \le r$, is rotated about the x axis.

SOLUTION

```
y = √(r² - x²);
```

$$\pi \int_{-r}^r y^2 \, dx$$

$$\frac{4\,\pi r^3}{3}$$

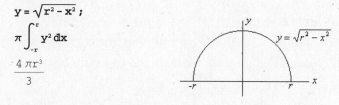

9.9 Compute the volume of a frustum of a cone with height h and radii r and R, and use this to derive the formula for the volume of a cone of radius R and height h.

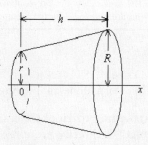

SOLUTION

The frustum is generated by rotating the region bounded by the line segment connecting $(0, r)$ and (h, R) and the x axis, about the x axis. The equation of the line segment is $y = \dfrac{R - r}{h}x + r$. The volume is $\pi \displaystyle\int_0^h y^2 \, dx$.

$$y = \frac{R - r}{h} x + r;$$

$$\pi \int_0^h y^2 \, dx \, // \text{Factor}$$

$$\frac{1}{3} h \pi \, (r^2 + r R + R^2)$$

$$\% \, / . \, r \to 0 \qquad\qquad \leftarrow \text{Let } r = 0 \text{ for a cone.}$$

$$\frac{1}{3} h \pi R^2$$

9.10 The arc length of a curve represented by $f(x)$, $a \le x \le b$, is given by $L = \int_a^b \sqrt{1 + [f'(x)]^2} \, dx$. Compute the approximate length of arc of one "arch" of a sine curve.

SOLUTION

One arch of the curve is generated for $0 \le x \le \pi$.

```
f[x_] = Sin[x];
NIntegrate[Sqrt[1+f'[x]^2], {x, 0, Pi}]
3.8202
```

9.11 The Mean Value Theorem for integrals says that if f is continuous on a closed interval $[a, b]$, there exists a number, c, between a and b, such that $\int_a^b f(x)dx = f(c)(b - a)$. Find the value of c which satisfies the Mean Value Theorem for $f(x) = \ln x$ on the interval $[1, 2]$.

SOLUTION

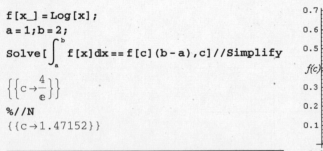

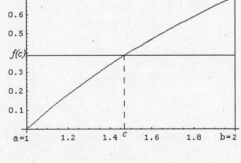

The area below the curve, above the x axis, is equal to the area enclosed by the rectangle determined by c.

9.12 The work done in moving an object from a to b by a variable force, $f(x)$, is $\int_a^b f(x) \, dx$.

According to Hooke's law, the force required to hold a spring stretched beyond its natural length is directly proportional to the displaced distance. If the natural length of a spring is 10 cm, and the force that is required to hold the spring 5 cm beyond this length is 40 newtons, how much work is done in stretching the spring from 10 to 15 cm?

SOLUTION

Hooke's law states that $f(x) = kx$ where x represents the distance beyond the spring's natural length. Since a force of 40 newtons is required to hold the spring 5 cm (0.05 m) beyond its natural length, $40 = 0.05k$.

```
k = 40/0.05;
f[x_] = kx;
```

$$\texttt{work} = \int_0^{.05} \texttt{f[x] dx}$$

> The work done is 1 joule.

```
1
```

9.3 FUNCTIONS DEFINED BY INTEGRALS

If f is continuous on $[a,b]$, we can define a new function

$$F(x) = \int_a^x f(t)\,dt$$

Intuitively, if $f(t) \geq 0$, $F(x)$ represents the area bounded by $f(t)$ and the t axis from a to x, if $x \geq a$, and the negative of this area if $x < a$. The (Second) Fundamental Theorem of Calculus tells us that (i) F is differentiable on (a,b) and (ii) $F'(x) = f(x)$ for all $x \, \varepsilon \, (a,b)$.

EXAMPLE 14

Let $f(x) = 1/x$, $x > 0$. The shaded area in the diagram below represents $F(x)$, assuming $x \geq 1$.

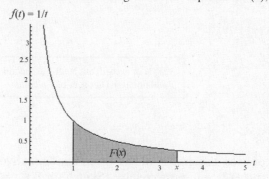

Students of calculus will recognize that this integral defines $F(x)$ to be the natural logarithm function. *Mathematica* knows this also.

```
f[x_] = 1/x;
```

$$\texttt{F[x\_]} = \int_1^x \texttt{f[t] dt}$$

```
Log[x]
F[2] //N
0.693147
F[1/2] //N
-0.693147
```

EXAMPLE 15

The function $f(x) = x^x$ has an antiderivative, but it cannot be put into "closed" form in terms of elementary functions. However, *Mathematica* can deal with it as a function defined by an integral. Since all antiderivatives of $f(x)$ differ by a constant, we define $F(x)$ to be the antiderivative for which $F(0) = 0$. Let us plot this antiderivative for $0 \leq x \leq 4$.

```
f[x_] = x^x;

       x
F[x_] = ∫ f[t] dt          ←By making the lower limit 0, we force F(0) = 0.
       0

Plot[F[x], {x, 0, 4}]
```

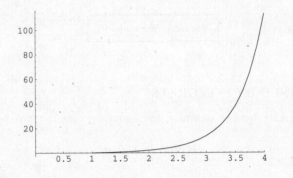

SOLVED PROBLEMS

9.13   Let $F(x) = \displaystyle\int_1^x e^{\sin t}\, dt.$ Find $F'(x)$.

SOLUTION

```
       x
F[x_] = ∫ Exp[Sin[t]] dt;
       1

F'[x]
 Sin[x]
e
```

> This is in accordance with the second
> Fundamental Theorem of Calculus.

9.14   Sketch, on one set of axes, the graphs of the three antiderivatives of $f(x) = e^{\sin x}$, for which $F(0) = 0$, $F(1) = 0$, and $F(2) = 0$, respectively.   Sketch for $0 \le x \le 2\pi$.

SOLUTION

```
f[x_] = Exp[Sin[x]];

        x
F1[x_] = ∫ f[t] dt;
        0

        x
F2[x_] = ∫ f[t] dt;
        1

        x
F3[x_] = ∫ f[t] dt;
        2

Plot[{F1[x],F2[x],F3[x]}, {x, 0, 2π}];
```

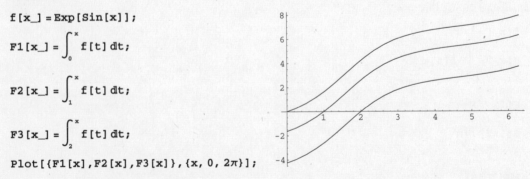

9.15   Consider the semicircle $x^2 + y^2 = 16$, $y \ge 0$ shown on the next page. Find the height, h, so that the shaded area is half the area of the semicircle.

SOLUTION

We solve for x as a function of y:

`Solve[x² + y² == 16, x]`

$$\left\{ \left\{ x \to -\sqrt{16 - y^2} \right\}, \left\{ x \to \sqrt{16 - y^2} \right\} \right\}$$

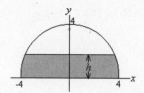

By subdividing the y axis and taking advantage of symmetry, we obtain the following representation for $A(h)$, the shaded area:

$$A(h) = 2\int_0^h x(y)\,dy \text{ where } x(y) = \sqrt{16 - y^2}.$$

`x[y_] = √16 - y² ;`

`A[h_] = 2 ∫₀ʰ x[y] dy ;`

We compute the total area inside the semicircle. (Since we know a formula for the area of a circle, this is a good check for errors.)

`A[4]`
8π

To approximate the solution, draw a graph of $A(h)$:

`Plot[A[h], {h, 0, 4}];`

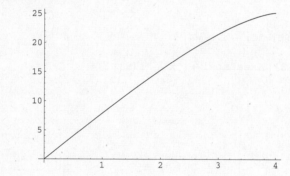

It appears that half the semicircular area, $4\pi \approx 12.5$, corresponds to a value of h near 1.5.   We finish the job with `FindRoot`.

`FindRoot[A[h] == 4π, {h, 1.5}]`
`{h → 1.61589}`

9.16   The curve below is the parabola $y = 9 - x^2$.   Find h so that the shaded area is 2/3 the total area bounded by the curve and the x axis.

SOLUTION

`Solve[y == 9 - x², x]`

$$\left\{ \left\{ x \to -\sqrt{9 - y} \right\}, \left\{ x \to \sqrt{9 - y} \right\} \right\}$$

`x[y_] = √9 - y ;`

`A[h_] = 2 ∫₀ʰ x[y] dy ;`

`totalarea = A[9]`

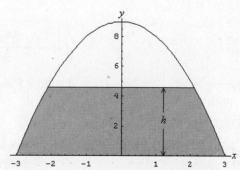

36

```
Plot[A[h], {h, 0, 9}];
```

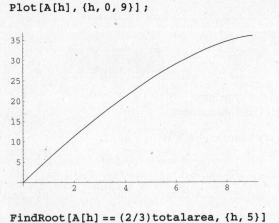

2/3 of the total area, 24, appears to correspond to a value of h near 5.

```
FindRoot[A[h] == (2/3)totalarea, {h, 5}]
{h → 4.67325 }
```

9.17   What point on the parabola $y = x^2$ is 5 units away from the origin along the curve?

SOLUTION

The length of arc of a function, $f(x)$, from a to b is $L = \int_a^b \sqrt{1 + [f'(x)]^2}\,dx$

```
f[x_] = x²;

l[x_] = ∫₀ˣ √(1 + f'[t]²) dt;

Plot[l[x], {x, 0, 3}, AxesLabel → {"x", "l(x)"}];
```

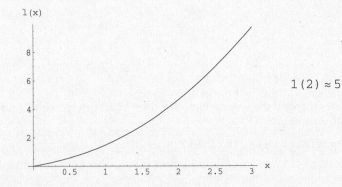

$1(2) \approx 5$

```
FindRoot[l[x] == 5, {x, 2}]
{x → 2.08401}
{%[[1, 2]], f[%[[1, 2]]]}
{2.08401, 4.34308}
```

The point is (2.08401, 4.34308).

9.18   A mixing bowl is a hemisphere of radius 5 inches.   Determine the height of 100 cubic inches of liquid.

SOLUTION

The equation of the hemisphere in three dimensions is $x^2 + y^2 + z^2 = 25$, $z \le 0$.   Its intersection with the plane $z = z_0$ is the circle $x^2 + y^2 = 25 - z_0^2$, whose radius is $\sqrt{25 - z_0^2}$ and whose area is $\pi(25 - z_0^2)$.

Integrating with respect to z, the volume of the shaded region is $V(h) = \pi \int_{-5}^{-5+h} (25 - z^2)dz$.

v[h_] = π $\int_{-5}^{-5+h}$ **(25 - z²) dz**

$$\left(5h^2 - \frac{h^3}{3}\right)\pi$$

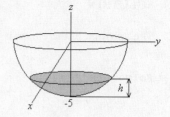

As a check, v(5) should give the volume of the hemisphere.

v[5]

$\dfrac{250\pi}{3}$

┌───┐
│ The volume of a hemisphere of radius r is $\frac{2}{3}\pi r^3$. │
└───┘

We can plot v as a function of h.

Plot[v[h], {h, 0, 5}, AxesLabel→{"h", "v[h]"}];

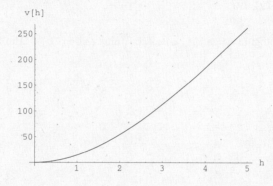

Since v(h) is a polynomial function, we can use NSolve to determine the approximate solution to the problem. (From the graph, it looks like h is near 3.)

NSolve[v[h] == 100]

{{h→-2.34629}, {h→2.79744}, {h→14.5489}}

Obviously, the only realistic solution is $h = 2.79744$ inches.

9.19 An underground fuel tank is in the shape of an elliptical cylinder. The tank has length $l = 20$ feet, semi-major axis $a = 10$ feet, and semi-minor axis $b = 5$ feet. To measure the amount of fuel in the tank, we insert a stick through the center of the cylinder until it touches the bottom of the tank and measure how high the fuel level is on the stick. How far from the end of the stick should a mark be placed to indicate that only 500 cubic feet of fuel is left?

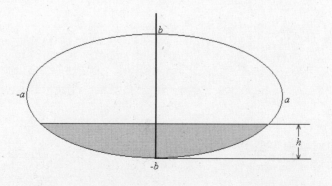

SOLUTION

The equation of the ellipse is $\dfrac{x^2}{a^2} + \dfrac{y^2}{b^2} = 1$. We first define x as a function of y.

```
Solve[ x²/a² + y²/b² == 1, x ]
```

$$\left\{ \left\{ x \to -a\,\sqrt{1 - \frac{y^2}{b^2}} \right\},\ \left\{ x \to a\,\sqrt{1 - \frac{y^2}{b^2}} \right\} \right\}$$

Next we obtain an integral representing the cross-sectional area of the tank. We take the positive solution and double the area, taking advantage of symmetry.

```
a = 10; b = 5;
```

$$x[y\_] = a\,\sqrt{1 - \frac{y^2}{b^2}}\,;$$

$$area[h\_] = 2 \int_{-b}^{-b+h} x[y]\,dy;$$

As a check, we can compute `area[0]`, `area[b]`, and `area[2b]`. The area enclosed by the ellipse $\dfrac{x^2}{a^2} + \dfrac{y^2}{b^2} = 1$ is $\pi a b$.

```
area[0]
```
0
```
area[b]
```
25π
```
area[2b]
```
50π

Since the tank has a uniform cross section, its volume = length × cross-sectional area.

```
l = 20;
v[h_] = l area[h];
```

To approximate the location on the stick which corresponds to 500 cubic feet, we draw the graph of v(h). Then we use FindRoot to obtain a more accurate value. (We cannot use **NSolve**, as in the previous problem, since v(h) is a nonalgebraic function.)

```
Plot[v[h], {h, 0, 2b}, AxesLabel→{"h", "v[h]"}];
```

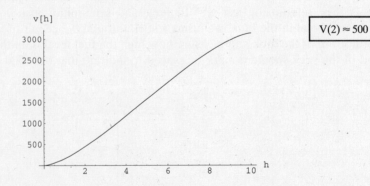

$$V(2) \approx 500$$

```
FindRoot[v[h] == 500, {h, 2}]
```
{h→2.1623}

9.20　An underground fuel tank is in the shape of an ellipsoid with semi-axes 6, 10, and 6 feet.　(This problem, although more difficult than the previous problem, is somewhat more realistic.) To measure the amount of fuel in the tank, we insert a stick through the center of the ellipsoid until it touches the bottom of the tank and measure how high the fuel level is on the stick.　How far from the end of the stick should a mark be placed to indicate that only 500 cubic feet of fuel is left?

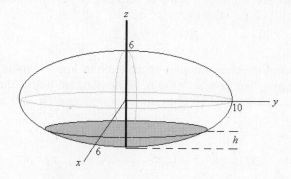

SOLUTION

The equation of this ellipsoid is

$\dfrac{x^2}{a^2} + \dfrac{y^2}{b^2} + \dfrac{z^2}{c^2} = 1$ with $a = 6$, $b = 10$, and $c = 6$.　The intersection of the ellipsoid with the plane $z = z_0$ is

the ellipse $\dfrac{x^2}{a^2} + \dfrac{y^2}{b^2} = 1 - \dfrac{z_0^2}{c^2}$ whose area can be computed as a function of z_0.　We then integrate with respect to z to obtain the volume.

To determine the area of the ellipse, we take advantage of the fact that the area enclosed by an ellipse

is π times the product of its semi-major and semi-minor axes.　If we rewrite $\dfrac{x^2}{a^2} + \dfrac{y^2}{b^2} = 1 - \dfrac{z_0^2}{c^2}$ in the

equivalent form $\dfrac{x^2}{a^2\left(1 - \dfrac{z_0^2}{c^2}\right)} + \dfrac{y^2}{b^2\left(1 - \dfrac{z_0^2}{c^2}\right)} = 1$ we see that the semi-axes are $a\sqrt{1 - \dfrac{z_0^2}{c^2}}$ and $b\sqrt{1 - \dfrac{z_0^2}{c^2}}$.

The elliptical area is then $\pi ab\left(1 - \dfrac{z_0^2}{c^2}\right)$.　For our values of a, b, and c, this becomes $60\pi\left(1 - \dfrac{z_0^2}{6^2}\right) =$

$\dfrac{5\pi}{3}(36 - z_0^2)$.　The integral representing the volume of fuel as a function of h is then

$$V = \frac{5\pi}{3}\int_{-6}^{-6+h}(36 - z^2)\,dz.$$

```
v[h_] = 5π/3 ∫_{-6}^{-6+h} (36 - z²) dz;

Plot [v[h], {h, 0, 12},
    AxesLabel → {"h", "v[h]"}];
NSolve[v[h] == 500]
{{h → - 3.63858}, {h → 4.62871},
 {h → 17.0099}}
```

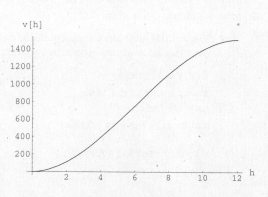

The mark should be placed approximately 4.62871 feet from the end of the stick.　The other solutions are extraneous.

9.4   RIEMANN SUMS

A partition, P, of the interval $I = [a, b]$ is a collection of subintervals,

$$[x_0, x_1], [x_1, x_2], \ldots, [x_{n-1}, x_n]$$

where $x_0 = a$ and $x_n = b$.   If we let x_i^* be any point in the ith subinterval, $\Delta x_i = x_i - x_{i-1}$, represent the length of the ith subinterval, and then the Riemann sum of f over I with respect to

$$P \text{ is } \sum_{i=1}^{n} f(x_i^*) \Delta x_i$$

If $f(x) \geq 0$ for $a \leq x \leq b$, the Riemann sum represents an approximation of the area under the graph of $f(x)$, above the x axis, from $x = a$ to $x = b$.   The diagram below shows the Riemann sum of the function $f(x) = x^2$, over the interval $[1, 2]$ as the area enclosed by four approximating rectangles of equal width.

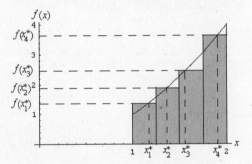

The Riemann sum, represented by the gray area enclosed by the rectangles, offers only an approximation to the area under the curve.   However, as the width of each rectangle approaches 0, the approximation gets better, and the *exact* area under the curve is approached as a limit.

The definite integral of $f(x)$ over $[a, b]$ is defined in most calculus texts by

$$\int_a^b f(x)\,dx = \lim_{\|P\| \to 0} \sum_{i=1}^{n} f(x_i^*) \Delta x_i \quad \text{where} \quad \|P\| = \max_{i \leq i \leq n} \Delta x_i$$

The condition $\|P\| \to 0$ guarantees that all subintervals shrink towards 0 length as we take more and more subintervals.   If all subintervals are of equal length, this condition is equivalent to $n \to \infty$.   For convenience we shall only consider subintervals of equal length.   However, in theory, this need not be the case.

EXAMPLE 16

We will consider the function $f(x) = \sin x$ on the interval $[0, \pi/2]$.   Because $\displaystyle\int_0^{\pi/2} \sin x\,dx = 1$, this is a good example for comparative purposes.

(a)   We use 100 subintervals and choose x_i^* to be the left endpoint of each subinterval.   Since $\sin x$ increases over this interval, we expect an underapproximation.

```
f[x_] = Sin[x];
a = 0; b = π/2; n = 100;
Δx = (b - a)/n;                    ← Since each Δx has the same value, subscripts are not necessary.
xstar[i_] = a + (i - 1)Δx;

 n
 ∑  f[xstar[i]] Δx //N
i=1
0.992125
```

(b) We choose the value of x_i^* to be the *right* endpoint of each subinterval. This time we expect an overapproximation.

```
xstar[i_] = a + i Δx;

 n
 ∑ f[xstar[i]] Δx //N
i=1
1.00783
```

(c) To improve the accuracy of the approximation offered above we can choose the value of x_i^* to be the midpoint of each subinterval. This leads to an approximation method called the *midpoint rule*.

```
xstar[i_] = a + (i-.5) Δx;

 n
 ∑ f[xstar[i]] Δx //N
i=1
1.00001
```

As expected, the accuracy of the approximation further improves.

Another simple approximation method, called the *trapezoidal rule*, improves accuracy by connecting the points on the curve corresponding to the points of subdivision with line segments, forming trapezoidal approximations of the area in place of rectangular approximations.

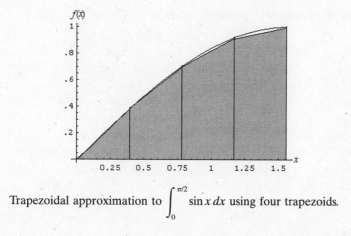

Trapezoidal approximation to $\displaystyle\int_0^{\pi/2} \sin x \, dx$ using four trapezoids.

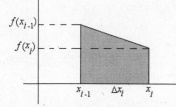

The area enclosed by a trapezoid with base Δx and sides A and B is $\dfrac{\Delta x}{2}(A+B)$. Thus the area enclosed by the trapezoid constructed in the ith interval, $[x_{i-1}, x_i]$, is $\dfrac{\Delta x_i}{2}[f(x_{i-1}) + f(x_i)]$.

The total trapezoidal area, obtained by adding the individual areas, is

$$\frac{\Delta x_1}{2}[f(x_0) + f(x_1)] + \frac{\Delta x_2}{2}[f(x_1) + f(x_2)] + \frac{\Delta x_3}{2}[f(x_2) + f(x_3)] + \ldots + \frac{\Delta x_n}{2}[f(x_{n-1}) + f(x_n)]$$

If the intervals are all the same length, Δx, this reduces to

$$\frac{\Delta x}{2}\Big[[f(x_0) + f(x_1)] + [f(x_1) + f(x_2)] + [f(x_2) + f(x_3)] + \ldots + [f(x_{n-1}) + f(x_n)]\Big]$$

or

$$\frac{\Delta x}{2}\Big[f(x_0) + 2f(x_1) + 2f(x_2) + 2f(x_3) + \ldots + 2f(x_{n-1}) + f(x_n)\Big]$$

EXAMPLE 17

```
f[x_] = Sin[x];
a = 0; b = π/2; n = 100;
Δx = (b - a)/n;
x[i_] = a + i * Δx;

approximation = Δx/2 (f[a] + 2 ∑_{i=1}^{n-1} f[x[i]] + f[b]) //N

0.999979
```

SOLVED PROBLEMS

9.21   Compute the Riemann sums of $f(x) = x \ln x$ over the interval $[1, 2]$ using

(a)   the left endpoint of each subinterval.
(b)   the right endpoint of each subinterval.
(c)   the midpoint of each subinterval.

Compare with *Mathematica*'s approximation.

SOLUTION

```
f[x_] = x Log[x];
a = 1; b = 2;

∫_a^b f[x] dx //N

0.636294                              ← Mathematica's approximation

n = 100;
Δx = (b - a) / n;
xstar[i_] = a + (i - 1) Δx;

∑_{i=1}^{n} f[xstar[i]] Δx //N
0.629369                              ← left endpoint approximation

xstar[i_] = a + i Δx;

∑_{i=1}^{n} f[xstar[i]] Δx //N
0.643232                              ← right endpoint approximation

xstar[i_] = a + (i - .5) Δx;

∑_{i=1}^{n} f[xstar[i]] Δx //N
0.636291                              ← midpoint approximation
```

9.22   Compute the lower and upper Riemann sums for the function $f(x) = x^2$ on the interval $[0, 1]$ for $n = 2, 4, 8, 16, \ldots, 2^{20}$ subintervals.   Explain the behavior of the approximations in terms of the integral $\int_0^1 x^2 \, dx$.

SOLUTION

```
f[x_] = x²;
a = 0; b = 1;
n = 2^m;
Δx = (b - a) / n;
nn = PaddedForm[n, 10];
```

$$temp1 = PaddedForm\left[N\left[\sum_{i=1}^{n} f[a + (i - 1) \Delta x] \Delta x\right], \{8, 6\}\right];$$

$$temp2 = PaddedForm\left[N\left[\sum_{i=1}^{n} f[a + i \Delta x] \Delta x\right], \{8, 6\}\right];$$

```
lst = Table[{nn, temp1, temp2}, {m, 1, 20}];
TableForm[lst, TableHeadings →
    {None, {"        n", "   lower", "    upper"}}]
```

n	lower	upper
2	0.125000	0.625000
4	0.218750	0.468750
8	0.273438	0.398438
16	0.302734	0.365234
32	0.317871	0.349121
64	0.325562	0.341187
128	0.329437	0.337250
256	0.331383	0.335289
512	0.332357	0.334311
1024	0.332845	0.333822
2048	0.333089	0.333578
4096	0.333211	0.333455
8192	0.333272	0.333394
16384	0.333303	0.333364
32768	0.333318	0.333349
65536	0.333326	0.333341
131072	0.333330	0.333337
262144	0.333331	0.333335
524288	0.333332	0.333334
1048576	0.333333	0.333334

> As n gets larger, the lower sums increase, approaching a limit of 1/3, and the upper sums decrease, approaching a limit of 1/3.
>
> $$\int_0^1 x^2 \, dx = \frac{1}{3}$$

9.23 Compute an approximation of $\int_0^1 e^{x^2} \, dx$ using the trapezoidal rule with 10, 50, and 100 subintervals. Compare with *Mathematica*'s approximation.

SOLUTION

```
f[x_] = Exp[x²];
a = 0; b = 1;
```

$$\int_a^b f[x] \, dx \, //N$$

```
1.46265
Clear[n];
Δx = (b - a)/n;
x[i_] = a + i * Δx;
n = 10;
```

$$\texttt{approximation} = \frac{\Delta x}{2}\left(\texttt{f[a]} + 2\sum_{i=1}^{n-1}\texttt{f[x[i]]} + \texttt{f[b]}\right) \texttt{//N}$$

1.46717

`n = 50;`

$$\texttt{approximation} = \frac{\Delta x}{2}\left(\texttt{f[a]} + 2\sum_{i=1}^{n-1}\texttt{f[x[i]]} + \texttt{f[b]}\right) \texttt{//N}$$

1.46283

`n = 100;`

$$\texttt{approximation} = \frac{\Delta x}{2}\left(\texttt{f[a]} + 2\sum_{i=1}^{n-1}\texttt{f[x[i]]} + \texttt{f[b]}\right) \texttt{//N}$$

1.4627

CHAPTER 10

Multivariate Calculus

10.1 PARTIAL DERIVATIVES

The commands **D**, **∂**, and **Derivative** discussed in Chapter 8 are actually commands for computing *partial derivatives*. Of course, if there is only one variable present in a function, the partial derivative becomes an ordinary derivative. If two or more variables are present, however, all variables other than the one specified are treated as constants.

In the following descriptions, **f** stands for a function of several variables.

- **D[f, x]** or **∂$_x$f** (on the BasicInput palette) returns $\partial f/\partial x$, the partial derivative of f with respect to x.
- **D[f, {x, n}]** or **∂$_{(x, n)}$f** returns $\partial^n f/\partial x^n$, the nth-order partial derivative of f with respect to x.
- **D[f, x$_1$, x$_2$, ..., x$_k$]** or **∂$_{x_1, x_2, ..., x_k}$f** returns the "mixed" partial derivative $\partial^k f/\partial x_1 \partial x_2 \ldots \partial x_k$.
- **D[f, {x$_1$, n$_1$}, {x$_2$, n$_2$}, ..., {x$_k$, n$_k$}]** or **∂$_{(x_1, n_1), (x_2, n_2), ..., (x_k, n_k)}$f** returns the partial derivative $\dfrac{\partial^n f[x_1, x_2, \ldots, x_k]}{\partial_{x_1}^{n_1} \partial_{x_2}^{n_2} \ldots \partial_{x_k}^{n_k}}$ where $n_1 + n_2 + \ldots + n_k = n$.

For convenience, an *invisible* comma may be used to separate variables in the partial derivative symbol. An invisible comma is entered by the three-key sequence [ESC] + , + [ESC]. An invisible comma works like an ordinary comma, but is hidden from the display.

EXAMPLE 1

```
D[x² y³ z⁴, x]
2 x y³ z⁴
∂_y (x² y³ z⁴)
3 x² y² z⁴
D[x² y³ z⁴, {z, 2}]
12 x² y³ z²
∂_{x, y} (x² y³ z⁴)
6 x y² z⁴
```

←The parentheses are important here. Why?

239

EXAMPLE 2

Let $f(x, y) = x^5 y^7$. We wish to compute $f_{xxxyyyy}$.

```
f[x_, y_] = x⁵ y⁷;
D[f[x, y], {x, 3}, {y, 4}]
```
$50400 \, x^2 y^3$
$$\partial_{(x, 3), (y, 4)} f[x, y]$$
$50400 \, x^2 y^3$

The **Derivative** command can also be used to construct partial derivatives. Suppose f is a function of k variables, $x_1, x_2, \ldots, x_k$.

- **Derivative[n₁, n₂, ..., n_k] [f]** gives the partial derivative $\dfrac{\partial^n}{\partial_{x1}^{n_1} \partial_{x2}^{n_2} \ldots \partial_{x_k}^{n_k}}$ where $n_1 + n_2 + \ldots + n_k = n$. It returns a pure function which may then be evaluated at $[x_1, x_2, \ldots x_k]$.

EXAMPLE 3

```
f[x_, y_] = x⁵ y⁷;
g = Derivative[3, 4][f];
g[x, y]
```
$50400 \, x^2 y^3$

Although the D command can be used to evaluate partial derivatives at a given point, Derivative is perhaps a bit more convenient.

EXAMPLE 4

Let $f(x, y) = x^3 \sin y$. Evaluate f_{xy} at the point $(2, \pi)$.

```
f[x_, y_] = x³ Sin[y];
D[f[x, y], x, y] /. {x→2, y→π}
```
-12
```
Derivative[1, 1][f][2, π]
```
-12

SOLVED PROBLEMS

10.1   Compute the first- and second-order partial derivatives of $f(x, y) = x e^{xy}$.

SOLUTION

```
f[x_, y_] = x Exp[xy];
D[f[x, y], x]
```
$e^{xy} + e^{xy} x y$
```
D[f[x, y], y]
```
$e^{xy} x^2$
```
D[f[x, y], {x, 2}]
```
$2 e^{xy} y + e^{xy} x y^2$
```
D[f[x, y], {y, 2}]
```
$e^{xy} x^3$
```
D[f[x, y], x, y]
```
$2 e^{xy} x + e^{xy} x^2 y$

10.2   The partial derivatives of $f(x, y)$ are defined by the following limits:

$$f_x(x, y) = \lim_{h \to 0} \frac{f(x + h, y) - f(x, y)}{h}$$

$$f_y(x, y) = \lim_{h \to 0} \frac{f(x, y + h) - f(x, y)}{h}$$

Compute the derivatives of $f(x, y) = \ln(x^2 + y^3)$ using the definition and verify using the *Mathematica* D command.

SOLUTION

```
f[x_, y_] = Log[x² + y³];
Limit[ f[x+h, y] - f[x, y] / h , h→0]
```
$$\frac{2x}{x^2 + y^3}$$
```
D[f[x,y], x]
```
$$\frac{2x}{x^2 + y^3}$$
```
Limit[ f[x, y+h] - f[x, y] / h , h→0]
```
$$\frac{3y^2}{x^2 + y^3}$$
```
D[f[x,y], y]
```
$$\frac{3y^2}{x^2 + y^3}$$

10.3   Let $z = e^{xy}$.   Compute $\dfrac{\partial^3 z}{\partial^2 x\, \partial y}$.

SOLUTION

```
z = Exp[x y];
D[z, {x, 2}, y]  or  ∂_{x, 2}, y z
2 eˣʸ y + eˣʸ x y²
```

10.4   Verify that $u = e^{-a^2 k^2 t} \sin kx$ is a solution of the heat equation: $\dfrac{\partial u}{\partial t} = a^2 \dfrac{\partial^2 u}{\partial x^2}$.

SOLUTION

```
Clear[a, k]
u = Exp[-a² k² t] Sin[k x];
lhs = D[u, t]
-a² e⁻ᵃ²ᵏ²ᵗ k² Sin[k x]
rhs = a² D[u, {x, 2}]
-a² e⁻ᵃ²ᵏ²ᵗ k² Sin[k x]
lhs == rhs
True
```

10.5 A function of three variables, $f(x, y, z)$, is said to be harmonic if it satisfies Laplace's equation: $\dfrac{\partial^2 f}{\partial x^2} + \dfrac{\partial^2 f}{\partial y^2} + \dfrac{\partial^2 f}{\partial z^2} = 0$. Let $f(x, y, z) = \dfrac{1}{\sqrt{x^2 + y^2 + z^2}}$. Compute f_{xx}, f_{yy}, and f_{zz} and show that f is harmonic.

SOLUTION

```
f[x_,y_,z_] = 1/√(x² + y² + z²);
∂{x, 2} f[x,y,z] //Together
```

$$\frac{2x^2 - y^2 - z^2}{(x^2 + y^2 + z^2)^{5/2}}$$

```
∂{x, 2} f[x,y,z] //Together
```

$$\frac{-x^2 + 2y^2 - z^2}{(x^2 + y^2 + z^2)^{5/2}}$$

```
∂{x, 2} f[x,y,z] //Together
```

$$\frac{-x^2 - y^2 + 2z^2}{(x^2 + y^2 + z^2)^{5/2}}$$

```
%%% + %% + %//Together
```

0

10.6 The plane tangent to the surface defined by $z = f(x, y)$ at the point (x_0, y_0, z_0) is

$$z = z_0 + f_x(x_0, y_0)(x - x_0) + f_y(x_0, y_0)(y - y_0)$$

Determine the equation of the plane tangent to the paraboloid $z = 10 - x^2 - 2y^2$ at the point $(1, 2)$. Sketch the paraboloid and its tangent plane.

SOLUTION

```
f[x_, y_] = 10 - x² - 2y²;
z = f[1, 2] + Derivative[1, 0][f][1, 2] (x - 1)
          + Derivative[0, 1][f][1, 2] (y - 2)//Expand
19 - 2x - 8y
```

The tangent plane has equation $z = 19 - 2x - 8y$.

```
g1 = Plot3D[f[x, y], {x, -5, 5}, {y, -5, 5},
         DisplayFunction → Identity];
g2 = Plot3D[z, {x, -5, 5}, {y, -5, 5},
      DisplayFunction → Identity];
```

> `DisplayFunction → Identity`
> suppresses graphics output (see Chapter 4).

```
Show[g1, g2, ViewPoint → {2.330, -2.223, 1.040},
         DisplayFunction → $DisplayFunction];
```

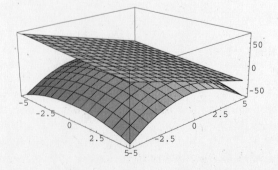

10.7   The plane tangent to the surface $f(x, y, z) = 0$ at the point (x_0, y_0, z_0) is

$$f_x(x_0, y_0, z_0)(x - x_0) + f_y(x_0, y_0, z_0)(y - y_0) + f_z(x_0, y_0, z_0)(z - z_0) = 0$$

Sketch the sphere $x^2 + y^2 + z^2 = 14$ and its tangent plane at the point $(1, 2, 3)$.

SOLUTION

The sphere is centered at the origin and has a radius of $\sqrt{14}$. Its equation is rewritten $x^2 + y^2 + z^2 - 14 = 0$. We can use the graphics primitive **Sphere** to construct its graph. (See Chapter 5.)

```
<<Graphics`Shapes`
f(x_, y_, z_) = x² + y² + z² - 14;
g1 = Graphics3D[Sphere[√14, 20, 15]];
a = Derivative[1, 0, 0][f][1, 2, 3];
b = Derivative[0, 1, 0][f][1, 2, 3];
c = Derivative[0, 0, 1][f][1, 2, 3];
Solve[a(x - 1) + b(y - 2) + c(z - 3) == 0, z]
```

$$\left\{ \left\{ z \rightarrow \frac{1}{3}(14 - x - 2y) \right\} \right\}$$

```
g2 = Plot3D[1/3 (14 - x - 2y), {x, -5, 5}, {y, -5, 5}, DisplayFunction→Identity];
Show[g1, g2, DisplayFunction→$DisplayFunction]
```

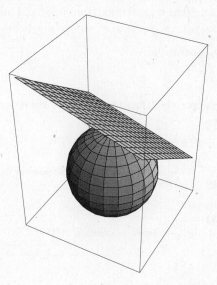

10.2   MAXIMUM AND MINIMUM VALUES

A function f has a local (or relative) maximum at (x_0, y_0) if there exists an open disk centered at (x_0, y_0) such that $f(x, y) \leq f(x_0, y_0)$ for all (x, y) in the disk. A similar definition (with the inequality reversed) holds for a local minimum. If f has either a local maximum or local minimum at (x_0, y_0), we say the f has a local *extremum* at (x_0, y_0).

A necessary condition for $f(x, y)$ to have a local extremum at (x_0, y_0) is $f_x(x_0, y_0) = f_y(x_0, y_0) = 0$. Such a point is called a *critical point* of the function f.

EXAMPLE 5

To find the critical point(s) for the function $f(x, y) = x^4 + y^4 - 4xy$, we compute the first-order partial derivatives, set them both equal to 0, and solve the resulting equations.

```
f[x_, y_] = x^4 + y^4 - 4 x y;
pdx = D[f[x, y], x]
4x³ - 4y

pdy = D[f[x, y], y]
-4x + 4y³

Solve[{pdx == 0, pdy == 0}, {x, y}]
{{x → -1, y → -1}, {x → 0, y → 0},
 {x → -i, y → i}, {x → i, y → -i}, {x → 1, y → 1},
 {x → -(-1)^(1/4), y → -(-1)^(3/4)}, {x → (-1)^(1/4), y → (-1)^(3/4)},
 {x → -(-1)^(3/4), y → -(-1)^(1/4)}, {x → (-1)^(3/4), y → (-1)^(1/4)}}
```

The only *real* critical points are $(-1, -1)$, $(0, 0)$, and $(1, 1)$.

Unfortunately, not all critical points turn out to be local extrema. To determine whether a function has a local extremum at a critical point, and if so, whether it is a maximum or minimum, we use the <u>Second Partial Derivatives Test:</u>

Let $D(x, y) = f_{xx}(x, y) f_{yy}(x, y) - [f_{xy}(x, y)]^2$ and let (x_0, y_0) be a critical point of f.

1. If $D(x_0, y_0) > 0$ and $f_{xx}(x_0, y_0) > 0$ then f has a local minimum at (x_0, y_0).
2. If $D(x_0, y_0) > 0$ and $f_{xx}(x_0, y_0) < 0$ then f has a local maximum at (x_0, y_0).
3. If $D(x_0, y_0) < 0$ then f has neither a local maximum nor local minimum at (x_0, y_0). We say that f has a *saddle point* at (x_0, y_0).

If $D(x_0, y_0) = 0$, the test is inconclusive.

EXAMPLE 6

Continuing with the previous example, we define $D(x, y)$. (We use **d** to avoid conflict with **D**, *Mathematica*'s derivative command.)

```
d[x_, y_] = ∂_{x,2} f[x,y] ∂_{y,2} f[x,y] - (∂_{x,y} f[x,y])²;
d[0, 0]
-16                          ←negative number; saddle point at (0,0)

d[1, 1]
128
∂_{x,2} f[x, y] /. {x → 1, y → 1}
12                           ←relative minimum at (1,1)

d[-1, -1]
128
∂_{x,2} f[x, y] /. {x → -1, y → -1}
12                           ←relative minimum at (-1,-1)
```

It is certainly worthwhile plotting this function. *Mathematica* makes it easy, although some experimentation with the options is necessary to show the details clearly.

```
Plot3D[f[x, y], {x, - 2, 2}, {y, - 2, 2},
        PlotRange → { - 2, 5},
        ViewPoint → {1.761, - 2.816, 0.647},
        PlotPoints → 30];
```

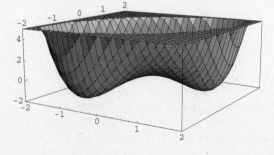

To find the maximum and minimum values of a function $f(x, y)$ subject to the constraint $g(x, y) = 0$, the method of *Lagrange multipliers* can be used. Geometrically, it can be shown that the maximum (minimum) value of f will occur where the level curves of f and the level curves of g share a common tangent line. At this point the gradient of f and the gradient of g will be parallel and $\nabla f(x, y) = \lambda \nabla g(x, y)$. It follows that

$$f_x(x, y) = \lambda g_x(x, y)$$
$$f_y(x, y) = \lambda g_y(x, y)$$

Using these equations, together with $g(x, y) = 0$, λ can be eliminated and the values of x and y corresponding to the maximum and minimum values of f can be determined. The next example illustrates the procedure.

EXAMPLE 7

Suppose we wish to find the maximum and minimum values of $f(x, y) = 2x^2 + 3y^2$ subject to the constraint $x^2 + y^2 = 4$. We define $g(x, y) = x^2 + y^2 - 4$ and eliminate λ.

```
f[x_, y_] = 2 x² + 3 y²; g[x_, y_] = x² + y² - 4;
conditions = Eliminate[{∂ₓf[x,y] == λ∂ₓg[x,y],
                        ∂ᵧf[x,y] == λ∂ᵧg[x,y], g[x,y] == 0}, λ]
```
x² == 4 - y² && xy == 0 && - 4y + y³ == 0

> **Eliminate** [*equations, variables*] eliminates *variables* between a set of simultaneous equations.

```
pts = Solve[conditions]
```
{{x → - 2, y → 0}, {x → 0, y → - 2}, {x → 0, y → 2}, {x → 2, y → 0}}

To determine the maximum and minimum value of f, we compute its values at these points.

```
functionvalues = f[x, y] /. pts
```
{8, 12, 12, 8}
```
Max[functionvalues]
```
12
```
Min[functionvalues]
```
8

The method of Lagrange multipliers can be extended to functions of three (or more) variables.

EXAMPLE 8

To find the maximum and minimum values of $f(x, y, z) = xyz$, subject to the constraint $x^2 + 2y^2 + 3z^2 = 6$ we define $g(x, y, z) = x^2 + 2y^2 + 3z^2 - 6$.

```
f[x_,y_,z_] = x y z;
g[x_,y_,z_] = x² + 2 y² + 3 z² - 6;
conditions = Eliminate [{∂ₓf[x,y,z] == λ∂ₓg[x,y,z],
 ∂ᵧf[x,y,z] == λ∂ᵧg[x,y,z], ∂_zf[x,y,z] == λ∂_zg[x,y,z],
 g[x,y,z] == 0}, λ]
```
x² == 6 - 2 y² - 3 z² && 4 y² z == z (6 - 3 z²) &&
  x (2 y² - 3 z²) == 0 && xy (- 2 + 3 z²) == 0 && xz (- 2 + 3 z²) == 0 &&
  yz (- 2 + 3 z²) == 0 && 4 z - 8 z³ + 3 z⁵ == 0 && y³ + y (- 3 + 3 z²) == 0
```
pts = Solve[conditions]
```
$$\left\{\left\{x \to 0, y \to 0, z \to -\sqrt{2}\right\}, \left\{x \to 0, y \to 0, z \to -\sqrt{2}\right\},\right.$$
$$\left\{x \to 0, y \to 0, z \to \sqrt{2}\right\}, \left\{x \to 0, y \to 0, z \to \sqrt{2}\right\},$$
$$\left\{x \to 0, y \to -\sqrt{3}, z \to 0\right\}, \left\{x \to 0, y \to -\sqrt{3}, z \to 0\right\},$$
$$\left.\left\{x \to 0, y \to \sqrt{3}, z \to 0\right\}, \left\{x \to 0, y \to \sqrt{3}, z \to 0\right\},\right.$$

$$\left\{x \to -\sqrt{2}, y \to -1, z \to -\sqrt{\frac{2}{3}}\right\}, \left\{x \to -\sqrt{2}, y \to -1, z \to \sqrt{\frac{2}{3}}\right\},$$

$$\left\{x \to -\sqrt{2}, y \to 1, z \to -\sqrt{\frac{2}{3}}\right\}, \left\{x \to -\sqrt{2}, y \to 1, z \to \sqrt{\frac{2}{3}}\right\},$$

$$\left\{x \to \sqrt{2}, y \to -1, z \to -\sqrt{\frac{2}{3}}\right\}, \left\{x \to \sqrt{2}, y \to -1, z \to \sqrt{\frac{2}{3}}\right\},$$

$$\left\{x \to \sqrt{2}, y \to 1, z \to -\sqrt{\frac{2}{3}}\right\}, \left\{x \to \sqrt{2}, y \to 1, z \to \sqrt{\frac{2}{3}}\right\},$$

$$\left\{x \to -\sqrt{6}, y \to 0, z \to 0\right\}, \left\{x \to -\sqrt{6}, y \to 0, z \to 0\right\},$$
$$\left\{x \to \sqrt{6}, y \to 0, z \to 0\right\}, \left\{x \to \sqrt{6}, y \to 0, z \to 0\right\}\right\}$$

functionvalues = f[x, y, z] /. pts

$$\left\{0, 0, 0, 0, 0, 0, 0, 0, -\frac{2}{\sqrt{3}}, \frac{2}{\sqrt{3}}, \frac{2}{\sqrt{3}}, -\frac{2}{\sqrt{3}}, \frac{2}{\sqrt{3}}, -\frac{2}{\sqrt{3}}, -\frac{2}{\sqrt{3}}, \frac{2}{\sqrt{3}}, 0, 0, 0, 0\right\}$$

Min[functionvalues]

$$-\frac{2}{\sqrt{3}}$$

Max[functionvalues]

$$\frac{2}{\sqrt{3}}$$

SOLVED PROBLEMS

10.8   Find all relative extrema of the function $f(x, y) = x^2 - y^2$.   Sketch the surface.

SOLUTION

```
f[x_, y_] = x² - y²;
Solve[{∂ₓf[x, y] == 0, ∂_y f[x, y] == 0}, {x, y}]
  {{x → 0, y → 0}}
d[x_, y_] = ∂_(x, 2) f[x, y] ∂_(y, 2) f[x, y] -
  (∂_(x,y) f[x,y])²;
d[0, 0]
  -4            ← negative number; saddle point at (0, 0)
Plot3D[f[x, y], {x, -5, 5}, {y, -5, 5},
     BoxRatios → {1, 1, 1}, PlotPoints → 25];
```

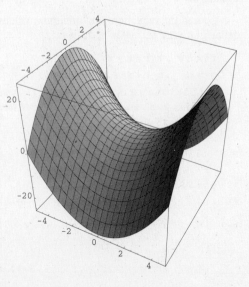

10.9 Find all relative extrema of the function $f(x, y) = xye^{-x^2-y^2}$.   Sketch the surface.

SOLUTION

```
f[x_, y_] = x y Exp[- x² - y²];
pdx = ∂ₓf[x,y]//Factor
```
$- e^{-x^2-y^2} (- 1 + 2 x^2) y$

```
pdy = ∂_y f[x,y]//Factor
```
$- e^{-x^2-y^2} x (- 1 + 2 y^2)$

If we try to use `Solve` to find where the partial derivatives are 0, we will get an error message due to the presence of the (nonalgebraic) exponential.   However, since $-e^{-x^2-y^2}$ cannot equal zero, we can ignore its presence.

```
Solve[{-1 + 2x²)y == 0, x(-1 + 2y²) == 0}, {x, y}]
```
$$\left\{ \{x \to 0, y \to 0\}, \left\{x \to -\frac{1}{\sqrt{2}}, y \to -\frac{1}{\sqrt{2}}\right\}, \left\{x \to \frac{1}{\sqrt{2}}, y \to \frac{1}{\sqrt{2}}\right\}, \right.$$
$$\left. \left\{x \to \frac{1}{\sqrt{2}}, y \to -\frac{1}{\sqrt{2}}\right\}, \left\{x \to \frac{1}{\sqrt{2}}, y \to \frac{1}{\sqrt{2}}\right\}\right\}$$

```
d[x_,y_] = ∂_{x, 2} f[x,y] ∂_{y, 2} f[x,y] - (∂_{x,y} f[x,y])²;
d[0, 0]
```
-1 $\leftarrow$ negative number; no relative extremum

```
d[-1/√2, -1/√2]
```
$\dfrac{4}{e^2}$

```
∂_{x, 2} f[x,y] /. {x → -1 / √2, y → -1 / √2}
```
$-\dfrac{2}{e}$ $\leftarrow$ relative maximum at $\left(\dfrac{-1}{\sqrt{2}}, \dfrac{-1}{\sqrt{2}}\right)$

```
d[-1 / √2, 1 / √2]
```
$\dfrac{4}{e^2}$

```
∂_{x, 2} f[x,y] /. {x → -1 / √2, y → 1 / √2}
```
$\dfrac{2}{e}$ $\leftarrow$ relative minimum at $\left(\dfrac{-1}{\sqrt{2}}, \dfrac{1}{\sqrt{2}}\right)$

```
d[1 / √2, -1 / √2]
```
$\dfrac{4}{e^2}$

```
∂_{x, 2} f[x,y] /. {x → 1 / √2, y → -1 / √2}
```
$\dfrac{2}{e}$ $\leftarrow$ relative minimum at $\left(\dfrac{1}{\sqrt{2}}, \dfrac{-1}{\sqrt{2}}\right)$

```
d[1 / √2, 1 / √2]
```
$\dfrac{4}{e^2}$

```
∂_{x, 2} f[x,y] /. {x → 1 / √2, y → 1 / √2}
```
$-\dfrac{2}{e}$ $\leftarrow$ relative maximum at $\left(\dfrac{1}{\sqrt{2}}, \dfrac{1}{\sqrt{2}}\right)$

We sketch the surface showing two views.

```
Plot3D[f[x, y], {x, - 3, 3}, {y, - 3, 3}, PlotPoints → 30,
    ViewPoint → {1.391, - 3.001, 0.713}, PlotRange → All];
Plot3D[f[x, y], {x, - 3, 3}, {y, - 3, 3}, PlotPoints → 30,
    ViewPoint → {0.617, - 3.318, 0.245}, PlotRange → All];
```

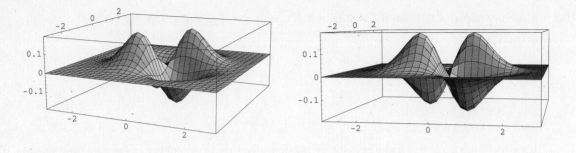

10.10 Use Lagrange multipliers to find the points on the circle $x^2 + y^2 - 2x - 4y = 0$ closest to and furthest from $P(4, 4)$.

SOLUTION

```
<<Graphics`ImplicitPlot`
circ = ImplicitPlot[x² + y² - 2x - 4y == 0, {x, -5, 5},
                DisplayFunction → Identity];
p = Graphics[{PointSize[.02], Point[{4, 4}]}];
Show[circ, p, DisplayFunction → $DisplayFunction];
```

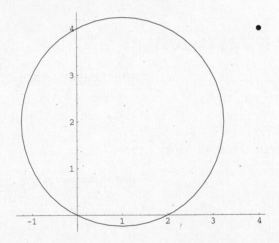

```
f[x_, y_] = (x - 4)² + (y - 4)²;        ← We minimize the square of the distance from P.
g[x_, y_] = x² + y² - 2x - 4y;
conditions = Eliminate[{∂ₓf[x,y] == λ∂ₓg[x,y],
   ∂ᵧf[x,y] == λ∂ᵧg[x,y], g[x,y] == 0},λ]
```

$2 x == -4 + 3 y \;\&\&\; -52 y + 13 y^2 == -32$

```
pts = Solve[conditions]
```

$$\left\{\left\{x \to \frac{1}{13}\left(13 - 3\sqrt{65}\right), y \to \frac{2}{13}\left(13 - \sqrt{65}\right)\right\},\right.$$

$$\left.\left\{x \to \frac{1}{13}\left(13 + 3\sqrt{65}\right), y \to \frac{2}{13}\left(13 + \sqrt{65}\right)\right\}\right\}$$

```
√f[x,y] / .pts //N
```

$\{5.84162, 1.36948\}$

> Based upon the computed distances, the first point is farthest from P and the second point is closest.

10.11 Find the points on the sphere $x^2 + y^2 + z^2 = 1$ that are closest to and farthest from $(1, 2, 3)$.

SOLUTION

```
f[x_, y_, z_] = (x - 1)² + (y - 2)² + (z - 3)²;
g[x_, y_, z_] = x² + y² + z² - 1;                          ← minimize the square of the distance from (1, 2, 3)
conditions = Eliminate [{∂ₓf[x,y,z] == λ∂ₓg[x,y,z],
  ∂ᵧf[x,y,z] == λ∂ᵧg[x,y,z], ∂_zf[x,y,z] == λ∂_zg[x,y,z],
  g[x,y,z] == 0}, λ]
3 x == z && 3 y == 2 z && 14 z² == 9
pts = Solve[conditions, {x, y, z}]
```

$$\left\{ \left\{ x \to -\frac{1}{\sqrt{14}}, \; y \to -\sqrt{\frac{2}{7}}, \; z \to -\frac{3}{\sqrt{14}} \right\}, \right. \qquad \text{← farthest point}$$

$$\left. \left\{ x \to \frac{1}{\sqrt{14}}, \; y \to \sqrt{\frac{2}{7}}, \; z \to \frac{3}{\sqrt{14}} \right\} \right\} \qquad \text{← closest point}$$

```
√f[x,y,z] /.pts //N
{4.74166, 2.74166}
```

10.3 THE TOTAL DIFFERENTIAL

The command D, discussed in Section 10.1, gives the partial derivative of a function of several variables. All variables other than the variable of differentiation are considered as constants. If f is a function, say, of two variables x and y, but y is a function of x, D will compute an incorrect derivative.

The command **Dt** returns the *total differential* of a function.

- **Dt[f[x, y]]** returns the total differential of f[x, y].
- **Dt[f[x, y], x]** returns the total derivative of f[x, y] with respect to x.

Of course, f may be a function of more that two variables and the independent variable, listed as x in the above description, can be any of the variables defining f.

EXAMPLE 9

```
f[x_, y_] = x²y³
Dt[f[x, y]]
2x y³ Dt[x] + 3x² y² Dt[y]
Dt[f[x, y], x]
2x y³ + 3 x² y² Dt[y, x]
Dt[f[x, y], y]
3x² y² + 2 x y³ Dt[x, y]
```

If $z = f(x, y)$, where y is a function of x, **D[f[x, y], x]** returns $\dfrac{\partial z}{\partial x}$ but **Dt[f[x, y], x]** returns $\dfrac{dz}{dx}$.

EXAMPLE 10

Suppose $f(x, y) = x^4 y^5$ where $y = x^3$.   The following sequence gives an *incorrect* derivative.

```
z = x⁴ y⁵;
D[z, x] /. y → x³
4x¹⁸
```

In reality $z = x^{19}$, so dz/dx should equal $19x^{18}$.

```
Dt[z, x] /. y → x³
19x¹⁸
```

In some expressions, constants represented by letters might cause confusion. The option **Constants** can be used to instruct *Mathematica* to treat a particular symbol as a constant.

- **Constants → *objectlist*** causes all symbols in the list *objectlist* to be treated as constant.

EXAMPLE 11

```
Dt[xⁿ]
n x⁻¹⁺ⁿ + xⁿ Dt[n, x] Log[x]
Dt[xⁿ, Constants → {n}]
n x⁻¹⁺ⁿ
```

SOLVED PROBLEMS

10.12 Let $z = \sin xy$.   Let $x = 1$, $y = 2$, $dx = \Delta x = 0.03$, and $dy = \Delta y = 0.02$.   Compute dz and compare it with the value of Δz.

SOLUTION

```
z = f[x_, y_] = Sin[xy];
Δz = f[x + Δx, y + Δy] - f[x, y];
Dt[z] /. {x → 1, y → 2, Dt[x] → 0.03, Dt[y] → 0.02}
- 0.0332917
Δz /. {x → 1, y → 2, Δx → 0.03, Δy → 0.02}
- 0.0364571
```

10.13 Use differentials to approximate $e^{0.1} \sqrt{4.01}$ and determine the percentage error of the estimate.

SOLUTION

```
f[x_, y_] = Exp[x] Sqrt[y];
approximation = f[0, 4] + Dt[f[x, y]] /. {x → 0, y → 4, Dt[x] → 0.1, Dt[y] → 0.01}
2.2025
exactvalue = f[0.1, 4.01]
2.2131
percenterror = Abs[approximation - exactvalue]/exactvalue * 100;
Print["The error is ", percenterror, "%"]
The error is 0.479103 %
```

10.14 Use differentials to approximate the amount of metal in a tin can with height 8 cm and radius 2 cm, if the thickness of the metal in the wall of the cylinder is 0.03 cm and the top and bottom are each 0.02 cm thick.

SOLUTION

> The change in height is the sum of the thicknesses of the top and bottom.

```
v = π r² h;
Dt[v] /. {h → 8, r → 2, Dt[h] → 0.04, Dt[r] → 0.03}
3.51858
```

The amount of metal is approximately 3.52 cm³.

10.15 If three resistors of resistance, R_1, R_2, and R_3 ohms are connected in parallel, their effective resistance is $\dfrac{1}{\frac{1}{R_1} + \frac{1}{R_2} + \frac{1}{R_3}}$ ohms. If a 20-ohm, 30-ohm, and 50-ohm resistor, each with maximum errors 5%, 2%, and 1%, respectively, are connected in parallel, what range of resistance is possible from this combination?

SOLUTION

```
f[R1_, R2_, R3_] = 1/(1/R₁ + 1/R₂ + 1/R₃);

Dt[f[R1, R2, R3]]
```

$$-\frac{-\frac{Dt[R1]}{R1^2} - \frac{Dt[R2]}{R2^2} - \frac{Dt[R3]}{R3^2}}{\left(\frac{1}{R1} + \frac{1}{R2} + \frac{1}{R3}\right)^2}$$

```
f[20, 30, 50]//N
9.67742
Dt[f[R1, R2, R3]] /. {R1 → 20, R2 → 30, R3 → 50, Dt[R1] → 1.0,
    Dt[R2] → 0.6, Dt[R3] → 0.5}
0.315297
```

The combined resistance is 9.67742 ± 0.315297 ohms.

10.4 MULTIPLE INTEGRALS

Multiple integrals, or more precisely iterated integrals, are invoked by the **Integrate** command and are an extension of the command for a function of one variable.

- **Integrate[f[x, y], {x, xmin, xmax}, {y, ymin, ymax}]** evaluates the double integral

$$\int_{x\min}^{x\max} \int_{y\min}^{y\max} f(x, y)\, dy\, dx.$$

- **Integrate[f[x, y, z], {x, xmin, xmax}, {y, ymin, ymax}, {z, zmin, zmax}]**

evaluates the triple integral $\displaystyle\int_{x\min}^{x\max} \int_{y\min}^{y\max} \int_{z\min}^{z\max} f(x, y, z)\, dz\, dy\, dx.$

Higher-order iterated integrals are evaluated in a similar manner. Note that the *rightmost* variable of integration in Integrate is the variable that is evaluated first.

As an alternative, the integral symbol from the BasicInput palette may be used repeatedly for the evaluation of multiple integrals.

EXAMPLE 12

To evaluate $\displaystyle\int_1^2 \int_1^x (x+y)\, dy\, dx$ we would type

`Integrate[x + y, {x, 1, 2}, {y, 1, x}]`

$\dfrac{3}{2}$

Obviously, if we reverse the order of the variables of integration, we get a *completely wrong answer*.

`Integrate[x + y, {y, 1, x}, {x, 1, 2}]`

$-2 + \dfrac{3x}{2} + \dfrac{x^2}{2}$

To resemble standard mathematical notation, the BasicInput palette can be used.

$\displaystyle\int_1^2 \int_1^x (x+y)\ \mathrm{dy}\ \mathrm{dx}$

$\dfrac{3}{2}$

EXAMPLE 13

To evaluate the triple integral $\displaystyle\int_0^2 \int_0^x \int_0^{xy} xyz\, dz\, dy\, dx$ we can either use

`Integrate[x y z, {x, 0, 2}, {y, 0, x}, {z, 0, x y}]`

4

or

$\displaystyle\int_0^2 \int_0^x \int_0^{xy} x\ y\ z\ \mathrm{dz}\ \mathrm{dy}\ \mathrm{dx}$

4

If the integral is such that its *exact* value cannot be determined, numerical integration can be used instead.

- `NIntegrate[f[x, y], {x, xmin, xmax}, {y, ymin, ymax}]` returns a numerical approximation of the double integral $\displaystyle\int_{x\min}^{x\max} \int_{y\min}^{y\max} f(x,y)\, dy\, dx$.

- `NIntegrate[f[x, y, z], {x, xmin, xmax}, {y, ymin, ymax}, {z, zmin, zmax}]` returns a numerical approximation the triple integral $\displaystyle\int_{x\min}^{x\max} \int_{y\min}^{y\max} \int_{z\min}^{z\max} f(x,y,z)\, dz\, dy\, dx$.

Higher-order iterated integrals are evaluated in a similar manner. If the BasicInput palette is used, the `N` command (or `//N` to the right of the integral) may be used. All of the options for `Nintegrate` as applied to functions of a single variable apply to multiple integrals.

EXAMPLE 14

`Integrate[Exp[x² y²], {x, 0, 1}, {y, 0, 1}]`

$\dfrac{1}{2}\sqrt{\pi} \displaystyle\int_0^1 \dfrac{\mathrm{Erfi}\,[x]}{x}\, \mathrm{dx}$.

```
NIntegrate[Exp[x² y²], {x, 0, 1}, {y, 0, 1}]
```
1.1351

$$\int_0^1 \int_0^1 \text{Exp}[x^2 \, y^2] \, dy \, dx \text{ //N}$$

1.1351

SOLVED PROBLEMS

10.16　Use a double integral to compute the area bounded by the parabola $y = x^2 - 2x + 2$ and the line $y = x + 1$.

SOLUTION

```
f[x_] = x² - 2x + 2;
g[x_] = x + 1;
Plot[{f[x], g[x]}, {x, -1, 3}];
```

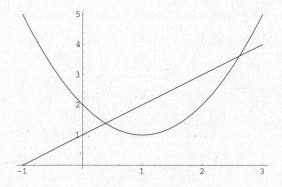

> It's always a good idea to sketch the region under consideration before integrating.

```
intersections = Solve[f[x] == g[x]]
```

$$\left\{ \left\{ x \to \frac{1}{2} \left(3 - \sqrt{5} \right) \right\}, \left\{ x \to \frac{1}{2} \left(3 + \sqrt{5} \right) \right\} \right\}$$

```
a = intersections[[1, 1, 2]];
b = intersections[[2, 1, 2]];
```

$$\int_a^b \int_{f[x]}^{g[x]} dy \, dx$$

$$\frac{5\sqrt{5}}{6}$$

10.17　Find the center of mass of the lamina bounded by the parabola $y = 9 - x^2$ and the x axis if the density at each point is proportional to its distance from the x axis.

SOLUTION

Let R be the region bounded by $y = 9 - x^2$ and the x axis.

```
Plot[9 - x², {x, -3, 3}];
```

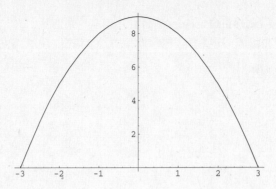

The graph intersects the x axis at -3 and 3. The coordinates of the center of mass, $(\bar{x}, \bar{y})$, are $\left(\dfrac{M_y}{M}, \dfrac{M_x}{M}\right)$, where the density function $\rho(x, y) = k\, y$.

$$M_y = \text{Moment about the } y \text{ axis} = \iint_R x\rho(x, y)\,dA$$

$$M_x = \text{Moment about the } x \text{ axis} = \iint_R y\rho(x, y)\,dA$$

$$M = \text{Mass of lamina} = \iint_R \rho(x, y)\,dA$$

```
ρ[x_, y_] = k y;

my = ∫₋₃³ ∫₀^{9-x²} x ρ[x,y] dy dx

0

mx = ∫₋₃³ ∫₀^{9-x²} y ρ[x,y] dy dx

23328 k
───────
  35

m = ∫₋₃³ ∫₀^{9-x²} ρ[x,y] dy dx

648 k
─────
  5

{my/m, mx/m}

{0, 36/7}
```

10.18 Compute the shaded area. The curve shown is the spiral of Archimedes, and has polar equation $r = \theta$. It is shown for $0 \le \theta \le 6\pi$.

SOLUTION

The area inside a polar region, R, is $\iint_R r\,dr\,d\theta$. The smaller arc of the shaded region is described by $r = \theta$,

$2\pi \leq \theta \leq 5\pi/2$, and the larger arc may be represented by $r = \theta + 2\pi$, $2\pi \leq \theta \leq 5\pi/2$. Hence the enclosed area can be expressed as $\displaystyle\int_{2\pi}^{5\pi/2} \int_{\theta}^{\theta+2\pi} r\,dr\,d\theta$.

$$\int_{2\pi}^{5\pi/2} \int_{\theta}^{\theta+2\pi} r\,dr\,d\theta$$

$$\frac{13\pi^3}{4}$$

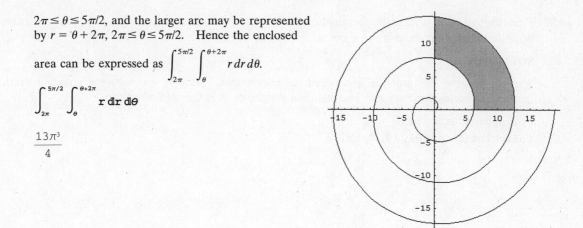

10.19 Compute the volume under the paraboloid $z = x^2 + y^2$, above the region in the *x-y* plane bounded by $y = x^2$ and $y = \sqrt{x+1}$.

SOLUTION

The volume bounded by a surface $z = f(x, y)$ and the *x-y* plane, above a region R, is $\displaystyle\iint_R f(x,y)\,dA$. First let's look at R.

$$\texttt{Plot}\left[\left\{\texttt{x}^2, \sqrt{\texttt{x+1}}\right\}, \{\texttt{x}, \texttt{-1}, \texttt{2}\}\right]$$

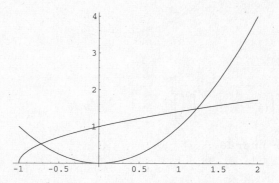

Next we find the points of intersection. Because of the complicated nature of the solution, we will use a numerical approximation.

$$\texttt{NSolve}[\texttt{x}^2 == \sqrt{\texttt{x+1}}]$$

$$\{\{\texttt{x} \to 1.22074\}, \{\texttt{x} \to -0.724492\}\}$$

Now we can express the volume as a double integral. Two solutions are shown.

$$\int_{-.724492}^{1.22074} \int_{\texttt{x}^2}^{\sqrt{\texttt{x+1}}} (\texttt{x}^2 + \texttt{y}^2)\,d\texttt{y}\,d\texttt{x} \,/\!/\texttt{N}$$

1.11738

$$\texttt{NIntegrate[x\^{}2 + y\^{}2, \{x, -0.724492, 1.22074\}, \{y, x\^{}2, Sqrt[x+1]\}]}$$

1.11738

10.20 Find the volume under the hemisphere $z = 4 - x^2 - y^2$ above the region in the x-y plane bounded by the cardioid $r = 1 - \cos\theta$.

SOLUTION

We will translate the problem into cylindrical coordinates. Since $r^2 = x^2 + y^2$, the equation of the hemisphere becomes $z = 4 - r^2$. The region of integration R is the cardioid shown.

```
<<Graphics`Graphics`
PolarPlot[1-Cos[θ], {θ, 0, 2π}];
```

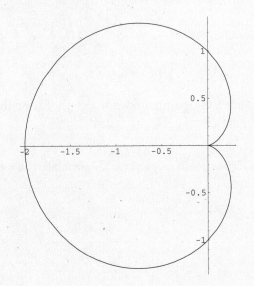

The volume $V = \iint\limits_{R} (4 - r^2)dA = \int_0^{2\pi} \int_0^{1-\cos\theta} (4 - r^2)\, r\, dr\, d\theta$.

$$\int_0^{2\pi} \int_0^{1-\text{Cos}[\theta]} (4 - r^2)\, r\, dr\, d\theta$$

$$\frac{61\pi}{16}$$

10.21 Find the volume of the solid that lies under the paraboloid $z = x^2 + y^2$, above the x-y plane, and inside the cylinder $(x - 1)^2 + y^2 = 1$.

SOLUTION

The cylinder $(x - 1)^2 + y^2 = 1$ is a cylinder of radius 1 whose axis is translated from the z axis by the vector $(1, 0, 0)$.

```
<<Graphics`Shapes`
s1 = Graphics3D[TranslateShape[Cylinder[1, 4, 30], {1, 0, 0}]];
s2 = Plot3D[x² + y², {x, -2, 2}, {y, -2, 2}, DisplayFunction → Identity];
Show[s1, s2, PlotRange → {0, 4},
    ViewPoint → {1.217, -3.125, 0.447}, DisplayFunction → $DisplayFunction];
```

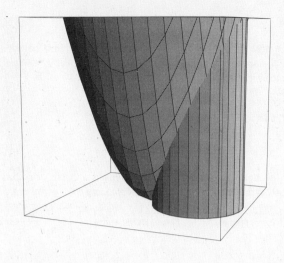

Now that we know what the region looks like, we must look at its projection in the *x-y* plane.

```
<< Graphics`ImplicitPlot`
ImplicitPlot[(x - 1)² + y² == 1, {x, 0, 2}];
```

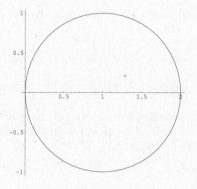

Although the problem can be solved in rectangular coordinates, it is easier to solve it in cylindrical coordinates. The equation of the circle can be expanded to become $x^2 + y^2 = 2x$, which is equivalent, in polar coordinates, to $r = 2 \cos \theta$. The complete circle is generated as θ varies from $-\pi/2$ to $\pi/2$. The paraboloid $z = x^2 + y^2$ becomes $z = r^2$. The required volume is expressed as a double integral.

$$\int_{-\pi/2}^{\pi/2} \int_0^{2\cos[\theta]} (r^2) \; r \; dr \; d\theta$$

$$\frac{3\pi}{2}$$

10.22 The area of the surface $z = f(x, y)$ above the region R in the x-y plane is $\iint\limits_R \sqrt{[f_x(x, y)]^2 + [f_y(x, y)]^2 + 1}$. Compute the surface area of a sphere of radius a.

SOLUTION

We compute the surface area of the portion of the sphere in the first octant and, by symmetry, multiply by 8. The equation of the sphere is $x^2 + y^2 + z^2 = a^2$. Solving for z, we get $f(x, y) = z = \sqrt{a^2 - x^2 - y^2}$ as the function representing the upper hemisphere.

```
f[x_,y_] = √(a² - x² - y²);
1 + (∂ₓf[x, y])² + (∂_yf[x,y])²//Together
```

$$-\frac{a^2}{-a^2 + x^2 + y^2}$$

Since the projection of the hemisphere onto the x-y plane is a circle of radius a centered at the origin, it is most convenient to use cylindrical coordinates. Replacing $x^2 + y^2$ by r^2,

$$\iint_R \sqrt{1 + (\partial_x f[x, y])^2 + (\partial_y f[x, y])^2}\, dA = 8 \int_0^{\pi/2} \int_0^a \sqrt{\frac{a^2}{a^2 - r^2}}\, r\, dr\, d\theta$$

$$8 \int_0^{\pi/2} \int_0^a \sqrt{\frac{a^2}{a^2 - r^2}}\, r\, dr\, d\theta$$

$$4a^2\pi$$

10.23 Find the volume of the "ice cream cone" bounded by the cone $z = 3\sqrt{x^2 + y^2}$ and the sphere $x^2 + y^2 + (z - 9)^2 = 9$.

SOLUTION

The required volume is represented by the triple integral $\iiint_G dV$.

Because of the nature of the bounding surfaces, this problem is done most conveniently using cylindrical coordinates. First, rewrite the equation of the sphere, solving for z in terms of x and y.

```
Solve[x² + y² + (z - 9)² == 9, z]
```

$$\left\{\left\{z \to 9 - \sqrt{9 - x^2 - y^2}\right\}, \left\{z \to 9 + \sqrt{9 - x^2 - y^2}\right\}\right\}$$

Using the second solution (corresponding to the upper hemisphere), and replacing $x^2 + y^2$ by r^2, the equation of the sphere becomes $z = 9 + \sqrt{9 - r^2}$. Now we can sketch the surfaces which form our region.

```
<<Graphics`ParametricPlot3D`
cone = CylindricalPlot3D[3r, {r, 0, 3}, {θ, 0, 2π},
                        DisplayFunction → Identity];
hemisphere = CylindricalPlot3D[9 + √(9 - r²), {r, 0, 3}, {θ, 0, 2π},
                        DisplayFunction → Identity];
Show[cone, hemisphere,
    DisplayFunction → $DisplayFunction,
    Axes → False, Boxed → False];
```

To compute the volume, we observe that the projection of the region onto the x-y plane is a circle. To determine its radius, we find the intersection of the cone and the hemisphere.

```
Solve[3r == 9 + √(9 - r²)]
```

```
{{r → 3}}
```

The projection onto the *x-y* plane is a circle of radius 3 centered at the origin.   The required volume is

$$\int_0^{2\pi} \int_0^3 \int_{3r}^{9+\sqrt{9-r^2}} r\, dz\, dr\, d\theta$$

45π

10.24 A "silo" is formed above the *x-y* plane by the intersection of a right circular cylinder of radius 3 and a sphere of radius 5.   Compute its volume.

SOLUTION

It is easiest to work in cylindrical coordinates.   The equation of the spherical cap is $z = \sqrt{25 - r^2}$.

```
<<Graphics`Shapes`
<<Graphics`ParametricPlot3D`
cyl = Graphics3D[Cylinder[3, 4, 30]];
cap = CylindricalPlot3D[√25 - r², {r, 0, 3}, {θ, 0, 2π},
    DisplayFunction → Identity];
Show[cyl, cap, PlotRange → {0, 5}, Boxed → False,
    DisplayFunction → $DisplayFunction];
```

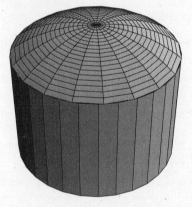

The projection of the solid is a circle of radius 3, centered at the origin.

$$\mathbf{volume} = \int_0^{2\pi} \int_0^3 \int_0^{\sqrt{25-r^2}} r\, dz\, dr\, d\theta$$

$\dfrac{122\,\pi}{3}$

10.25 Find the center of mass of a solid hemisphere of radius *a* if its density at each point is proportional to its distance above the *x-y* plane.

SOLUTION

The center of mass has coordinates $(\bar{x}, \bar{y}, \bar{x})$ where

$$\bar{x} = \frac{\iiint\limits_G x\,\sigma(x,y,z)\,dV}{\iiint\limits_G \sigma(x,y,z)\,dV},\ \bar{y} = \frac{\iiint\limits_G y\,\sigma(x,y,z)\,dV}{\iiint\limits_G \sigma(x,y,z)\,dV},\ \bar{z} = \frac{\iiint\limits_G z\,\sigma(x,y,z)\,dV}{\iiint\limits_G \sigma(x,y,z)\,dV}$$

The density function $\sigma(x, y, z) = kz$, where k is a constant of proportionality. The problem is most conveniently solved by using spherical coordinates:

$$x = \rho \sin \phi \cos \theta, \qquad y = \rho \sin \phi \sin \theta, \qquad z = \rho \cos \phi$$

$\mathbf{x = \rho \, Sin[\phi] \, Cos[\theta] \, ;}$

$\mathbf{y = \rho \, Sin[\phi] \, Sin[\theta] \, ;}$

$\mathbf{z = \rho \, Cos[\phi] \, ;}$

$\mathbf{\sigma = k \, z \, ;}$

$$\mathbf{mass = \int_0^{2\pi} \int_0^{\pi/2} \int_0^a \sigma \rho^2 \, Sin[\phi] \, d\rho \, d\phi \, d\theta}$$

$\dfrac{1}{4} a^4 k \pi$

$$\mathbf{centerofmass = \left\{ \frac{\int_0^{2\pi} \int_0^{\pi/2} \int_0^a x \, \sigma \rho^2 \, Sin[\phi] \, d\rho \, d\phi \, d\theta}{mass}, \frac{\int_0^{2\pi} \int_0^{\pi/2} \int_0^a y \, \sigma \rho^2 \, Sin[\phi] \, d\rho \, d\phi \, d\theta}{mass}, \right.}$$

$$\mathbf{\left. \frac{\int_0^{2\pi} \int_0^{\pi/2} \int_0^a z \, \sigma \rho^2 \, Sin[\phi] \, d\rho \, d\phi \, d\theta}{mass} \right\}}$$

$\left\{ 0, \, 0, \, \dfrac{8 \, a}{15} \right\}$

10.26 Find the moment of inertia of a solid hemisphere of radius a about the z axis if its density is proportional to the distance from the center of its base.

SOLUTION

The moment of inertia about the z axis is $\iiint\limits_G [\delta(x, y, z)]^2 \, \sigma(x, y, z) \, dV$ where $\delta(x, y, z)$ is the distance from the point (x, y, z) to the z axis and $\sigma(x, y, z)$ is the density at the point (x, y, z). In this problem, we should use a spherical coordinate system:

$$x = \rho \sin \phi \cos \theta, \qquad y = \rho \sin \phi \sin \theta, \qquad z = \rho \cos \phi$$

$$\delta(x, y, z) = \sqrt{x^2 + y^2} = \sqrt{(\rho \sin \phi \cos \theta)^2 + (\rho \sin \phi \sin \theta)^2} = \rho \sin \phi$$

$$\sigma(x, y, z) = \sqrt{x^2 + y^2 + z^2} = \rho$$

$$\int_0^{2\pi} \int_0^{\pi/2} \int_0^a (\rho \, Sin[\phi])^2 \, \rho \, (\rho^2 \, Sin[\phi]) \, d\rho \, d\phi \, d\theta$$

$\dfrac{2 a^6 \pi}{9}$

CHAPTER 11

Ordinary Differential Equations

11.1 ANALYTICAL SOLUTIONS

Simply put, a differential equation is an equation expressing a relationship between a function and one or more of its derivatives. A function which satisfies a differential equation is called a *solution*.

The *Mathematica* command **DSolve** is used to solve differential equations. As with algebraic or transcendental equations, a double equal sign, **==**, is used to separate the two sides of the equation.

- **DSolve[*equation*, y[x], x]** gives the general solution, y[x], of the differential equation, *equation*, whose independent variable is x.
- **DSolve[*equation*, y, x]** gives the general solution, y, of the differential equation expressed as a "pure" function (see Appendix) within a list. ReplaceAll (/.) may then be used to evaluate the solution. Alternatively, one may use Part or [[ ]] to extract the solution from the list.

EXAMPLE 1

To solve the first-order differential equation $\frac{dy}{dx} = x + y$, we simply type

```
DSolve[y'[x] == x + y[x], y[x], x]
```
$\{\{y[x] \rightarrow -1 - x + e^x C[1]\}\}$

EXAMPLE 2

To obtain the solution of $\frac{dy}{dx} = x + y$ as a pure function, we enter

```
solution = DSolve[y'[x] == x + y[x], y, x]
```
$\{\{y \rightarrow (-1 + e^{\#1} C[1] - \#1 \&)\}\}$

If we wish to evaluate the solution, we can type

`y[x] /. solution`

$\{ -1 - x + e^x\, C[1] \}$

Using the pure function, we can evaluate the derivatives of the solution. This would be clumsy using the solution of Example 1.

`y'[x] /. solution`

$\{ -1 + e^x\, C[1] \}$

`y''[x] + y'[x] /. solution`

$\{ -1 + 2\, e^x\, C[1] \}$

We can define a function f representing the solution

`f = solution[[1, 1, 2]]`

$-1 + e^{\#1}\, C[1] - \#1\, \&$

We can then directly evaluate f or any of its derivatives.

`f[x]`

$-1 - x + e^x\, C[1]$

`f'[x]`

$-1 + e^x\, C[1]$

`f''[x]`

$e^x\, C[1]$

It is *extremely important* that the unknown function be represented `y[x]`, not `y`, within the differential equation. Similarly, its derivatives must be represented `y'[x]`, `y''[x]`, etc. The next example illustrates some common errors.

EXAMPLE 3

`DSolve[y'[x] == x + y, y[x], x]`

$$\left\{\left\{ y[x] \to \frac{x^2}{2} + x\,y + C[1] \right\}\right\}$$ ← *Mathematica* treats *y* as a constant.

`DSolve[y' == x + y, y, x]`   ← The function and its derivative must be
specified as `y[x]` and `y'[x]`.

```
DSolve::nvld:
 The description of the equations appears to be ambiguous or invalid.
DSolve::deqx :
 Supplied equations are not differential equations of the given functions.
```

The solution of a first order differential equation *without* initial conditions involves an arbitrary constant labeled, by default, `C[1]`. Additional constants (for higher-order equations) are labeled `C[2]`, `C[3]`, ... If a different labeling is desired, the option `DSolveConstants` may be used.

- `DSolveConstants → constantlabel` specifies that the constants should be labeled `constantlabel[1]`, `constantlabel[2]`, etc.

EXAMPLE 4

`DSolve[y'[x] == x + y[x], y[x], x, DSolveConstants → mylabel]`

$\{\{ y[x] \to -1 - x + e^x\, mylabel[1] \}\}$

Higher-order differential equations are solved in a similar manner. The derivatives are represented `y'[x]`, `y''[x]`, `y'''[x]`, ... Alternatively, D, ∂, or `Derivative` may be used.

EXAMPLE 5

```
DSolve[y''[x] + y[x] == 0, y[x], x]
{{y[x] → C[2]Cos[x] - C[1]Sin[x]}}
DSolve[D[y[x], {x, 2}] + y[x] == 0, y[x], x]
{{y[x] → C[2]Cos[x] - C[1]Sin[x]}}
DSolve[∂_(x, 2) y[x] + y[x] == 0, y[x], x]
{{y[x] → C[2]Cos[x] - C[1]Sin[x]}}
DSolve[Derivative[2][y][x] + y[x] == 0, y[x], x]
{{y[x] → C[2]Cos[x] - C[1]Sin[x]}}
```

Complex differential equations are solved, if possible, using special functions. If *Mathematica* cannot solve the equation, it will either return the equation unsolved, or in terms of unevaluated integrals. In such cases a numerical solution (see Section 11.2) may be more appropriate.

EXAMPLE 6

$x^2\dfrac{d^2y}{dx^2} + x\dfrac{dy}{dx} + (x^2 - 4)y = 0$ is a special case of Bessel's equation. The solution is expressed in terms of Bessel functions of the first (`BesselJ`) and second (`BesselY`) kind.

```
DSolve[x²y''[x] + xy'[x] + (x² - 4)y[x] == 0, y[x], x]
{{y[x] → BesselJ[2, x] C[1] + BesselY[2, x] C[2]}}
```

EXAMPLE 7

$\dfrac{d^2y}{dx^2} + \dfrac{dy}{dx} + y^2 = 0$ is a nonlinear differential equation which *Mathematica* cannot solve.

```
DSolve[y''[x] + y'[x] + y[x]² == 0, y[x], x]
DSolve[[y[x]² + y'[x] + y''[x] == 0, y[x], x]
```

If values of *y*, and perhaps one or more of its derivatives, are specified along with the differential equation, the task of finding *y* is known as an *initial value problem*. The differential equation and the initial conditions are specified as a list within the `DSolve` command. A unique solution is returned, provided an appropriate number of initial conditions are supplied.

EXAMPLE 8

Solve the equation $\dfrac{dy}{dx} = x + y$ with initial condition $y(0) = 2$. Then plot the solution.

```
solution = DSolve[{y'[x] == x + y[x], y[0] == 2}, y[x], x]
{{y[x] → -1 + 3 eˣ - x}}
Plot[y[x] /. solution, {x, -5, 2}];
```

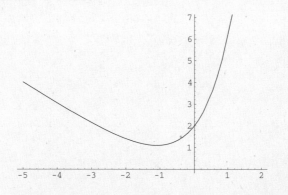

Here is another way the solution can be plotted:

```
solution = DSolve[{y'[x] == x + y[x], y[0] == 2}, y, x]
{{y → (-1 + 3 e^#1 - #1 &)}}
f = solution[[1, 1, 2]]
-1 + 3 e^#1 - #1 &
Plot[f[x], {x, -5, 2}];
```

A useful way of visualizing the solution of a first-order differential equation is to introduce the concept of a vector field. A vector field on $\mathbb{R}^2$ is a function $\mathbf{F}$ that assigns to each point (x, y) a two-dimensional vector $\mathbf{F}(x, y)$. By drawing the vectors $\mathbf{F}(x, y)$ for a (finite) subset of $\mathbb{R}^2$, one obtains a geometric interpretation of the behavior of $\mathbf{F}$.

- **PlotVectorField[{Fx, Fy}, {x, xmin, xmax}, {y, ymin, ymax}]** produces a vector field plot of the two-dimensional vector function **F**, whose components are **Fx** and **Fy**. The direction of the arrow is the direction of the vector field at the point (x, y). The magnitude of the arrow is proportional to the magnitude of the vector field.

PlotVectorField is contained within the package **Graphics`PlotField`** which must be loaded prior to its use.

EXAMPLE 9

Plot the vector field $\mathbf{F}(x, y) = -y\mathbf{i} + x\mathbf{j}$. By default, no axes are drawn so the option **Axes → Automatic** will be used.

```
<<Graphics`PlotField`
PlotVectorField[{-y, x}, {x, -5, 5}, {y, -5, 5}, Axes → Automatic];
```

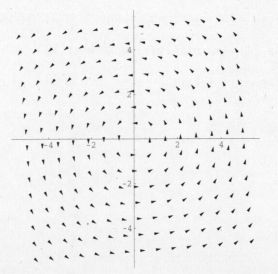

Any first-order differential equation can be used to define a vector field.   Indeed, the vector field $\mathbf{i} + f(x, y)\mathbf{j}$, corresponding to the equation $\dfrac{dy}{dx} = f(x, y)$, generates a field whose vectors are tangent to the solution at any point.   The next example, although simple, illustrates this nicely.

EXAMPLE 10

Plot the vector field of the solution of the equation $\dfrac{dy}{dx} = 2x$.   The solutions to this equation, parabolas $y = x^2 + c$, can be seen quite vividly.

```
<<Graphics`PlotField`
PlotVectorField[{1, 2 x}, {x, -1, 1}, {y, -1, 1}, Axes → Automatic];
```

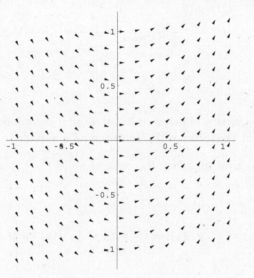

EXAMPLE 11

In this example we plot the vector field generated by the equation $\dfrac{dy}{dx} = 2x + y$.   The option **HeadLength → 0** suppresses the vector heads for a clearer picture.   Then the solutions with initial conditions $y(0) = -2, -1, 0, 1$, and 2 are plotted on the vector field for comparison.

```
<< Graphics`PlotField`
vf = PlotVectorField[{1, 2x + y}, {x, -2, 1}, {y, -4, 6},
    Axes → Automatic, HeadLength → 0,
    AspectRatio → 1/GoldenRatio,
    DisplayFunction → $DisplayFunction];
```

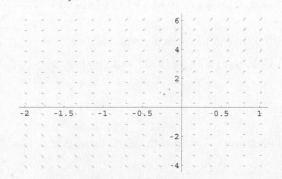

```
sols = Table[DSolve[{y'[x] == 2x + y[x], y[0] == k}, y[x], x], {k, -2, 2}]
```
$\{\{\{y[x] \rightarrow -2 - 2x\}\}, \{\{y[x] \rightarrow -2 + e^x - 2x\}\}, \{\{y[x] \rightarrow -2 + 2e^x - 2x\}\},$
$\quad \{\{y[x] \rightarrow -2 + 3e^x - 2x\}\}, \{\{y[x] \rightarrow -2 + 4e^x - 2x\}\}\}$
```
Do[g[k] =
 Plot[sols[[k, 1, 1, 2]], {x, -2, 1}, PlotRange→All,
 DisplayFunction→Identity], {k, 1, 5}]
Show[g[1], g[2], g[3], g[4], g[5], vf, DisplayFunction→$DisplayFunction];
```

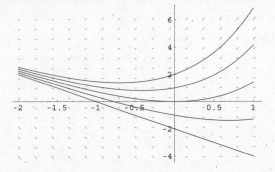

A *system* of differential equations consists of *n* differential equations involving $n + 1$ variables. Solving a system of differential equations with *Mathematica* is similar to solving a single equation.

The next example illustrates how to solve the system $\dfrac{dx}{dt} = t^2, \dfrac{dy}{dt} = t^3$ with initial conditions $x(0) = 2$, $y(0) = 3$.

EXAMPLE 12

```
solution = DSolve[{x'[t] == t², y'[t] == t³, x[0] == 2, y[0] == 3}, {x[t], y[t]}, t]
```
$$\left\{\left\{x[t] \rightarrow \frac{1}{3}(6 + t^3), y[t] \rightarrow \frac{1}{4}(12 + t^4)\right\}\right\}$$

Instead of specifying the values of *f* and its derivatives at a single point, values at two distinct points may be given. The problem of solving the differential equation then becomes known as a *boundary value problem*. However, unlike initial value problems, which can be shown to have unique solutions for a wide variety of cases, a boundary value problem may have no solution even for the simplest of equations.

EXAMPLE 13

Consider the equation $\dfrac{d^2 y}{dx^2} + y = 0$ with boundary conditions $y(0) = 0$, $y(\pi) = 1$. This equation has no solution.

```
DSolve[{y''[x] + y[x] == 0, y[0] == 0, y[π] == 1}, y[x], x]
{}
```

The same equation with $y(0) = 0$, $y(\pi/2) = 1$ has a unique solution.

```
DSolve[{y''[x] + y[x] == 0, y[0] == 0, y[π/2] == 1}, y[x], x]
{{y[x] → Sin[x]}}
```

SOLVED PROBLEMS

11.1   Solve the differential equation $\dfrac{dy}{dx} = xy$ with initial condition $y(1) = 2$ and graph the solution for $-2 \leq x \leq 2$.

SOLUTION

```
equation = DSolve[{y'[x] == x y[x], y[1] == 2}, y[x], x]
```

$$\left\{\left\{y[x] \to 2 \, e^{-\frac{1}{2} + \frac{x^2}{2}}\right\}\right\}$$

```
Plot[y[x] /. equation, {x, -2, 2}];
```

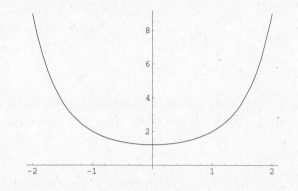

11.2   Plot the vector field for the differential equation of the previous example.

SOLUTION

```
<< Graphics`PlotField`
PlotVectorField[{1, x y}, {x, - 2, 2}, {y, - 10, 10},
              AspectRatio → 1/GoldenRatio, ScaleFactor → 2,
              HeadLength → 0.01, Axes → Automatic];
```

> **ScaleFactor → *length*** is an option that scales the length of the vectors so that the longest vector is of length *length*.   The default is Automatic.   At this setting, the vectors often appear too short to view their direction clearly.   Headlength adjusts the size of the arrowheads.

11.3   Plot the vector field for the equation $\dfrac{dy}{dx} = x^2 + y$ together with its solutions for $y(0) = 0, 1, 2, 3,$ and 4.

SOLUTION

```
<< Graphics`PlotField`
vf = PlotVectorField[{1, x² + y}, {x, 0, 1}, {y, 0, 12},
     Axes → Automatic, HeadLength → 0,
     ScaleFactor → 0.25, AspectRatio → 1/GoldenRatio,
     DisplayFunction → Identity];
sols = Table[DSolve[{y'[x] == x² + y[x], y[0] == k}, y[x], x], {k, 0, 4}];
Do[g[k] = Plot[sols[[k, 1, 1, 2]], {x, 0, 1}, PlotRange → All,
     DisplayFunction → Identity], {k, 1, 5}]
Show[g[1], g[2], g[3], g[4], g[5], vf, DisplayFunction → $DisplayFunction];
```

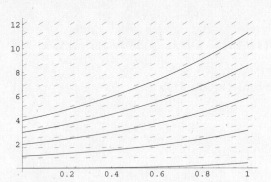

11.4 The *escape velocity* is the minimum velocity with which an object must be propelled in order to escape the gravitational field of a celestial body. Compute the escape velocity for the planet Earth.

SOLUTION

We shall assume that the initial velocity is in a radial direction away from the Earth's center. According to Newton's laws of motion, the acceleration of a particle is inversely proportional to the square of the distance of the particle from the center of the Earth. If r represents that distance, R the radius of the Earth (approximately 3,960 miles), v the velocity of the particle, and a its acceleration, then $a = \dfrac{dv}{dt} = \dfrac{k}{r^2}$.

At the Earth's surface ($r = R$), $a = -g$, where $g = 32.16$ ft/sec^2 = .00609 mi/sec^2. It follows that $k = -gR^2$ so $a = -\dfrac{gR^2}{r^2}$. Since $a = \dfrac{dv}{dt}$ and $v = \dfrac{dr}{dt}$, by the chain rule we have $a = \dfrac{dv}{dt} = \dfrac{dv}{dr}\dfrac{dr}{dt} = v\dfrac{dv}{dr}$. If v_0 re-

presents the escape velocity, we are led to the differential equation $v\dfrac{dv}{dr} = -\dfrac{gR^2}{r^2}$ with initial condition $v = v_0$ when $r = R$.

```
DSolve[{v[r]v'[r] == - g R^2/r^2, v[R] == v₀}, v[r], r]
```

$$\left\{\left\{v[r] \rightarrow -\sqrt{-2\,g\,R + \frac{2\,g\,R^2}{r} + v_0^2}\right\}, \left\{v[r] \rightarrow \sqrt{-2\,g\,R + \frac{2\,g\,R^2}{r} + v_0^2}\right\}\right\}$$

Since the velocity is positive at the surface of the Earth ($r = R$), and must remain positive for the duration of the flight, we reject the first solution. Furthermore, $v(r)$ will remain positive if and only if $-2gR + v_0^2 \geq 0$ so $v_0 \geq \sqrt{2gR}$.

```
√(2gR) /. {g → .00609, R → 3960}
6.94498
```

The escape velocity is 6.94498 miles per second.

11.5 According to Newton's law of cooling, the temperature of an object changes at a rate proportional to the difference in temperature between the object and the outside medium. If an object whose temperature is 70° Fahrenheit is placed in a medium whose temperature is 20°, and is found to be 40° after 3 minutes, what will its temperature be after 6 minutes?

SOLUTION

If $u(t)$ represents the temperature of the object at time t, $\dfrac{du}{dt} = k(u - 20)$. The initial condition is $u(0) = 70$.

```
DSolve[{u'[t] == k(u[t] - 20), u[0] == 70}, u[t], t]
{{u[t] → 20 + 50 e^(kt)}}

u[t_] = %[[1, 1, 2]]
20 + 50 e^(kt)
```

We determine k using the information about the temperature 3 minutes later.   Since we are using `Solve` for a transcendental function e^x, *Mathematica* supplies a warning which may be safely ignored.

```
Solve[u[3] == 40, k]
```
```
Solve::ifun: Inverse functions are being
    used by Solve, so some solutions may not be found.
```
$$\left\{\left\{k \to -\frac{1}{3} \text{Log}\left[\frac{5}{2}\right]\right\}\right\}$$
```
u[6] /. k → %[[1, 1, 2]]
28
```

The temperature 6 minutes later is 28° Fahrenheit.

11.6   A freely falling body falls with an acceleration g, which is approximately 32.16 ft/sec^2.   If air resistance is considered, its motion is changed dramatically.   If an object whose mass is 5 slugs is dropped from a height of 1000 feet, determine how long it will take to hit the ground (*a*) neglecting air resistance and (*b*) assuming that the force of air resistance is equal to the velocity of the object.

SOLUTION

Let $h(t)$ represent the height of the object at time t, $v(t)$ its velocity, and $a(t)$ its acceleration.   Recall that $v(t) = h'(t)$ and $a(t) = v'(t) = h''(t)$ and, by Newton's law, the sum of the external forces acting upon the object is equal to its mass times its acceleration: $m\,a(t) = \sum F$.   In what follows we take "up" to be the positive direction.

(*a*)   If air resistance is neglected, the only force acting on the object is gravity, so $m\,a(t) = -m\,g$.   We can divide by m and solve the differential equation $h''(t) = -g$ with initial conditions $h'(0) = 0$, $h(0) = 1000$.

```
g = 32.16;
solution = DSolve[{h''[t] == -g, h'[0] == 0, h[0] == 1000}, h[t], t];
height[t_] = solution[[1, 1, 2]]
1000. - 16.08 t²
```

When the object reaches the ground its height will be 0.

```
Solve[height[t] == 0, t]
{{t → -7.886}, {t → 7.886}}
```

It takes 7.886 seconds to reach the ground.

(*b*)   If air resistance is taken into account, there is an external force acting upon the object, in addition to gravity, equal to $-v(t)$.   The differential equation becomes

$$m\,a(t) = -m\,g - v(t)$$

or

$$m\,h''(t) = -m\,g - h'(t)$$

with initial conditions as in (*a*).

```
m = 5; g = 32.16;
solution = DSolve[{m h''[t] == -m g - h'[t], h'[0] == 0, h[0] == 1000}, h[t], t];
height[t_] = solution[[1, 1, 2]]
1804. - 804. e^{-0.2 t} - 160.8 t
```

```
FindRoot[height[t] == 0, {t, 10}]
{t → 10.6213}
```

It now takes 10.6213 seconds to reach the ground.

11.7   A baseball is hit with a velocity of 100 ft/sec at an angle of 30° with the horizontal.   The height
of the bat is 3 ft above the ground.   Neglecting air and wind resistance, will it clear a 35-ft-high
fence located 200 ft from home plate?

SOLUTION

```
g = 32; h = 3; θ = 30 Degree; v₀ = 100;
solution = DSolve[{x''[t] == 0, y''[t] == - g, x'[0] == v₀ Cos[θ],
   y'[0] == v₀ Sin[θ], x[0] == 0, y[0] == h}, {x[t], y[t]}, t];
horiz[t_] = solution[[1, 1, 2]]
```
$50\sqrt{3}\,t$
```
vert[t_] = solution[[1, 2, 2]]
```
$3 + 50\,t - 16\,t^2$
```
ParametricPlot[{horiz[t], vert[t]}, {t, 0, 3.2}, AxesLabel → {"x", "y"}];
```

> Assume g = 32 ft/sec.

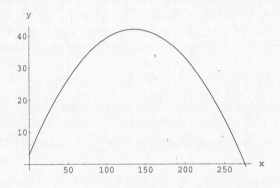

```
Solve[horiz[t] == 200]
```
$\left\{\left\{t \to \dfrac{4}{\sqrt{3}}\right\}\right\}$
```
vert[t] /. % // N
```
{33.1367}

> From the graph it is questionable whether $y \geq 35$
> when $x = 200$, so we compute precisely when the ball
> reaches the fence and then calculate its height at that
> instant.

Since the height of the ball is less than 35 ft, the ball will not go over the fence.

11.8   At what angle should the ball in Problem 11.7 be hit so that it goes over the fence?

SOLUTION

First we want to get a relationship between y and θ.

```
Clear[θ]
g = 32; h = 3; v₀ = 100;
solution = DSolve[{x''[t] == 0, y''[t] == - g, x'[0] == v₀ Cos[θ],
   y'[0] == v₀ Sin[θ], x[0] == 0, y[0] == h}, {x[t], y[t]}, t]
```
$\{\{x[t] \to 100\,t\,Cos[\theta],\ y[t] \to 3 - 16\,t^2 + 100\,t\,Sin[\theta]\}\}$
```
horiz[t_] = solution[[1, 1, 2]];
vert[t_] = solution[[1, 2, 2]];
temp = Solve[horiz[t] == 200, t]
```
$\{\{t \to 2\,Sec[\theta]\}\}$ ← Solve for t as a function of θ.
```
height[θ_] = vert[t] /. temp
```
$\{3 - 64\,Sec[\theta]^2 + 200\,Tan[\theta]\}$ ← Define a function representing
                                                    the height as a function of θ.

```
NSolve[height[θ Degree] == 35, θ]
Solve::ifun : Inverse functions are being
    used by Solve, so some solutions may not be found.
        {{θ → -149.364}, {θ → -111.545},
          {θ → 30.6357, {θ → 68.4546}}
```

← Find θ which gives a height of 35 ft.
The value of θ is expressed in degrees.

This warning may be safely disregarded.

The negative values of θ may be disregarded.   The ball will go over the fence if θ lies between 30.6357° and 68.4546°.   We conclude by sketching these two trajectories.   The vertical line represents the 35-ft fence located 200 ft from home plate.

```
θ = 30.6357 Degree;
horiz[t_] = solution[[1, 1, 2]];
vert[t_] = solution[[1, 2, 2]];
graph1 = ParametricPlot[{horiz[t], vert[t]}, {t, 0, 6}, DisplayFunction → Identity];
θ = 68.4546 Degree;
horiz[t_] = solution[[1, 1, 2]];
vert[t_] = solution[[1, 2, 2]];
graph2 = ParametricPlot[{horiz[t], vert[t]}, {t, 0, 6}, DisplayFunction → Identity];
graph3 = Graphics[Line[{{200, 0}, {200, 35}}]];
Show[graph1, graph2, graph3, PlotRange → {- 50, 150},
    AxesLabel → {"x", "y"}, DisplayFunction → $DisplayFunction];
```

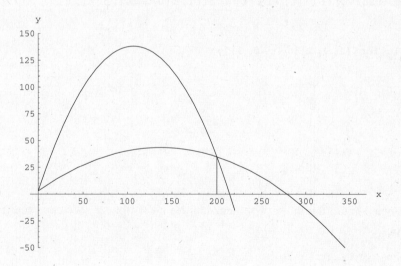

11.9   A culture of microorganisms grows at a rate proportional to the amount present at any given time.   If there are 500 bacteria present after 1 day and 1200 after 2 days, how many bacteria will be present after 4 days?

SOLUTION

The differential equation described by this situation is $\dfrac{dN}{dt} = kN$ where N is the number of bacteria present in the culture and k is a constant to be determined by the given information.   The initial condition is $N = 500$ when $t = 1$.

```
solution = DSolve[{n'[t] == k n[t], n[1] == 500}, n[t], t]
```
$\{\{n[t] \to 500\, e^{-k+k\,t}\}\}$

```
population[t_] = solution[[1, 1, 2]];
```

```
Solve[population[2] == 1200, k]
```
Solve::ifun : Inverse functions are being
   used by Solve, so some solutions may not be found.

$\left\{\left\{k \to \text{Log}\left[\dfrac{12}{5}\right]\right\}\right\}$

```
population[4] /. k → Log[12/5]
```
6912

11.10 The equation governing the amount of current I flowing through a simple resistance–inductance circuit when an EMF (voltage) E is applied is $L\dfrac{dI}{dt} + RI = E$.   The units for E, I, and L are, respectively, volts, amperes, and henries.   If $R = 10$ ohms, $L = 1$ henry, the EMF source is an alternating voltage whose equation is $E(t) = 10 \sin 5t$, and the current is initially 4 amperes, find an expression for the current at time t and plot the graph of the current for the first 3 seconds.

SOLUTION

Note: Care must be taken not to use **E** or **I** to represent voltage and current.

```
r = 10; l = 1; e[t_] = 10Sin[5t];
```

```
solution = DSolve[{l i'[t] + r i[t] == e[t], i[0] == 4}, i[t], t]
```

$\left\{\left\{i[t] \to -\dfrac{1}{5} e^{-10\,t}\,(-22 + 2\, e^{10\,t} \text{Cos}\,[5\,t] - 4\, e^{10\,t} \text{Sin}\,[5\,t])\right\}\right\}$

```
Plot[solution[[1, 1, 2]], {t, 0, 3}];
```

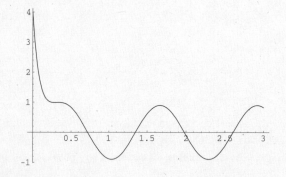

11.11 If a spring with mass m attached at one one end is suspended from its other end, it will come to rest in an equilibrium position.   If the system is then perturbed by releasing the mass with an initial velocity of v_0 at a distance y_0 below its equilibrium position, its motion satisfies the differential equation $m\dfrac{d^2y}{dt^2} + a\dfrac{dy}{dt} + ky = 0$, $y'(0) = v_0$, $y(0) = y_0$. a is a damping constant (determined experimentally) due to friction and air resistance, and k is the spring constant given in Hooke's law.   (We are assuming that the positive direction is downward).

A mass of $\frac{1}{4}$ slug is attached to a spring with a spring constant k of 6 lb/ft.   The mass is pulled downward from its equilibrium position 1 ft and then released.   Assuming a damping constant a of $\frac{1}{2}$, determine the motion of the mass and sketch its graph for the first 5 seconds of motion.

SOLUTION

```
m = 1/4; y₀ = 1; v₀ = 0; a = 1/2; k = 6;
solution = DSolve[{m y''[t] + a y'[t] + k y[t] == 0, y'[0] == v₀, y[0] == y₀}, y[t], t]
```

$$\left\{ \left\{ y[t] \to e^{-t} \left(Cos\left[\sqrt{23}\, t \right] + \frac{Sin\left[\sqrt{23}\, t \right]}{\sqrt{23}} \right) \right\} \right\}$$

```
Plot[solution[[1, 1, 2]], {t, 0, 5}];
```

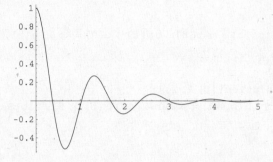

11.12 If a cable of uniform cross section is suspended between two supports, the cable will sag forming a curve called a catenary. If we assume the lowest point on the curve to lie on the y axis, a distance y_0 above the origin, the differential equation governing its shape can be shown to be $\dfrac{d^2 y}{dx^2} = \dfrac{1}{a} \sqrt{1 + \left(\dfrac{dy}{dx} \right)^2}$, $y(0) = y_0$, $y'(0) = 0$, where a is a positive constant depending upon the physical properties of the cable. Find an equation of the catenary and sketch its graph.

SOLUTION

```
solution = DSolve[{y''[x] == 1/a √(1 + y'[x]²), y'[0] == 0, y[0] == y₀}, y[x], x]
```

```
Solve::ifun : Inverse functions are being
   used by Solve, so some solutions may not be found.
```

$$\left\{ y[x] \to - a + a\, Cosh\left[\frac{x}{a} \right] + y_0,\ y[x] \to a + a\, Cosh\left[\frac{x}{a} \right] + y_0 \right\}$$

Since $y_0 > 0$, the first solution applies. We take $y_0 = a = 1$ and plot its graph.

```
catenary[x_] = solution[[1, 2]] /. {a → 1, y₀ → 1};
Plot[catenary[x], {x, -1, 1}, Ticks → {Automatic, {0, 1, 2}}, PlotRange → {0, 2}];
```

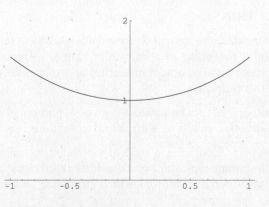

11.13 The logistic equation for population growth, $\dfrac{dp}{dt} = ap - bp^2$, was discovered in the mid-nineteenth century by the biologist Pierre Verhulst. The constant b is generally small in comparison to a so that for small population size p the quadratic term in p will be negligible and the population will grow approximately exponentially. For large p, however, the quadratic term serves to slow down the rate of growth of the population. Solve the logistic equation and sketch the solution for $a = 2$, $b = .005$, and an initial population $p_0 = 1$ (thousand). Then determine the limiting value of the population as $t \to \infty$.

SOLUTION

```
solution = DSolve[{p'[t] == a p[t] - b p[t]², p[0] == p₀}, p[t], t]
```

$$\left\{ \left\{ p[t] \to \frac{a\, e^{a\,t}\, p_0}{a - b\, p_0 + b\, e^{a\,t}\, p_0} \right\} \right\}$$

```
population[t_] = solution[[1, 1, 2]];
Plot[population[t] /. {p₀ → 1, a → 2, b → .05}, {t, 0, 5}, PlotRange → All];
```

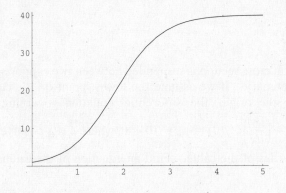

```
Limit[population[t] /. {p₀ → 1, a → 2, b → .05}, t → ∞]
40.
```

11.14 Solve the boundary value problem $\dfrac{d^2 y}{dx^2} + 4\pi^2 y = 0$, $y(0) = y(1) = 0$.

SOLUTION

```
DSolve[{y''[x] + 4 π² y[x] == 0, y[0] == 0, y[1] == 0}, y[x], x]
{{y[x] → -C[1] Sin[2 π x]}}
```

11.2 NUMERICAL SOLUTIONS

Although certain differential equations can be solved analytically in terms of elementary functions, the vast majority of equations that arise in applications cannot. Even if unique solutions can be shown to exist, it may only be possible to obtain numerical approximations. The command **NDSolve** is designed specifically for this purpose.

- **NDSolve[***equations*, **y, {x, xmin, xmax}]** gives a numerical approximation to the solution, **y**, of the differential equation with initial conditions, *equations*, whose independent variable x satisfies xmin ≤ x ≤ xmax.

Because NDSolve yields a numerical solution to a differential equation, or system of differential equations, an appropriate set of initial conditions which guarantee uniqueness *must* be specified.

EXAMPLE 14

In this example we consider the differential equation $\frac{dy}{dx} = x^2 + \sqrt{y}$ with initial condition $y(0) = 1$. Although this equation has a unique solution, it cannot be found in terms of elementary functions using `DSolve`.

```
DSolve[{y'[x] == x² + √(y[x]), y[0] == 1}, y[x], x]
```

```
DSolve[{y'[x] == x² + √(y[x]), y[0] == 1}, y[x], x]
```

We can only obtain a numerical approximation to the solution of this equation. Because numerical techniques construct approximations at a finite number of points, *Mathematica* interpolates, i.e., constructs a smooth function passing through these points and returns the solution as an `InterpolatingFunction` object.

EXAMPLE 15

```
temp = NDSolve[{y'[x] == x² + √(y[x]), y[0] == 1}, y, {x, 0, 1}]
{{y → InterpolatingFunction[{{0., 1.}}, <>]}}
```

The actual `InterpolatingFunction` can now be extracted from this expression:

```
solution = temp[[1, 1, 2]]
InterpolatingFunction[{{0., 1.}}, <>]
```

Only the *domain* of an `InterpolatingFunction` object is printed explicitly. The remaining elements are represented as < >. To see the data used in its construction, enter the command `FullForm[solution]`. Using the interpolated solution, `solution`, we can compute the solution at one or more points, and we can even plot it. One must be careful, however, to stay within the domain of the `InterpolatingFunction` or a warning will be generated.

```
solution[0.5]
1.60643
```

```
solution[1.1]
InterpolatingFunction::dmval :
   Input value {1.1} lies outside the range of data in the
      interpolating function. Extrapolation will be used.
```

2.89132   ← An extrapolated value is not as reliable as an interpolated value in terms of accuracy.

```
Table[{x, solution[x]}, {x, 0, 1, .1}] // TableForm
```

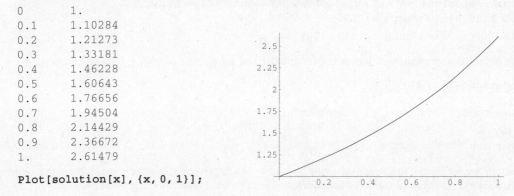

0	1.
0.1	1.10284
0.2	1.21273
0.3	1.33181
0.4	1.46228
0.5	1.60643
0.6	1.76656
0.7	1.94504
0.8	2.14429
0.9	2.36672
1.	2.61479

```
Plot[solution[x], {x, 0, 1}];
```

Although the default settings for `NDSolve` work nicely for most differential equations, *Mathematica* provides some options that can be used to set parameters to handle abnormal situations.

- **WorkingPrecision** specifies the number of significant digits used internally in computations. The default, **WorkingPrecision → $MachinePrecision**, typically yields 16 significant digits.
- **PrecisionGoal** specifies the number of digits of precision sought in the final answer. The default is **PrecisionGoal → Automatic**, which is WorkingPrecision - 10.
- **AccuracyGoal** specifies the number of digits of accuracy (number of digits to the right of the decimal point). The default is **AccuracyGoal → Automatic**, which is WorkingPrecision - 10. AccuracyGoal should be set larger, or to Infinity, if the value of the solution is close to 0.
- **MaxSteps** is the maximum number of steps to take in obtaining the solution. The default is **MaxSteps → Automatic** which, for ordinary differential equations, is 1000.
- **MaxStepSize** specifies the maximum size of each step in the iteration. The default is **MaxStepSize → Infinity**.
- **StartingStepSize** specifies the initial step size. The default is **StartingStepSize → Automatic** (*Mathematica* automatically determines the best step size for the given equation).
- **InterpolationPrecision** sets the precision of the InterpolatingFunction object which represents the solution of the equation. The default is **InterpolationPrecision → Automatic** which uses the value of WorkingPrecision.

EXAMPLE 16

The differential equation $\dfrac{d^2 y}{dx^2} + y = 0$ with initial conditions $y(0) = 0$, $y'(0) = 1$ has unique solution $y = \sin x$. We attempt to solve it for $0 \le x \le 1000$.

```
equation = NDSolve[{y''[x] + y[x] == 0, y[0] == 0, y'[0] == 1}, y, {x, 0, 1000}]
```
```
NDSolve::mxst : Maximum number of 1000
    steps reached at the point x == 199.22439653359802`.
```
```
{{y → InterpolatingFunction [{{0., 199.224}}, <>]}}
```

Because of the wide interval, [0, 1000], over which the solution is to be obtained, more than 1000 steps are necessary.

```
equation = NDSolve[{y''[x] + y[x] == 0, y[0] == 0, y'[0] == 1}, y,
    {x, 0, 1000}, MaxSteps → 10000]
```
```
{{y → InterpolatingFunction[{{0, 1000.}}, <>]}}
```

Having obtained a solution, we check it for accuracy. The solution at $x = (4k + 1)\dfrac{\pi}{2}$ should be 1.

```
f = equation[[1, 1, 2]];
```
```
f[π/2]
```
```
0.999997
```
```
f[633π/2]
```
```
0.999407
```

The further we get from 0 the larger the error becomes. To improve the accuracy, we increase the value of WorkingPrecision. This automatically increases PrecisionGoal and AccuracyGoal.

```
equation = NDSolve[{y''[x] + y[x] == 0, y[0] == 0, y'[0] == 1}, y,
    {x, 0, 1000}, MaxSteps → 10000, WorkingPrecision → 20]
```
```
{{y → InterpolatingFunction[{{0, 1000.0000000000000000}}, <>]}}
```
```
f2 = equation[[1, 1, 2]];
```

```
f2[π/2]
0.9999999999317246612
f2[633π/2]
1.000000004055726494
```

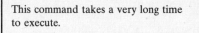

> This command takes a very long time to execute.

SOLVED PROBLEMS

11.15 Solve the differential equation $\dfrac{dy}{dx} = 1 + \dfrac{1}{2}y^2 = 0$, $y(0) = 1$, $0 \le x \le 1$, using `DSolve` and `NDSolve` and compare the results.

SOLUTION

```
equation1 = DSolve[{y'[x] == 1 + 1/2 y[x]², y[0] == 1}, y[x], x]
```

$$\left\{\left\{y[x] \to \sqrt{2}\, \text{Tan}\left[\frac{1}{2}\left(\sqrt{2}\, x + 2\, \text{ArcTan}\left[\frac{1}{\sqrt{2}}\right]\right)\right]\right\}\right\}$$

```
solution1[x_] = equation1[[1, 1, 2]];

equation2 = NDSolve[{y'[x] == 1 + 1/2 y[x]², y[0] == 1}, y[x], {x, 0, 1}]

{{y[x] → InterpolatingFunction[{{0., 1.}}, <>] [x]}}
solution2[x_] = equation2[[1, 1, 2]];
tabledata = Table[{x, solution1[x], solution2[x]}, {x, 0, 1, .1}];
TableForm[tabledata, TableHeadings → {None, {"x", "analytic", "numerical"}}]
```

x	analytic	numerical
0	1	1.
0.1	1.15817	1.15817
0.2	1.33582	1.33583
0.3	1.53895	1.53896
0.4	1.77601	1.77602
0.5	2.05935	2.05936
0.6	2.40786	2.40788
0.7	2.85196	2.85199
0.8	3.44406	3.44411
0.9	4.28301	4.28309
1.	5.58016	5.58031

> The numerical and analytic solutions agree to at least three decimal places. The numerical error increases as x gets further from 0.

11.16 Plot the solution to the differential equation $\dfrac{d^2 y}{dt^2} + \left(\dfrac{dy}{dt} + 1\right)^2 \dfrac{dy}{dt} + y = 0$, $y(0) = 1$, $y'(0) = 0$ for $0 \le t \le 10$.

SOLUTION

```
solution1 =
    NDSolve[{y''[t] + (y'[t] + 1)²
            y'[t] + y[t] == 0, y[0] == 1,
            y'[0] == 0}, y[t], {t, 0, 10}]
{{y[t] → InterpolatingFunction[{{0., 10.}},
    <>][t]}}
Plot[y[t] /. solution1, {t, 0, 10},
    PlotRange → All]
```

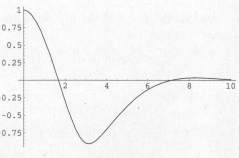

11.17 Plot the (five) solutions to $\dfrac{d^2 y}{dx^2} + 0.3\dfrac{dy}{dx} + \sin y = 0$ for $0 \le x \le 30$ using initial conditions $y'(0) = 0$, $y(0) = -2, -1, 0, 1$, and 2.

SOLUTION

```
Do[
  {solution =
     NDSolve[{y''[x] + 0.3 y'[x] + Sin[y[x]] == 0, y[0] == i,
       y'[0] == 1}, y[x], {x, 0, 30}];
  f[x_] = solution[[1, 1, 2]];
  graph[i] = Plot[f[x], {x, 0, 30}, PlotStyle → Hue[.2 i + .5],
       DisplayFunction → Identity]}, {i, -2, 2}]
  Show[graph[-2], graph[-1], graph[0], graph[1], graph[2],
    DisplayFunction → $DisplayFunction, PlotRange → All];
```

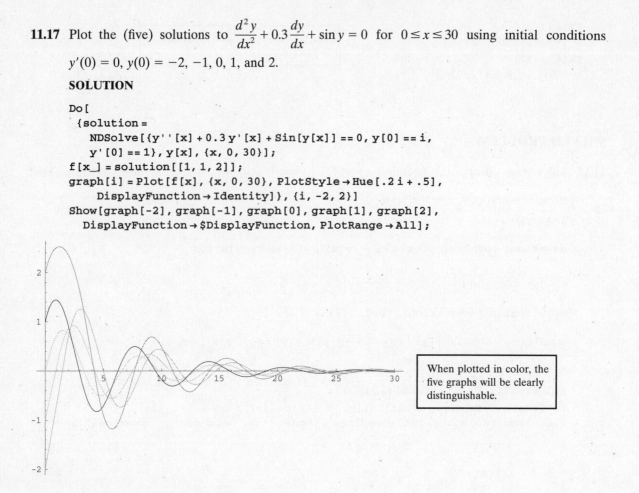

When plotted in color, the five graphs will be clearly distinguishable.

11.3 LAPLACE TRANSFORMS

In this section we describe an ingenious method for solving differential equations. Although the procedure can be used in a wide variety of problems, its real power lies in its ability to solve differential equations whose "right-hand sides" are either discontinuous or are zero except on a very short interval when its value is large. Because most of these types of problems arise within the context of time as the independent variable, it is convenient to express y and its derivatives as functions of t. We shall discuss Laplace transforms heuristically, and shall not concern ourselves with conditions sufficient for existence.

If f is defined on the interval $[0, \infty)$, the Laplace transform of $f(t)$ is defined

$$\mathcal{L}\{f(t)\} = \int_0^\infty e^{-st} f(t)\, dt$$

Its usefulness lies in the following properties which we list without proof:

$$\mathcal{L}\{1\} = \frac{1}{s} \qquad\qquad \mathcal{L}\{\sinh\{bt\}\} = \frac{b}{s^2 - b^2}$$

$$\mathcal{L}\{t\} = \frac{1}{s^2} \qquad\qquad \mathcal{L}\{\cosh(bt)\} = \frac{s}{s^2 - b^2}$$

$$\mathcal{L}\{t^n\} = \frac{n!}{s^{n+1}} \text{ for positive integers, } n. \qquad \mathcal{L}\{e^{at}\sinh(bt)\} = \frac{b}{(s-a)^2 - b^2}$$

$$\mathcal{L}\{e^{at}\} = \frac{1}{s-a}$$

$$\mathcal{L}\{\sin(bt)\} = \frac{b}{s^2+b^2}$$

$$\mathcal{L}\{\cos(bt)\} = \frac{s}{s^2+b^2}$$

$$\mathcal{L}\{e^{at}\sin(bt)\} = \frac{b}{(s-a)^2+b^2}$$

$$\mathcal{L}\{e^{at}\cos(bt)\} = \frac{s-a}{(s-a)^2+b^2}$$

$$\mathcal{L}\{e^{at}\cosh(bt)\} = \frac{s-a}{(s-a)^2-b^2}$$

$$\mathcal{L}\{f'(t)\} = s\mathcal{L}\{f(t)\} - f(0)$$

$$\mathcal{L}\{f''(t)\} = s^2\mathcal{L}\{f(t)\} - sf(0) - f'(0)$$

$$\mathcal{L}\{af(t) + bg(t)\} = a\mathcal{L}\{f(t)\} + b\mathcal{L}\{g(t)\}$$

If $F(s) = \mathcal{L}\{f(t)\}$, then $\mathcal{L}\{e^{at}f(t)\} = F(s-a)$

Mathematica computes the Laplace transform of a function *f* by the invocation of the command
`LaplaceTransform`.

- `LaplaceTransform[f[var1], var1, var2]` computes the Laplace transform of the function *f*, with independent variable *var1* and expresses it as a function of *var2*.

> Note: In versions 4 and 5 of *Mathematica*, the commands
> `LaplaceTransform`, `InverseLaplaceTransform`, `UnitStep`, and
> `DiracDelta`, which are discussed in the remainder of this section, are
> included in the kernel. In version 3, however, these commands are located
> in the packages `Calculus` `LaplaceTransform'` and `Calculus`
> `DiracDelta'`. Readers running version 3 must load these packages
> prior to using these commands.

EXAMPLE 17

The following agree with the properties listed above.

`LaplaceTransform[Exp[2t]Sin[3t], t, s]`

$$\frac{3}{13 - 4s + s^2}$$

`LaplaceTransform[Exp[2t]Cos[3t], t, s]`

$$\frac{-2 + s}{13 - 4s + s^2}$$

`LaplaceTransform[a f[t] + b g[t], t, s]`
`a LaplaceTransform[f[t], t, s] + b LaplaceTransform[g[t], t, s]`
`LaplaceTransform[f'[t], t, s]`
`- f[0] + s LaplaceTransform[f[t], t, s]`
`LaplaceTransform[f''[t], t, s]`
`- s f[0] + s² LaplaceTransform[f[t], t, s] - f'[0]`

The power of the Laplace transform is derived from the fact that there is a one-to-one
correspondence between $f(t)$ and $\mathcal{L}\{f(t)\}$. This means that if $\mathcal{L}\{f(t)\}$ is known, then $f(t)$ is uniquely
determined. If $F(s) = \mathcal{L}\{f(t)\}$, then $f(t) = \mathcal{L}^{-1}\{F(s)\}$. $\mathcal{L}^{-1}$ is called the *inverse* Laplace transform.

- `InverseLaplaceTransform[F[var1], var1, var2]` computes the inverse Laplace transform of the function F, with independent variable *var1*, and expresses it as a function of *var2*.

EXAMPLE 18

```
InverseLaplaceTransform[ 1 ,  s, t]
                        s - 3
```
e^{3t}

```
InverseLaplaceTransform[ 1 ,  s, t]
                        s³ - 8
```

$\frac{1}{12} e^{-t} \left(e^{3t} - \text{Cos}\left[\sqrt{3}\, t\right] - \sqrt{3}\, \text{Sin}\left[\sqrt{3}\, t\right] \right)$

> Traditionally, one would factor the denominator, expand into partial fractions, and find the inverse transformation separately for each term. *Mathematica* does it all automatically.

The next example illustrates how the Laplace transform can be used to solve a simple differential equation.

EXAMPLE 19

Solve the differential equation $\dfrac{d^2 y}{dt^2} - 3\dfrac{dy}{dt} + 2y = t^2$, $y'(0) = 1$, $y(0) = 2$.

First we compute the Laplace transform of both sides of the equation. This can be done in one step.

```
equation = y''[t] - 3y'[t] + 2y[t] == t²;
temp = LaplaceTransform[equation, t, s]
2 LaplaceTransform[y[t], t, s] + s² LaplaceTransform[y[t], t, s] -

   3 (s LaplaceTransform[y[t], t, s] - y[0]) - s y[0] - y'[0] == 2
                                                                   s³
```

Then we solve for the Laplace transform satisfying the given initial conditions.

```
temp2 = Solve[temp, LaplaceTransform[y[t], t, s]] /. {y'[0] → 1, y[0] → 2}
```

$\left\{ \left\{ \text{LaplaceTransform}[y[t], t, s] \to -\dfrac{5 - \frac{2}{s^3} - 2\,s}{2 - 3\,s + s^2} \right\} \right\}$

Next we extract the transform as a function of *s*.

```
temp3 = temp2[[1, 1, 2]]
```

$\dfrac{5 - \frac{2}{s^3} - 2\,s}{2 - 3\,s + s^2}$

Finally, we compute the inverse Laplace transform to get the solution of the equation.

```
InverseLaplaceTransform[temp3, s, t]
```

$\dfrac{1}{4} \left(7 + 4\,e^t - 3\,e^{2t} + 6\,t + 2\,t^2 \right)$

As indicated at the beginning of this section, Laplace transforms are the ideal tool to use when dealing with equations involving discontinuous "right-hand sides." In this context we shall find it convenient to introduce the Heaviside, or unit step, function.

- **UnitStep[x]** returns a value of 0 if $x < 0$ and 1 if $x \geq 0$.

The unit step function, which we represent as $u(t)$, offers a convenient way to define *piecewise-defined* functions.

EXAMPLE 20

Plot the graph of $g(x) = \begin{cases} x & \text{if } x < 1 \\ x^3 & \text{if } x \geq 1 \end{cases}$ for $0 \leq x \leq 2$.

```
g[x_] = UnitStep[1 - x]x + UnitStep[x - 1]x³;
Plot[g[x], {x, 0, 2}];
```

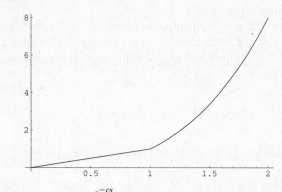

It is easily shown that $\mathcal{L}\{u(t-c)\} = \dfrac{e^{-cs}}{s}$ and, if $F(s) = \mathcal{L}\{f(t)\}$, then $\mathcal{L}\{u(t-c)f(t-c)\} = e^{-cs}F(s)$.

These properties make it convenient to solve differential equations involving piecewise continuous functions.

EXAMPLE 21

Solve $\dfrac{d^2y}{dt^2} - 3\dfrac{dy}{dt} + 2y = g(t)$, $y(0) = y'(0) = 0$ where $g(t) = \begin{cases} 1 \text{ if } 0 \le t < 1 \\ 0 \text{ if } t > 1 \end{cases}$

Plot the solution for $0 \le t \le 2$.

For $t \ge 0$, $g(t) =$ UnitStep $[1-t]$

```
temp = LaplaceTransform[y''[t] - 3y'[t] + 2y[t] == UnitStep[1 - t], t, s]
```

2 LaplaceTransform[y[t], t, s] + s² LaplaceTransform[y[t], t, s] -

$\qquad$ 3 (s LaplaceTransform[y[t], t, s] - y[0]) - s y[0] - y'[0] == $\dfrac{1 - e^{-s}}{s}$

```
temp2 = Solve[temp, LaplaceTransform[y[t], t, s]] /. {y'[0] → 0, y[0] → 0}
```

$\left\{\left\{\text{LaplaceTransform}[y[t], t, s] \to \dfrac{1 - e^{-s}}{s\ (2 - 3\ s + s^2)}\right\}\right\}$

```
temp3 = temp2[[1, 1, 2]]
```

$\dfrac{1 - e^{-s}}{s\ (2 - 3\ s + s^2)}$

```
f[t_] = InverseLaplaceTransform[temp3, s, t]
```

$\dfrac{-\ (e - e^t)^2\ \text{UnitStep}[-1 + t] + e^2\ (-1 + e^t)^2\ \text{UnitStep}[t]}{2\ e^2}$

```
Plot[f[t], {t, 0, 2}];
```

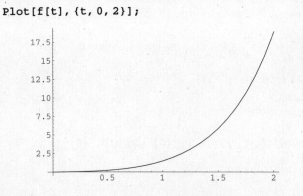

> Observe that the solution is
> continuous, even though the equation
> involves a discontinuous function.

In physical and biological applications, we are often led to differential equations whose right-hand side, $f(t)$, is a function of an *impulsive* nature; that is, $f(t)$ has zero value everywhere except over a short interval of time where its value is positive.

The Dirac delta function is a idealized impulse function.   Although not a true function in the classical sense, its validity is justified by the *theory of distributions*, developed by Laurent Schwartz in the mid-twentieth century.   It is defined by the following pair of conditions:

$$\delta(t - t_0) = 0 \ \text{ if } \ t \neq t_0$$

$$\int_{-\infty}^{\infty} \delta(t - t_0)\,dt = 1$$

An immediate consequence of the definition is the result that $\int_{-\infty}^{\infty} f(t)\delta(t - t_0)\,dt = f(t_0)$.   It follows, therefore, that $\mathcal{L}\{\delta(t - t_0)\} = \int_{0}^{\infty} e^{-st}\,\delta(t - t_0)\,dt = e^{-st_0}$, provided that $t_0 \geq 0$.   Otherwise its value is 0.

- **DiracDelta[t]** returns $\delta(t)$, the Dirac delta function which satisfies

  $$\delta(t) = 0 \ \text{ if } \ t \neq 0, \int_{-\infty}^{\infty} \delta(t)\,dt = 1.$$

EXAMPLE 22

$$\int_{-\infty}^{\infty} \texttt{DiracDelta[t]}\ \texttt{dt}$$

1

$$\int_{-\infty}^{\infty} \texttt{f[t]DiracDelta[t - a]}\ \texttt{dt}$$

f[a]

EXAMPLE 23

```
LaplaceTransform[DiracDelta[t - a], t, s]
e⁻ᵃˢUnitStep[a]
LaplaceTransform[DiracDelta[t - 3], t, s]
e⁻³ˢ
LaplaceTransform[DiracDelta[t + 3], t, s]
0
```

← Since *Mathematica* does not know whether *a* is negative or non-negative, UnitStep[a] is included in the Laplace transform.

Since we know the Laplace transform of the Dirac delta function, we can solve differential equations involving impulses much the same way as described in Example 19.   The following example illustrates the method.

EXAMPLE 24

Find the solution of the differential equation $\dfrac{d^2 y}{dt^2} - 2\dfrac{dy}{dt} + y = \delta(t - 1)$, $y(0) = y'(0) = 0$.

```
equation = y''[t] - 2y'[t] + y[t] == DiracDelta[t - 1];
temp = LaplaceTransform[equation, t, s]
LaplaceTransform[y[t], t, s] + s² LaplaceTransform[y[t], t, s] -
   2 (s LaplaceTransform[y[t], t, s] - y[0]) - s y[0] - y'[0] == e⁻ˢ
temp2 = Solve[temp, LaplaceTransform[y[t], t, s]] /. {y'[0] → 0, y[0] → 0}
```

$$\left\{\left\{\texttt{LaplaceTransform[y[t], t, s]} \to \frac{e^{-s}}{(-1 + s)^2}\right\}\right\}$$

```
temp3 = temp2[[1, 1, 2]]
```

$$\frac{e^{-s}}{(-1+s)^2}$$

```
solution[t_] = InverseLaplaceTransform[temp3, s, t]
```

$e^{-1+t}(-1+t)\,\text{UnitStep}[-1+t]$

```
Plot[solution[t], {t, 0, 2}, PlotStyle→Thickness[.01]];
```

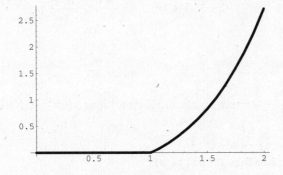

Laplace transforms can be used to solve systems of differential equations. The technique is similar to that of a single equation, except that a different transform is defined for each dependent variable. The next example illustrates the method for solving a system of two first-order equations. It generalizes in a natural way to larger and higher-order systems.

EXAMPLE 25

Solve the system $\begin{cases} \dfrac{dx}{dt} + y = t \\[2mm] 4x + \dfrac{dy}{dt} = 0 \end{cases}$ with initial condition $x(0) = 1,\ y(0) = -1$.

```
system = {x'[t] + y[t] == t, 4x[t] + y'[t] == 0};
temp = LaplaceTransform[system, t, s]
```

$\Big\{ s\,\text{LaplaceTransform}[x[t], t, s] + \text{LaplaceTransform}[y[t], t, s] -$

$x[0] == \dfrac{1}{s^2},\ 4\,\text{LaplaceTransform}[x[t], t, s] + s\,\text{LaplaceTransform}[y[t], t, s] - y[0] == 0 \Big\}$

```
temp2 = Solve[temp, {LaplaceTransform[x[t], t, s],
          LaplaceTransform[y[t], t, s]}] /. {x[0] → 1, y[0] → -1}
```

$\Big\{ \Big\{ \text{LaplaceTransform}[x[t], t, s] \to -\dfrac{-1-s-s^2}{s\,(-4+s^2)},$

$\text{LaplaceTransform}[y[t], t, s] \to -\dfrac{4+4s^2+s^3}{s^2\,(-4+s^2)} \Big\} \Big\}$

```
temp3a = temp2[[1, 1, 2]]
```

$-\dfrac{-1-s-s^2}{s\,(-4+s^2)}$

```
temp3b = temp2[[1, 2, 2]]
```

$\dfrac{4+4s^2+s^3}{s^2\,(-4+s^2)}$

```
InverseLaplaceTransform[temp3a, s, t]
```

$\frac{1}{8} \left(-2 + 3\, e^{-2\, t} + 7\, e^{2\, t} \right)$

```
InverseLaplaceTransform[temp3b, s, t]
```

$\frac{3\, e^{-2\, t}}{4} - \frac{7\, e^{2\, t}}{4} + t$

The solution to the system is $x = \frac{1}{8}(-2 + 3e^{-2t} + 7e^{2t})$, $y = \frac{3e^{-2t}}{4} - \frac{7e^{2t}}{4} + t$.

SOLVED PROBLEMS

11.18 Solve the equation $\dfrac{d^2 y}{dt^2} + y = \sin t$ with initial conditions $y(0) = 0$, $y'(0) = 2$.

SOLUTION

```
equation = y''[t] + y[t] == Sin[t];
temp = LaplaceTransform[equation, t, s]
LaplaceTransform[y[t], t, s] +
```

$\quad$ s² LaplaceTransform[y[t], t, s] $-$ s y[0] $-$ y'[0] $==\dfrac{1}{1 + s^2}$

```
temp2 = Solve[temp, LaplaceTransform[y[t], t, s]] /. {y'[0] → 2, y[0] → 0}
```

$\left\{ \left\{ \text{LaplaceTransform}[\text{y}[\text{t}], \text{t}, \text{s}] \to - \dfrac{-2 - \dfrac{1}{1 + s^2}}{1 + s^2} \right\} \right\}$

```
temp3 = temp2[[1, 1, 2]];
InverseLaplaceTransform[temp3, s, t]
```

$\dfrac{1}{2} \left(-t \operatorname{Cos}[t] + 5 \operatorname{Sin}[t] \right)$

11.19 Solve $\dfrac{d^2 y}{dt^2} + \dfrac{dy}{dt} + y = e^t$, $y(0) = 3$, $y'(0) = 2$.

SOLUTION

```
equation = y''[t] + y'[t] + y[t] == Exp[t];
temp = LaplaceTransform[equation, t, s]
LaplaceTransform[y[t], t, s] + s LaplaceTransform[y[t], t, s] +
```

$\quad$ s² LaplaceTransform[y[t], t, s] $-$ y[0] $-$ s y[0] $-$ y'[0] $==\dfrac{1}{-1 + s}$

```
temp2 = Solve[temp, LaplaceTransform[y[t], t, s]] /. {y'[0] → 2, y[0] → 3}
```

$\left\{ \left\{ \text{LaplaceTransform}[\text{y}[\text{t}], \text{t}, \text{s}] \to - \dfrac{-5 - \dfrac{1}{-1 + s} - 3\, s}{1 + s + s^2} \right\} \right\}$

```
temp3 = temp2[[1, 1, 2]];
InverseLaplaceTransform[temp3, s, t]
```

$\dfrac{1}{3} e^{-t/2} \left(e^{3\, t/2} + 8 \operatorname{Cos}\left[\dfrac{\sqrt{3}\, t}{2} \right] + 6 \sqrt{3} \operatorname{Sin}\left[\dfrac{\sqrt{3}\, t}{2} \right] \right)$

11.20 Solve the equation $\dfrac{d^2 y}{dt^2} - 2 \dfrac{dy}{dt} + y = g(t)$, $y(0) = y'(0) = 0$ where $g(t) = \begin{cases} t & \text{if } 0 \le t \le 1 \\ t^2 & \text{if } t > 1 \end{cases}$

and plot the solution for $0 \le x \le 4$.

SOLUTION

```
equation = y''[t] - 2 y'[t] + y[t] == t UnitStep[1 - t] + t² UnitStep[t - 1];
temp = LaplaceTransform[equation, t, s]
```
LaplaceTransform[y[t], t, s] + s² LaplaceTransform[y[t], t, s] -
  2 (s LaplaceTransform[y[t], t, s] - y[0]) - s y[0] - y'[0] ==

$$\frac{e^{-s} (-1 + e^s - s)}{s^2} + \frac{e^{-s} (2 + 2 s + s^2)}{s^3}$$

```
temp2 = Solve[temp, LaplaceTransform[y[t], t, s]] /. {y'[0] → 0, y[0] → 0}
```

$$\left\{\left\{ \text{LaplaceTransform}[y[t], t, s] \to \frac{e^{-s} (2 + s + e^s s)}{(-1 + s)^2 s^3} \right\}\right\}$$

```
temp3 = temp2[[1, 1, 2]] // Expand
```

$$\frac{2 e^{-s}}{(-1 + s)^2 s^3} + \frac{1}{(-1 + s)^2 s^2} + \frac{e^{-s}}{(-1 + s)^2 s^2}$$

> *Mathematica* 4 has trouble computing the inverse unless the Laplace transform is expanded.

```
f[t_] = InverseLaplaceTransform[temp3, s, t]
```
2 + e^t (-2 + t) + t +

$$2 \left(\frac{1}{2} (6 + 4 (-1 + t) + (-1 + t)^2) + e^{-1+t} (-4 + t) \right) \text{UnitStep}[-1 + t] +$$

(1 + e^{-1+t} (-3 + t) + t) UnitStep[-1 + t]

```
Plot[f[t], {t, 0, 4}];
```

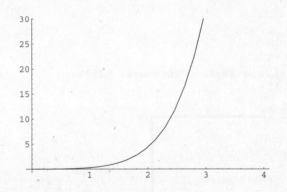

11.21 Solve the system

$$\begin{cases} \dfrac{dx}{dt} + y = t \sin t \\[2mm] x + \dfrac{dy}{dt} = t \cos t \end{cases} \qquad x(0) = y(0) = 0$$

SOLUTION

```
system = {x'[t] + y[t] == t Sin[t], x[t] + y'[t] == t Cos[t]};
temp = LaplaceTransform[system, t, s]
```

$$\left\{ s \text{ LaplaceTransform}[x[t], t, s] + \text{LaplaceTransform}[y[t], t, s] - \right.$$

$$x[0] == \frac{2 s}{(1 + s^2)^2}, \text{LaplaceTransform}[x[t], t, s] +$$

$$\left. s \text{ LaplaceTransform}[y[t], t, s] - y[0] == \frac{-1 + s^2}{(1 + s^2)^2} \right\}$$

```
temp2 = Solve[temp, {LaplaceTransform[x[t], t, s],
   LaplaceTransform[y[t], t, s]}] /. {x[0] → 0, y[0] → 0}
```

$$\left\{\left\{\text{LaplaceTransform}[x[t], t, s] \to \frac{1}{(-1+s^2)(1+s^2)},\right.\right.$$

$$\left.\left.\text{LaplaceTransform}[y[t], t, s] \to -\frac{\frac{2s}{(1+s^2)^2} - \frac{s(-1+s^2)}{(1+s^2)^2}}{-1+s^2}\right\}\right\}$$

```
temp3a = temp2[[1, 1, 2]];
temp3b = temp2[[1, 2, 2]];
InverseLaplaceTransform[temp3a, s, t]
```

$$\frac{1}{4}(-e^{-t} + e^t - 2\sin[t])$$

```
InverseLaplaceTransform[temp3b, s, t]
```

$$\frac{1}{4}(-e^{-t} - e^t + 2\cos[t] + 4t\sin[t])$$

The solution of the equation is $x = \frac{1}{4}(-e^{-t} + e^t - 2\sin t)$, $y = \frac{1}{4}(-e^{-t} - e^t + 2\cos t + 4t\sin t)$.

11.22 The equation governing the amount of current I flowing through a simple resistance–inductance circuit when an EMF (voltage) E is applied is $L\frac{dI}{dt} + RI = E$. The units for E, I, and L are, respectively, volts, amperes, and henries. Suppose $L = 1$ and $R = 10$. If a direct current of 1 volt is applied at time $t = 0$ and removed 1 second later, plot the current in the circuit during the first 2 seconds.

SOLUTION

```
e[t_] = UnitStep[1 - t];
Plot[e[t], {t, 0, 2}, PlotStyle → Thickness[.01]];
```

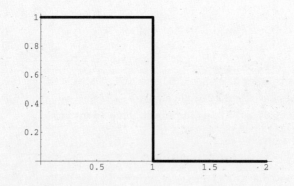

```
l = 1; r = 10;
equation = l i'[t] + r i[t] == e[t];
temp = LaplaceTransform[equation, t, s]
```

$$-i[0] + 10\,\text{LaplaceTransform}[i[t], t, s] + s\,\text{LaplaceTransform}[i[t], t, s] == \frac{1 - e^{-s}}{s}$$

```
temp2 = Solve[temp, LaplaceTransform[i[t], t, s]] /. i[0] → 0
```

$$\left\{\left\{\text{LaplaceTransform}[i[t], t, s] \to \frac{e^{-s}(-1+e^s)}{s(10+s)}\right\}\right\}$$

```
temp3 = temp2[[1, 1, 2]] // Expand
```

$$\frac{1}{s\,(10+s)} - \frac{e^{-s}}{s\,(10+s)}$$

```
f[t_] = InverseLaplaceTransform[temp3, s, t]
```

$$\frac{1}{10}\,(1 - e^{-10\,t}) - \frac{1}{10}\,(1 - e^{-10\,(-1+t)})\,\text{UnitStep}[-1 + t]$$

```
Plot[f[t], {t, 0, 2},
    PlotStyle → Thickness[.01]];
```

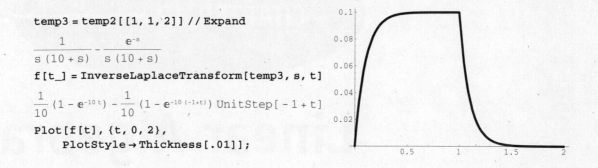

11.23 A particle of mass m is attached to one end of a spring and allowed to come to rest in an equilibrium position. If an external force $f(t)$ is then applied to the particle, its motion is described by the equation $m\dfrac{d^2 y}{dt^t} + a\dfrac{dy}{dt} + ky = f(t)$, $y'(0) = 0$, $y(0) = 0$ where a is a damping constant and k is the spring's stiffness constant. Assuming $m = 1$, $a = 2$, $k = 1$, and $f(t) = e^{-t}$, describe the motion of the spring. Then determine the motion of the spring if an impulse of 1 lb-sec is applied after 1 second. Plot both graphs on one set of axes.

SOLUTION

```
m = 1; a = 2; k = 1;
equation = m y''[t] + a y'[t] + k y[t] == Exp[-t]; (* without impulse *)
temp = LaplaceTransform[equation, t, s];
temp2 = Solve[temp, LaplaceTransform[y[t], t, s]] /. {y'[0] → 0, y[0] → 0};
temp3 = temp2[[1, 1, 2]] // Expand;
Print["Solution Without Impulse (dashed)"]
f1[t_] = InverseLaplaceTransform[temp3, s, t]
g1 = Plot[f1[t], {t, 0, 10}, PlotStyle → Dashing[{.01}],
        DisplayFunction → Identity];
equation = m y''[t] + a y'[t] + k y[t] ==
    Exp[-t] + DiracDelta[t - 1] (* with impulse *)
temp = LaplaceTransform[equation, t, s];
temp2 = Solve[temp, LaplaceTransform[y[t], t, s]] /. {y'[0] → 0, y[0] → 0};
temp3 = temp2[[1, 1, 2]] // Expand;
Print["Solution With Impulse (solid)"]
f2[t_] = InverseLaplaceTransform[temp3, s, t]
g2 = Plot[f2[t], {t, 0, 10}, DisplayFunction → Identity];
```

Solution Without Impulse (dashed)

$$\frac{1}{2}\,e^{-t}\,t^2$$

Solution With Impulse (solid)

$$\frac{1}{2}\,e^{-t}\,t^2 + e^{1-t}\,(-1 + t)\,\text{UnitStep}[-1 + t]$$

```
Show[g1, g2,
    DisplayFunction → $DisplayFunction];
```

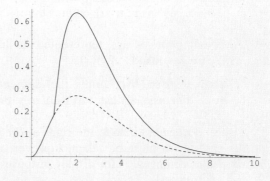

CHAPTER 12

Linear Algebra

12.1 VECTORS AND MATRICES

Vectors and matrices are represented as lists (Chapter 3) in *Mathematica*. A vector is a simple list and a matrix is a list of vectors.

The elements of a vector or a matrix may be entered manually as a list or, more conveniently, by the use of built-in commands.

Vectors

- **Table** [*expression*, {i, n}] constructs an *n*-dimensional vector whose elements are the values of *expression* for $i = 1, 2, 3, \ldots, n$.
- **Array** [f, n] generates an *n*-dimensional vector whose elements are f[1], f[2], ..., f[n]. f is a function of one variable.

Matrices

- **Table** [*expression*, {i, m}, {j, n}] constructs an $m \times n$ matrix whose elements are the values of *expression* for $(i, j) = (1, 1), \ldots, (m, n)$.
- **Array** [f, {m, n}] generates an $m \times n$ matrix whose elements are f[1, 1], ..., f[m,n]. f is a function of two variables.
- **DiagonalMatrix** [*list*] creates a diagonal matrix whose diagonal elements are the elements of *list*, a one-dimensional list.
- **IdentityMatrix** [n] creates the $n \times n$ identity matrix.

Although matrices are represented as lists, they may be *viewed* as matrices by using the **MatrixForm** command.

- **MatrixForm** [*list*] prints the elements of *list* in a rectangular array. If *list* is a simple (one-dimensional) list, MatrixForm prints as a column vector (i.e. an $n \times 1$ matrix).

Using **//MatrixForm** to the right of *list* is equivalent to MatrixForm[*list*] and is a bit more convenient. Care must be taken, however, not to use //MatrixForm in the definition of the matrix. This command is for display purposes only. (See Problem 12.3).

288

EXAMPLE 1

```
m = {{1, 1}, {1, 2}};
m // MatrixForm
```

$$\begin{pmatrix} 1 & 1 \\ 1 & 2 \end{pmatrix}$$

Additionally, a matrix can be introduced via the menu Input ⇒ Create Table/Matrix/Palette.

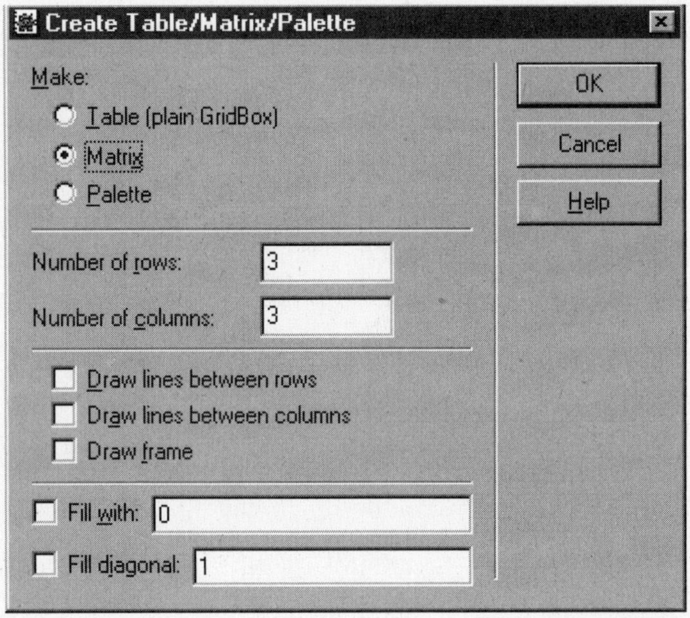

This produces a grid as shown to the right. Once the grid has been set up, you can conveniently enter the numbers, using the [TAB] key to go from cell to cell. Options for filling with 0's and 1's are particularly convenient for large, sparse matrices.

$$\begin{pmatrix} \square & \square & \square \\ \square & \square & \square \\ \square & \square & \square \end{pmatrix}$$

EXAMPLE 2

To generate a vector whose entries are the squares of the first five consecutive integers, we could simply enter them by hand.

```
squares = {1, 4, 9, 16, 25}
{1, 4, 9, 16, 25}
```

More conveniently, however, we can use the **Table** or **Array** command.

```
squares = Table[i², {i, 5}]
{1, 4, 9, 16, 25}
f[i_] = i²;
squares = Array[f, 5]
{1, 4, 9, 16, 25}
```

To view as a vector,

`MatrixForm[squares]` or `squares //MatrixForm`

$$\begin{pmatrix} 1 \\ 4 \\ 9 \\ 16 \\ 25 \end{pmatrix}$$

EXAMPLE 3

We will construct a 7×4 matrix whose entries are the sum of its row and column positions. For example, $a_{2,3} = 5$. We could, of course, input the entries directly using the tool in Input $\Rightarrow$ Create Table/Matrix/Palette, but it is certainly preferable to use one of the standard *Mathematica* commands. Here are two ways it can be done.

```
matrix = Table[i + j, {i, 5}, {j, 7}]
```
{{2, 3, 4, 5, 6, 7, 8}, {3, 4, 5, 6, 7, 8, 9}, {4, 5, 6, 7, 8, 9, 10},
  {5, 6, 7, 8, 9, 10, 11}, {6, 7, 8, 9, 10, 11, 12}}
```
f[i_, j_] = i + j;
```
```
matrix = Array[f, {5, 7}]
```
{{2, 3, 4, 5, 6, 7, 8}, {3, 4, 5, 6, 7, 8, 9}, {4, 5, 6, 7, 8, 9, 10},
  {5, 6, 7, 8, 9, 10, 11}, {6, 7, 8, 9, 10, 11, 12}}

Either way we can view the generated array as a matrix.

```
matrix //MatrixForm
```

$$\begin{pmatrix} 2 & 3 & 4 & 5 & 6 & 7 & 8 \\ 3 & 4 & 5 & 6 & 7 & 8 & 9 \\ 4 & 5 & 6 & 7 & 8 & 9 & 10 \\ 5 & 6 & 7 & 8 & 9 & 10 & 11 \\ 6 & 7 & 8 & 9 & 10 & 11 & 12 \end{pmatrix}$$

EXAMPLE 4

Submatrices can be constructed from a given matrix by careful implementation of `[[ ]]` (see the `Part` command, Chapter 3). First we construct a 5×5 matrix of consecutive integers.

```
matrix = Table[5i + j, {i, 0, 4}, {j, 1, 5}];
matrix //MatrixForm
```

$$\begin{pmatrix} 1 & 2 & 3 & 4 & 5 \\ 6 & 7 & 8 & 9 & 10 \\ 11 & 12 & 13 & 14 & 15 \\ 16 & 17 & 18 & 19 & 20 \\ 21 & 22 & 23 & 24 & 25 \end{pmatrix}$$

We can obtain a particular element of the matrix, say the element in row 3, column 4.

```
matrix[[3, 4]]
```
14

If we want the entire fourth row, we extract the fourth sublist from `matrix`.

matrix[[4]]

{16, 17, 18, 19, 20}

The entire fourth column can be obtained using the **All** directive.

matrix[[All, 4]]

{4, 9, 14, 19, 24}

We can obtain the submatrix whose elements are in rows 1, 3, and 5 and columns 2 and 4.

matrix[[{1, 3, 5}, {2, 4}]] // MatrixForm

$$\begin{pmatrix} 2 & 4 \\ 12 & 14 \\ 22 & 24 \end{pmatrix}$$

Or we can obtain the 3×5 matrix consisting of `matrix` with rows 2 and 4 deleted.

matrix[[{1, 3, 5}, All]] // MatrixForm

$$\begin{pmatrix} 1 & 2 & 3 & 4 & 5 \\ 11 & 12 & 13 & 14 & 15 \\ 21 & 22 & 23 & 24 & 25 \end{pmatrix}$$

With careful use of the `Take` command (Chapter 3) we can even construct the submatrix of `matrix` consisting of those elements in rows 2 through 4 and columns 3 through 5.

Take[matrix, {2, 4}, {3, 5}] // MatrixForm

$$\begin{pmatrix} 8 & 9 & 10 \\ 13 & 14 & 15 \\ 18 & 19 & 20 \end{pmatrix}$$

SOLVED PROBLEMS

12.1　　Construct a 10-dimensional vector of powers of 2.

　　　　SOLUTION

　　　　powersof2 = Table[2^k, {k, 1, 10}]

　　　　{2, 4, 8, 16, 32, 64, 128, 256, 512, 1024}

　　　　powersof2 // MatrixForm

$$\begin{pmatrix} 2 \\ 4 \\ 8 \\ 16 \\ 32 \\ 64 \\ 128 \\ 256 \\ 512 \\ 1024 \end{pmatrix}$$

12.2   Construct a 5×5 matrix of random digits.

SOLUTION

```
Table[Random[Integer, {0, 9}], {i, 5}, {j, 5}] // MatrixForm
```

$$\begin{pmatrix} 5 & 7 & 9 & 9 & 4 \\ 2 & 8 & 6 & 9 & 7 \\ 1 & 0 & 8 & 2 & 2 \\ 7 & 8 & 8 & 8 & 1 \\ 9 & 2 & 2 & 3 & 4 \end{pmatrix}$$

12.3   What happens if `//MatrixForm` is included within the definition of a matrix?

SOLUTION

```
m = {{1, 1}, {1, 2}} //MatrixForm
```

$$\begin{pmatrix} 1 & 1 \\ 1 & 2 \end{pmatrix}$$

```
m + m
```

$$\begin{pmatrix} 1 & 1 \\ 1 & 2 \end{pmatrix} + \begin{pmatrix} 1 & 1 \\ 1 & 2 \end{pmatrix}$$ ← We do *not* get the sum of the two matrices.

   Mathematica cannot perform the indicated operation because m is not a list.   To see this more clearly, we use `FullForm`.

```
FullForm[m]
```
MatrixForm[List[List[1, 1], List[1, 2]]]

Now we do it correctly.

```
m = {{1, 1}, {1, 2}}
```
{{1, 1}, {1, 2}}

```
m + m //MatrixForm
```

$$\begin{pmatrix} 2 & 2 \\ 2 & 4 \end{pmatrix}$$

12.4   Construct a 10×10 diagonal matrix whose diagonal entries are the first 10 primes.

SOLUTION

```
primelist = Array[Prime, 10];
DiagonalMatrix[primelist] //MatrixForm
```

| | **Prime** is a built-in *Mathematica* function. |

$$\begin{pmatrix} 2 & 0 & 0 & 0 & 0 & 0 & 0 & 0 & 0 & 0 \\ 0 & 3 & 0 & 0 & 0 & 0 & 0 & 0 & 0 & 0 \\ 0 & 0 & 5 & 0 & 0 & 0 & 0 & 0 & 0 & 0 \\ 0 & 0 & 0 & 7 & 0 & 0 & 0 & 0 & 0 & 0 \\ 0 & 0 & 0 & 0 & 11 & 0 & 0 & 0 & 0 & 0 \\ 0 & 0 & 0 & 0 & 0 & 13 & 0 & 0 & 0 & 0 \\ 0 & 0 & 0 & 0 & 0 & 0 & 17 & 0 & 0 & 0 \\ 0 & 0 & 0 & 0 & 0 & 0 & 0 & 19 & 0 & 0 \\ 0 & 0 & 0 & 0 & 0 & 0 & 0 & 0 & 23 & 0 \\ 0 & 0 & 0 & 0 & 0 & 0 & 0 & 0 & 0 & 29 \end{pmatrix}$$

12.5   Construct a 5×5 upper triangular matrix of 1's with 0's below the main diagonal.

SOLUTION

```
m=Table[If[i≤j,1,0], {i, 5}, {j, 5}];
m // MatrixForm
```

$$\begin{pmatrix} 1 & 1 & 1 & 1 & 1 \\ 0 & 1 & 1 & 1 & 1 \\ 0 & 0 & 1 & 1 & 1 \\ 0 & 0 & 0 & 1 & 1 \\ 0 & 0 & 0 & 0 & 1 \end{pmatrix}$$

12.6   Construct a 7×7 tridiagonal matrix with 2's on the main diagonal, 1's on the diagonals adjacent to the main diagonal, and 0's elsewhere.

SOLUTION

```
m=Table[If[Abs[i-j]==1, 1, If[i==j, 2, 0, 0]], {i, 1, 7}, {j, 1, 7}];
m // MatrixForm
```

$$\begin{pmatrix} 2 & 1 & 0 & 0 & 0 & 0 & 0 \\ 1 & 2 & 1 & 0 & 0 & 0 & 0 \\ 0 & 1 & 2 & 1 & 0 & 0 & 0 \\ 0 & 0 & 1 & 2 & 1 & 0 & 0 \\ 0 & 0 & 0 & 1 & 2 & 1 & 0 \\ 0 & 0 & 0 & 0 & 1 & 2 & 1 \\ 0 & 0 & 0 & 0 & 0 & 1 & 2 \end{pmatrix}$$

12.7   Let m be the 6×6 matrix containing the integers 1 through 36.   Construct the 3×3 matrix m2 consisting of the elements in the odd rows and even columns of m.

SOLUTION

```
m = Table[6i + j, {i, 0, 5}, {j, 1, 6}];
m // MatrixForm
```

$$\begin{pmatrix} 1 & 2 & 3 & 4 & 5 & 6 \\ 7 & 8 & 9 & 10 & 11 & 12 \\ 13 & 14 & 15 & 16 & 17 & 18 \\ 19 & 20 & 21 & 22 & 23 & 24 \\ 25 & 26 & 27 & 28 & 29 & 30 \\ 31 & 32 & 33 & 34 & 35 & 36 \end{pmatrix}$$

```
m2 = m[[{1, 3, 5}, {2, 4, 6}]];
m2 // MatrixForm
```

$$\begin{pmatrix} 2 & 4 & 6 \\ 14 & 16 & 18 \\ 26 & 28 & 30 \end{pmatrix}$$

12.2 MATRIX OPERATIONS

Since vectors and matrices are stored as lists in *Mathematica*, all list operations described in Chapter 3 apply.   In addition, there are some specialized commands which are applicable specifically to matrices.   Since n-dimensional vectors can be considered to be $n \times 1$ matrices, many of these commands apply to vectors as well.   In the following descriptions m, m1, and m2 denote matrices and v1 and v2 denote vectors.

- **m1 + m2** computes the sum of two matrices.
- **m1 - m2** computes the difference of two matrices
- **c m** multiplies each element of m by the scalar c.
- **m1 . m2** computes the matrix product of m1 and m2.   **v1.v2** computes the dot product of v1 and v2.   For matrices, the operation returns a list; for vectors a single number is returned.
- **Cross[v1, v2]** returns the cross product of **v1** and **v2**.   (This applies to three-dimensional vectors only.)   The cross product symbol, ×, can be inserted into the calculation by typing (without spaces) the key sequence [Esc]**cross**[Esc].

EXAMPLE 5

First we generate two 3×3 "random" matrices as lists.

m1 = Table[Random[Integer, {0, 9}], {i, 1, 3}, {j, 1, 3}]
{{9, 4, 2}, {2, 9, 3}, {0, 1, 4}}
m2 = Table[Random[Integer, {0, 9}], {i, 1, 3}, {j, 1, 3}]
{{2, 8, 1}, {8, 3, 4}, {6, 4, 0}}

Now we look at them in matrix form.

m1 // MatrixForm

$$\begin{pmatrix} 9 & 4 & 2 \\ 2 & 9 & 3 \\ 0 & 1 & 4 \end{pmatrix}$$

m2 // MatrixForm

$$\begin{pmatrix} 2 & 8 & 1 \\ 8 & 3 & 4 \\ 6 & 4 & 0 \end{pmatrix}$$

The next operation multiplies each element of m1 by 5.

5 m1

$$\begin{pmatrix} 45 & 20 & 10 \\ 10 & 45 & 15 \\ 0 & 5 & 20 \end{pmatrix}$$

Next we compute their sum, difference, and product.

m1 + m2 // MatrixForm

$$\begin{pmatrix} 11 & 12 & 3 \\ 10 & 12 & 7 \\ 6 & 5 & 4 \end{pmatrix}$$

m1 - m2 // MatrixForm

$$\begin{pmatrix} 7 & -4 & 1 \\ -6 & 6 & -1 \\ -6 & -3 & 4 \end{pmatrix}$$

    `m1.m2 // MatrixForm`

$$\begin{pmatrix} 62 & 92 & 25 \\ 94 & 55 & 38 \\ 32 & 19 & 4 \end{pmatrix}$$

Care must be taken not to use `*` between the matrices to be multiplied, as this simply multiplies corresponding entries of the matrices, in accordance with list conventions.

    `m1 * m2 // MatrixForm`

$$\begin{pmatrix} 18 & 32 & 2 \\ 16 & 27 & 12 \\ 0 & 4 & 0 \end{pmatrix}$$

EXAMPLE 6

    `v1 = {1, 2, 3};`
    `v2 = {4, 5, 6};`
    `v1.v2`
    32   ← *Mathematica* expresses the dot product as a
    `Cross[v1, v2]`   number rather than as a list containing a number.
    { - 3 , 6 , - 3 }

Mathematica makes no distinction between row and column vectors. Therefore, if v is an n-dimensional vector and m is an $n \times n$ matrix, both $v.m$ and $m.v$ are defined (although they generally yield different results). Furthermore, if $v1$ is an $n \times 1$ matrix (column vector) and $v2$ is a $1 \times n$ matrix (row vector), $v1.v2$ should be an $n \times n$ matrix, but *Mathematica* still computes a dot product. The command `Outer` can be used to compute the "outer" product of two vectors.

    ■ `Outer[Times, v1, v2]` computes the outer product of $v1$ and $v2$.

EXAMPLE 7

    `v1 = {1, 2, 3};`
    `v2 = {4, 5, 6};`
    `m = {{1, 2, 2}, {2, 3, 3}, {3, 1, 2}};`
    `m // MatrixForm`

$$\begin{pmatrix} 1 & 2 & 2 \\ 2 & 3 & 3 \\ 3 & 1 & 2 \end{pmatrix}$$

    `m.v1`
    {11, 17, 11}
$$\begin{pmatrix} 1 & 2 & 2 \\ 2 & 3 & 3 \\ 3 & 1 & 2 \end{pmatrix}\begin{pmatrix} 1 \\ 2 \\ 3 \end{pmatrix} = \begin{pmatrix} 11 \\ 17 \\ 11 \end{pmatrix}$$

    `v1.m`
    {14, 11, 14}
$$(1 \quad 2 \quad 3)\begin{pmatrix} 1 & 2 & 2 \\ 2 & 3 & 3 \\ 3 & 1 & 2 \end{pmatrix} = (14, 11, 14)$$

    `Outer[Times, v1, v2] // MatrixForm`

$$\begin{pmatrix} 4 & 5 & 6 \\ 8 & 10 & 12 \\ 12 & 15 & 18 \end{pmatrix}$$
$$\begin{pmatrix} 1 \\ 2 \\ 3 \end{pmatrix}(4 \quad 5 \quad 6) = \begin{pmatrix} 4 & 5 & 6 \\ 8 & 10 & 12 \\ 12 & 15 & 18 \end{pmatrix}$$

- **Inverse[*matrix*]** computes the inverse of *matrix*.
- **Det[*matrix*]** computes the determinant of *matrix*.
- **Transpose[*matrix*]** computes the transpose of *matrix*.
- **Tr[*matrix*]** computes the trace of *matrix*.
- **MatrixPower[*matrix*, n]** computes the nth power of *matrix*.
- **Minors [*matrix*]** produces a matrix whose (i, j)th entry is the determinant of the submatrix obtained from *matrix* by deleting row $n - i + 1$ and column $n - j + 1$ (*matrix* must be square).
- **Minors[*matrix*, k]** produces the matrix whose entries are the determinants of all possible $k \times k$ submatrices of *matrix*.   (*matrix* need not be square.)

EXAMPLE 8

$$m1 = \begin{pmatrix} 1 & 2 & 2 \\ 2 & 3 & 3 \\ 3 & 4 & 5 \end{pmatrix};$$

$$m2 = \begin{pmatrix} 1 & 2 & 3 & 4 \\ 5 & 6 & 7 & 8 \\ 9 & 10 & 11 & 12 \end{pmatrix};$$

These, and subsequent examples, were created using Create Table/Matrix/Palette in the Input menu.

Inverse[m1]//MatrixForm

$$\begin{pmatrix} -3 & 2 & 0 \\ 1 & 1 & -1 \\ 1 & -2 & 1 \end{pmatrix}$$

Tr[m1]
9

MatrixPower[m1, 3] // MatrixForm

$$\begin{pmatrix} 97 & 142 & 160 \\ 151 & 221 & 249 \\ 231 & 338 & 381 \end{pmatrix}$$

Transpose[m2]//MatrixForm

$$\begin{pmatrix} 1 & 5 & 9 \\ 2 & 6 & 10 \\ 3 & 7 & 11 \\ 4 & 8 & 12 \end{pmatrix}$$

Tr[m2]
18   The matrix does not have to be square in order for its trace to be defined.

EXAMPLE 9

m = Table[a[i, j], {i, 1, 3}, {j, 1, 3}];
m//MatrixForm

$$\begin{pmatrix} a[1,1] & a[1,2] & a[1,3] \\ a[2,1] & a[2,2] & a[2,3] \\ a[3,1] & a[3,2] & a[3,3] \end{pmatrix}$$

```
Minors[m] //MatrixForm
```

$$\begin{pmatrix} -a[1,1]a[2,1]+a[1,1]a[2,2] & -a[1,3]a[2,1]+a[1,1]a[2,3] & -a[1,3]a[2,2]+a[1,2]a[2,3] \\ -a[1,2]a[3,1]+a[1,1]a[3,2] & -a[1,3]a[3,1]+a[1,1]a[3,3] & -a[1,3]a[3,2]+a[1,2]a[3,3] \\ -a[2,2]a[3,1]+a[2,1]a[3,2] & -a[2,3]a[3,1]+a[2,1]a[3,3] & -a[2,3]a[3,2]+a[2,2]a[3,3] \end{pmatrix}$$

SOLVED PROBLEMS

12.8 The (Euclidean) norm of a vector is the square root of the sum of the squares of its components. Compute the norm of the vector (1, 3, 5, 7, 9, 11, 13, 15).

SOLUTION

```
v = Table[2k - 1, {k, 1, 8}]
{1, 3, 5, 7, 9, 11, 13, 15}
norm = √v.v
2√170
```

12.9 Prove that the cross product of two vectors in $\mathbb{R}^3$ is orthogonal to each of the vectors which form it.

SOLUTION

Let $\mathbf{u} = (u1, u2, u3)$ and $\mathbf{v} = (v1, v2, v3)$ and compute $\mathbf{w} = \mathbf{u} - \mathbf{v}$. Then verify that $\mathbf{w} \perp \mathbf{u}$ and $\mathbf{w} \perp \mathbf{v}$. Two vectors are orthogonal ($\perp$) if their dot product is 0.

```
u = {u1, u2, u3};
v = {v1, v2, v3};
w = Cross[u, v]
{ - u3 v2 + u2 v3, u3 v1 - u1 v3, - u2 v1 + u1 v2}
u.w // Expand
0
v.w // Expand
0
```

12.10 It can be shown that the volume of a parallelepiped formed by $\mathbf{u}$, $\mathbf{v}$, and $\mathbf{w}$ is $|\mathbf{u} \cdot (\mathbf{v} \times \mathbf{w})|$. Compute the volume of the parallelepiped formed by $\mathbf{i} + 2\mathbf{j} - 3\mathbf{k}, 2\mathbf{i} - 5\mathbf{j} + \mathbf{k}$, and $3\mathbf{i} + \mathbf{j} + 2\mathbf{k}$. (The quantity $\mathbf{u} \cdot (\mathbf{v} \times \mathbf{w})$ is called the scalar triple product.)

SOLUTION

```
u = {1, 2, -3};
v = {2, -5, 1};
w = {3, 1, 2};
volume = Abs[u.Cross[v, w]]
64
```

12.11 Let $\mathbf{u} = (u1, u2, u3)$, $\mathbf{v} = (v1, v2, v3)$, $\mathbf{w} = (w1, w2, w3)$. Prove that the scalar triple product

$$\mathbf{u} \cdot (\mathbf{v} \times \mathbf{w}) = \begin{vmatrix} u1 & u2 & u3 \\ v1 & v2 & v3 \\ w1 & w2 & w3 \end{vmatrix}.$$

SOLUTION

```
u = {u1, u2, u3};
v = {v1, v2, v3};
w = {w1, w2, w3};
matrix = {u, v, w}
lhs = u.Cross[v, w] // Expand;
rhs = Det[matrix] // Expand;
lhs == rhs
True
```

12.12 The Hilbert matrix is a square matrix whose element in row i column j is $\dfrac{1}{i+j-1}$. Construct the Hilbert matrix of order 6, and compute its determinant and its inverse.

SOLUTION

```
f[i_, j_] = 1/(i + j - 1);
hilbert = Array[f, {6, 6}];
hilbert // MatrixForm
```

$$\begin{pmatrix} 1 & \frac{1}{2} & \frac{1}{3} & \frac{1}{4} & \frac{1}{5} & \frac{1}{6} \\ \frac{1}{2} & \frac{1}{3} & \frac{1}{4} & \frac{1}{5} & \frac{1}{6} & \frac{1}{7} \\ \frac{1}{3} & \frac{1}{4} & \frac{1}{5} & \frac{1}{6} & \frac{1}{7} & \frac{1}{8} \\ \frac{1}{4} & \frac{1}{5} & \frac{1}{6} & \frac{1}{7} & \frac{1}{8} & \frac{1}{9} \\ \frac{1}{5} & \frac{1}{6} & \frac{1}{7} & \frac{1}{8} & \frac{1}{9} & \frac{1}{10} \\ \frac{1}{6} & \frac{1}{7} & \frac{1}{8} & \frac{1}{9} & \frac{1}{10} & \frac{1}{11} \end{pmatrix}$$

```
Det[hilbert]
```

$$\frac{1}{186313420339200000}$$

```
Inverse[hilbert]//MatrixForm
```

$$\begin{pmatrix} 36 & -630 & 3360 & -7560 & 7560 & -2772 \\ -630 & 14700 & -88200 & 211680 & -220500 & 83160 \\ 3360 & -88200 & 564480 & -1411200 & 1512000 & -582120 \\ -7560 & 211680 & -1411200 & 3628800 & -3969000 & 1552320 \\ 7560 & -220500 & 1512000 & -3969000 & 4410000 & -1746360 \\ -2772 & 83160 & -582120 & 1552320 & -1746360 & 698544 \end{pmatrix}$$

12.13 Construct a table which shows the determinant of the Hilbert matrices (see previous problem) of orders 1 through 10.

SOLUTION

```
Clear[hilbert]
f[i_, j_] = 1/(i + j - 1);
Do[hilbert[k] = Array[f, {k, k}], {k, 1, 10}];
TableForm[Table[{k, Det[hilbert[k]] // N}, {k, 1, 10}],
          TableHeadings → {None, {"k", "determinant"}}]
```

```
k      determinant
1      1.
2      0.833333
3      0.000462963
4      1.65344 × 10⁻⁷
5      3.7493 × 10⁻¹²
6      5.3673 × 10⁻¹⁸
7      4.8358 × 10⁻²⁵
8      2.73705 × 10⁻³³
9      9.72023 × 10⁻⁴³
10     2.16418 × 10⁻⁵³
```

	Since the determinants are nonzero, each Hilbert matrix is invertible. The Hilbert matrix is a classic example of an *ill-conditioned* matrix.

12.14 Let $M = \begin{pmatrix} \frac{1}{10} & \frac{2}{10} & \frac{7}{10} \\ \frac{3}{10} & \frac{3}{10} & \frac{4}{10} \\ \frac{5}{10} & \frac{4}{10} & \frac{1}{10} \end{pmatrix}$ (M is a *stochastic* matrix).   Compute $\lim_{n\to\infty} M^n$.

SOLUTION

$m = \frac{1}{10} \begin{pmatrix} 1 & 2 & 7 \\ 3 & 3 & 4 \\ 5 & 4 & 1 \end{pmatrix}$;

```
Clear[n]
Limit[MatrixPower[m, n], n → ∞] // MatrixForm
```

$\begin{pmatrix} \frac{47}{150} & \frac{23}{75} & \frac{19}{50} \\ \frac{47}{150} & \frac{23}{75} & \frac{19}{50} \\ \frac{47}{150} & \frac{23}{75} & \frac{19}{50} \end{pmatrix}$

12.15 Let $A = \begin{pmatrix} 1 & 2 & -1 & -2 & 3 \\ 2 & 1 & 2 & -2 & 0 \\ 0 & 1 & -2 & 3 & -1 \\ 1 & -1 & 1 & 2 & -3 \\ -2 & -2 & 1 & 1 & 2 \end{pmatrix}$ and $f(x) = x^5 + 2x^4 - x^3 + x^2 - 3x + 2$.   Compute $f(A)$.

SOLUTION

```
a = ⎛ 1   2  -1  -2   3 ⎞
    ⎜ 2   1   2  -2   0 ⎟
    ⎜ 0   1  -2   3  -1 ⎟ ;
    ⎜ 1  -1   1   2  -3 ⎟
    ⎝-2  -2   1   1   2 ⎠

MatrixPower[a, 5] + 2 MatrixPower[a, 4] - MatrixPower[a, 3] +
    MatrixPower[a, 2] - 3a + 2 IdentityMatrix[5] // MatrixForm
```

$\begin{pmatrix} -496 & -948 & -189 & 1776 & -1695 \\ -726 & -862 & 288 & 714 & -66 \\ -117 & 399 & -103 & -648 & 1233 \\ -174 & 324 & 315 & -1216 & 1875 \\ 1419 & 1068 & -267 & -702 & -1069 \end{pmatrix}$

12.16 It can be shown that the complex number $a + bi$ and the matrix $\begin{pmatrix} a & b \\ -b & a \end{pmatrix}$ have the same algebraic properties. Compute $(2 + 3i)^5$ using matrices and verify using complex arithmetic that this value is correct.

SOLUTION

```
a = ( 2   3 );
    (-3   2)
```

```
MatrixPower[a, 5]//MatrixForm
```

$$\begin{pmatrix} 122 & -597 \\ 597 & 122 \end{pmatrix}$$
        ← This represents the number $122 - 597\,i$.

```
(2 + 3I)^5
```

```
122 - 597 i
```

12.17 Compute the determinants

$$\begin{vmatrix} 1 & 1 & 1 & . & . & 1 \\ x_1 & x_2 & x_3 & . & . & x_n \\ x_1^2 & x_2^2 & x_3^2 & . & . & x_n^2 \\ . & . & . & . & & . \\ . & . & . & & . & . \\ x_1^{n-1} & x_2^{n-1} & x_3^{n-1} & . & . & x_n^{n-1} \end{vmatrix}$$

for $n = 2, 3, 4,$ and 5. Can you determine a pattern? These are known as *Vandermonde determinants*.

SOLUTION

```
m[n_] := Table[x[i]^j, {j, 0, n-1}, {i, 1, n}];
m[2]//MatrixForm
```

$$\begin{pmatrix} 1 & 1 \\ x[1] & x[2] \end{pmatrix}$$

```
m[3]//MatrixForm
```

$$\begin{pmatrix} 1 & 1 & 1 \\ x[1] & x[2] & x[3] \\ x[1]^2 & x[2]^2 & x[3]^2 \end{pmatrix}$$

```
Det[m[2]] // Factor
```

```
- x[1] + x[2]
```

```
Det[m[3]] // Factor
```

```
- (x[1] - x[2]) (x[1] - x[3]) (x[2] - x[3])
```

```
Det[m[4]] // Factor
```

```
(x[1] - x[2]) (x[1] - x[3]) (x[2] - x[3]) (x[1] - x[4]) (x[2] - x[4]) (x[3] - x[4])
```

```
Det[m[5]] // Factor
```

```
(x[1] - x[2]) (x[1] - x[3]) (x[2] - x[3]) (x[1] - x[4]) (x[2] - x[4]) (x[3] - x[4])
    (x[1] - x[5]) (x[2] - x[5]) (x[3] - x[5]) (x[4] - x[5])
```

In general, $\det\Big[m[n]\Big] = \prod_{i<j} \Big(x[j] - x[i]\Big)$.

12.18 A theorem in linear algebra says that the determinant of a matrix is the sum of the products of each entry of any row or column by its corresponding cofactor. (The cofactor, C_{ij}, of $a_{i,j}$ is $(-1)^{i+j} M_{ij}$ where $M_{i,j}$ is the corresponding minor.) Use this to compute the determinant of a randomly generated 5×5 matrix and verify its value.

SOLUTION

```
n = 5;
a = Table[Random[Integer, {0, 9}], {i, 1, n}, {j, 1, n}];
a // MatrixForm
```

$$\begin{pmatrix} 4 & 6 & 5 & 3 & 3 \\ 5 & 3 & 0 & 5 & 6 \\ 1 & 6 & 9 & 7 & 7 \\ 7 & 9 & 7 & 1 & 2 \\ 0 & 9 & 4 & 5 & 6 \end{pmatrix}$$

```
signs = Table[(-1)^(i + j), {i, 1, n}, {j, 1, n}];
matrixofminors = Minors[a];
matrixofminors // MatrixForm
```

$$\begin{pmatrix} 171 & 159 & -174 & -283 & -216 \\ -1350 & -1584 & -270 & -140 & -48 \\ 669 & 549 & 342 & 293 & -78 \\ -561 & -339 & 96 & -143 & -168 \\ -3231 & -3333 & -438 & -497 & 246 \end{pmatrix}$$

```
cofactors = matrixofminors * signs;
cofactors // MatrixForm
```

$$\begin{pmatrix} 171 & -159 & -174 & 283 & -216 \\ 1350 & -1584 & 270 & -140 & 48 \\ 669 & -549 & 342 & -293 & -78 \\ 561 & -339 & -96 & -143 & 168 \\ -3231 & 3333 & -438 & 497 & 246 \end{pmatrix}$$

```
i = 3; (* we expand using the third row *)
```

$$\text{determinant} = \sum_{j=1}^{n} a[[n-i+1, n-j+1]] * \text{cofactors}[[i, j]]$$

```
2082
Det[a]
2082
```

12.19 Let $\mathbf{x} = \begin{bmatrix} 1 \\ 2 \\ 3 \\ 4 \\ 5 \end{bmatrix}$. Compute $\mathbf{x}^T \mathbf{x} = \begin{bmatrix} 1 & 2 & 3 & 4 & 5 \end{bmatrix} \begin{bmatrix} 1 \\ 2 \\ 3 \\ 4 \\ 5 \end{bmatrix}$ and $\mathbf{x} \mathbf{x}^T = \begin{bmatrix} 1 \\ 2 \\ 3 \\ 4 \\ 5 \end{bmatrix} \begin{bmatrix} 1 & 2 & 3 & 4 & 5 \end{bmatrix}$.

SOLUTION

$\mathbf{x}^T\mathbf{x}$ is the dot product of the vector $\mathbf{x}$ with itself. $\mathbf{x}\mathbf{x}^T$, however, is a 5×5 matrix.

```
x.x
55
Outer[Times, x, x] // MatrixForm
```

$$\begin{pmatrix} 1 & 2 & 3 & 4 & 5 \\ 2 & 4 & 6 & 8 & 10 \\ 3 & 6 & 9 & 12 & 15 \\ 4 & 8 & 12 & 16 & 20 \\ 5 & 10 & 15 & 20 & 25 \end{pmatrix}$$

12.3 MATRIX MANIPULATION

Mathematica offers a package, `LinearAlgebra`MatrixManipulation``, which offers a variety of matrix manipulation commands which are quite useful when working problems in linear algebra.   Of course, as with any package, it must be loaded prior to its use.

The following commands enlarge a matrix:

- `AppendColumns[matrix1, matrix2, ...]` forms a new matrix from *matrix1*, *matrix2*, ... by joining their columns.
- `AppendRows[matrix1, matrix2, ...]` forms a new matrix from *matrix1*, *matrix2*, ... by joining their rows.
- `BlockMatrix[blocks]` forms a matrix of matrices by arranging them in block form.   *blocks* is a list of matrices whose construction specifies the block arrangement.

EXAMPLE 10

```
<< LinearAlgebra`MatrixManipulation`
```

$$m1 = \begin{pmatrix} a & b & c \\ d & e & f \\ g & h & i \end{pmatrix};$$

$$m2 = \begin{pmatrix} aa & bb & cc \\ dd & ee & ff \\ gg & hh & ii \end{pmatrix};$$

```
AppendColumns[m1, m2] //MatrixForm
```

$$\begin{pmatrix} a & b & c \\ d & e & f \\ g & h & i \\ aa & bb & cc \\ dd & ee & ff \\ gg & hh & ii \end{pmatrix}$$

> Corresponding columns of m1 and m2 are combined.

```
AppendRows[m1, m2] //MatrixForm
```

$$\begin{pmatrix} a & b & c & aa & bb & cc \\ d & e & f & dd & ee & ff \\ g & h & i & gg & hh & ii \end{pmatrix}$$

> Corresponding rows of m1 and m2 are combined.

BlockMatrix[{{m1,m2}, {m2,m1}}] // MatrixForm

$$
\begin{pmatrix}
a & b & c & aa & bb & cc \\
d & e & f & dd & ee & ff \\
g & h & i & gg & hh & ii \\
aa & bb & cc & a & b & c \\
dd & ee & ff & f & e & f \\
gg & hh & ii & g & h & i
\end{pmatrix}
$$

> The arrangement of the lists indicates that the two blocks on top, left to right, are m1 and m2. The bottom blocks are m2 and m1.

The following commands allow you to form submatrices:

- **TakeRows[*matrix*, n]** forms the submatrix consisting of the first n rows of *matrix*. If n is negative, the last n rows are taken.
- **TakeRows[*matrix*, {m, n}]** forms the submatrix consisting of rows m through n of *matrix*.
- **TakeColumns[*matrix*, n]** forms the submatrix consisting of the first n columns of *matrix*. If n is negative, the last n columns are taken.
- **TakeColumns[*matrix*, {m, n}]** forms the submatrix consisting of columns m thru n of *matrix*.
- **TakeMatrix[*matrix*, {i1, j1}, {i2, j2}]** forms the submatrix of *matrix* between elements (i1, j1) and (i2, j2).
- **SubMatrix[*matrix*, {i,j}, {m,n}]** takes the submatrix of *matrix* of dimension m × n starting at position {i,j}.

EXAMPLE 11

<< LinearAlgebra`MatrixManipulation`

$$
m = \begin{pmatrix}
1 & 2 & 3 & 4 & 5 \\
6 & 7 & 8 & 9 & 10 \\
11 & 12 & 13 & 14 & 15 \\
16 & 17 & 18 & 19 & 20
\end{pmatrix};
$$

TakeRows[m, 3] // MatrixForm

$$
\begin{pmatrix}
1 & 2 & 3 & 4 & 5 \\
6 & 7 & 8 & 9 & 10 \\
11 & 12 & 13 & 14 & 15
\end{pmatrix}
$$

TakeColumns[m, - 3] // MatrixForm

$$
\begin{pmatrix}
3 & 4 & 5 \\
8 & 9 & 10 \\
13 & 14 & 15 \\
18 & 19 & 20
\end{pmatrix}
$$

TakeRows[m, {2, 3}] // MatrixForm

$$
\begin{pmatrix}
6 & 7 & 8 & 9 & 10 \\
11 & 12 & 13 & 14 & 15
\end{pmatrix}
$$

TakeColumns[m, {2, 3}] // MatrixForm

$$
\begin{pmatrix}
2 & 3 \\
7 & 8 \\
12 & 13 \\
17 & 18
\end{pmatrix}
$$

```
TakeMatrix[m, {2, 2}, {3, 4}] // MatrixForm
```

$$\begin{pmatrix} 7 & 8 & 9 \\ 12 & 13 & 14 \end{pmatrix}$$

```
SubMatrix[m, {2, 3}, {2, 2}] // MatrixForm
```

$$\begin{pmatrix} 8 & 9 \\ 13 & 14 \end{pmatrix}$$

Although many matrices can be created using `Table` or `Array`, as discussed in the previous section, the matrix manipulation package offers some commands for constructing certain specialized matrices.

- **UpperDiagonalMatrix[f, n]** creates an $n \times n$ upper triangular matrix with elements `f[i, j]` on or above the main diagonal (zeros below).
- **LowerDiagonalMatrix[f, n]** creates an $n \times n$ lower triangular matrix with elements `f[i, j]` on or below the main diagonal (zeros above).
- **ZeroMatrix[n]** creates an $n \times n$ matrix of zeros; **ZeroMatrix[m, n]** creates an $m \times n$ matrix of zeros.
- **HilbertMatrix[n]** creates an $n \times n$ Hilbert matrix; **HilbertMatrix[m, n]** creates an $m \times n$ Hilbert matrix.
- **HankelMatrix[n]** creates a Hankel matrix, whose first row (and column) is $\{1, 2, 3, \ldots, n\}$; **HankelMatrix[n, *list*]** creates a Hankel matrix whose first row (and column) is *list*.

EXAMPLE 12

```
<< LinearAlgebra`MatrixManipulation`
f[i_, j_] = i + j;
UpperDiagonalMatrix[f, 5] //MatrixForm
```

$$\begin{pmatrix} 2 & 3 & 4 & 5 & 6 \\ 0 & 4 & 5 & 6 & 7 \\ 0 & 0 & 6 & 7 & 8 \\ 0 & 0 & 0 & 8 & 9 \\ 0 & 0 & 0 & 0 & 10 \end{pmatrix}$$

```
LowerDiagonalMatrix[f, 5] //MatrixForm
```

$$\begin{pmatrix} 2 & 0 & 0 & 0 & 0 \\ 3 & 4 & 0 & 0 & 0 \\ 4 & 5 & 6 & 0 & 0 \\ 5 & 6 & 7 & 8 & 0 \\ 6 & 7 & 8 & 9 & 10 \end{pmatrix}$$

```
ZeroMatrix[3, 5] //MatrixForm
```

$$\begin{pmatrix} 0 & 0 & 0 & 0 & 0 \\ 0 & 0 & 0 & 0 & 0 \\ 0 & 0 & 0 & 0 & 0 \end{pmatrix}$$

```
HilbertMatrix[5] //MatrixForm
```

$$\begin{pmatrix} 1 & \frac{1}{2} & \frac{1}{3} & \frac{1}{4} & \frac{1}{5} \\ \frac{1}{2} & \frac{1}{3} & \frac{1}{4} & \frac{1}{5} & \frac{1}{6} \\ \frac{1}{3} & \frac{1}{4} & \frac{1}{5} & \frac{1}{6} & \frac{1}{7} \\ \frac{1}{4} & \frac{1}{5} & \frac{1}{6} & \frac{1}{7} & \frac{1}{8} \\ \frac{1}{5} & \frac{1}{6} & \frac{1}{7} & \frac{1}{8} & \frac{1}{9} \end{pmatrix}$$

```
HankelMatrix[{a, b, c, d, e}]//MatrixForm
```

$$\begin{pmatrix} a & b & c & d & e \\ b & c & d & e & 0 \\ c & d & e & 0 & 0 \\ d & e & 0 & 0 & 0 \\ e & 0 & 0 & 0 & 0 \end{pmatrix}$$

SOLVED PROBLEMS

12.20 Construct a 10×10 upper triangular matrix whose nonzero entries are random digits. Show that its determinant is equal to the product of the entries on its main diagonal.

SOLUTION

```
<< LinearAlgebra`MatrixManipulation`
f[i_, j_] := Random[Integer, {0, 9}]
m = UpperDiagonalMatrix[f, 10];
m // MatrixForm
```

$$\begin{pmatrix} 3 & 1 & 6 & 7 & 0 & 6 & 4 & 8 & 9 & 1 \\ 0 & 1 & 7 & 5 & 1 & 9 & 9 & 0 & 5 & 8 \\ 0 & 0 & 5 & 6 & 9 & 5 & 3 & 3 & 3 & 5 \\ 0 & 0 & 0 & 1 & 5 & 7 & 2 & 5 & 3 & 4 \\ 0 & 0 & 0 & 0 & 6 & 8 & 4 & 0 & 9 & 0 \\ 0 & 0 & 0 & 0 & 0 & 5 & 2 & 0 & 5 & 0 \\ 0 & 0 & 0 & 0 & 0 & 0 & 3 & 5 & 5 & 7 \\ 0 & 0 & 0 & 0 & 0 & 0 & 0 & 3 & 6 & 8 \\ 0 & 0 & 0 & 0 & 0 & 0 & 0 & 0 & 3 & 7 \\ 0 & 0 & 0 & 0 & 0 & 0 & 0 & 0 & 0 & 4 \end{pmatrix}$$

```
Det[m] == ∏(i=1 to 10) m[[i,i]]
```

$$\text{Det}[m] == \prod_{i=1}^{10} m[[i,i]]$$

```
True
```

12.21 Construct a 9×9 block diagonal matrix with a 2×2 block of 2's, a 3×3 block of 3's, and a 4×4 block of 4's. (A block diagonal matrix is a square partitioned matrix whose diagonal matrices are square and all others are zero matrices.)

SOLUTION

```
<< LinearAlgebra`MatrixManipulation`
m1 = Table[2, {2}, {2}];
m2 = Table[3, {3}, {3}];
m3 = Table[4, {4}, {4}];

BlockMatrix[{{m1, ZeroMatrix[2, 7]},
             {ZeroMatrix[3, 2], m2, ZeroMatrix[3, 4]},
             {ZeroMatrix[4, 5], m3}}] // MatrixForm
```

$$\begin{pmatrix} 2 & 2 & 0 & 0 & 0 & 0 & 0 & 0 & 0 \\ 2 & 2 & 0 & 0 & 0 & 0 & 0 & 0 & 0 \\ 0 & 0 & 3 & 3 & 3 & 0 & 0 & 0 & 0 \\ 0 & 0 & 3 & 3 & 3 & 0 & 0 & 0 & 0 \\ 0 & 0 & 3 & 3 & 3 & 0 & 0 & 0 & 0 \\ 0 & 0 & 0 & 0 & 0 & 4 & 4 & 4 & 4 \\ 0 & 0 & 0 & 0 & 0 & 4 & 4 & 4 & 4 \\ 0 & 0 & 0 & 0 & 0 & 4 & 4 & 4 & 4 \\ 0 & 0 & 0 & 0 & 0 & 4 & 4 & 4 & 4 \end{pmatrix}$$

12.22 Let $M = \begin{pmatrix} a & b & c & d & e \\ f & g & h & i & j \\ k & l & m & n & o \\ p & q & r & s & t \\ u & v & w & x & y \end{pmatrix}$.

Find the matrix P obtained from M by deleting its third column and fourth row.

SOLUTION

```
<< LinearAlgebra`MatrixManipulation`
temp = CharacterRange["a", "y"];          ← generates a list of alphabet letters
m = Partition[temp, 5];                    ← forms five sublists of five letters each
m // MatrixForm
```

$$\begin{pmatrix} a & b & c & d & e \\ f & g & h & i & j \\ k & l & m & n & o \\ p & q & r & s & t \\ u & v & w & x & y \end{pmatrix}$$

```
temp1 = TakeColumns[m, {1, 2}];           ← columns 1–2 of m
temp2 = TakeColumns[m, {4, 5}];           ← columns 4–5 of m
temp3 = AppendRows[temp1, temp2];         ← combines columns 1–2 with columns 3–4
temp3 //MatrixForm
```

$$\begin{pmatrix} a & b & d & e \\ f & g & i & j \\ k & l & n & o \\ p & q & s & t \\ u & v & x & y \end{pmatrix}$$ ← We print the matrix with column 3 removed.   Next we remove row 4.

```
temp4 = TakeRows[temp3, {1, 3}];          ← rows 1–3 of temp3
temp5 = TakeRows[temp3, -1];              ← row 5 of temp3
p = AppendColumns[temp4, temp5];          ← combines rows 1 to 3 with 5
p //MatrixForm
```

$$\begin{pmatrix} a & b & d & e \\ f & g & i & j \\ k & l & n & o \\ u & v & x & y \end{pmatrix}$$

12.4 LINEAR SYSTEMS OF EQUATIONS

Mathematica offers a number of ways to solve systems of linear equations. Solve, discussed in Chapter 6, offers one alternative, but it is somewhat clumsy and inefficient for use on large systems. In this section we discuss a number of other procedures for solving systems of linear equations.

■ LinearSolve[a, b] produces vectors x, such that a.x = b.

Here **a** is the matrix of coefficients of the unknowns, and **b** is the "right-hand side" of the linear system. If a is invertible, LinearSolve will produce a unique solution to the linear system. If a is singular, either no solution exists, or there are an infinite number of solutions.

If a system has a unique solution, *Mathematica* returns the solution. If no solution exists, *Mathematica* returns an error message.

EXAMPLE 13

The system

$2x + y + z = 7$, $x - 4y + 3z = 2$, $3x + 2y + 2z = 13$ has a unique solution.
$2x + y + z = 7$, $x - 4y + 3z = 2$, $3x - 3y + 4z = 10$ has no solution.

$$a1 = \begin{pmatrix} 2 & 1 & 1 \\ 1 & -4 & 3 \\ 3 & 2 & 2 \end{pmatrix};$$

$$a2 = \begin{pmatrix} 2 & 1 & 1 \\ 1 & -4 & 3 \\ 3 & -3 & 4 \end{pmatrix};$$

$$b1 = \begin{pmatrix} 7 \\ 2 \\ 13 \end{pmatrix};$$

$$b2 = \begin{pmatrix} 7 \\ 2 \\ 10 \end{pmatrix};$$

LinearSolve[a1, b1]

{{1}, {2}, {3}}

LinearSolve[a2, b2]

LinearSolve::nosol :
Linear equation encountered which has no solution.

LinearSolve[{{2, 1, 1}, {1, -4, 3}, {3, -3, 4}}, {{7}, {2}, {10}}]

If the system a.x = b has an infinite number of solutions, the treatment is a bit more complicated. In this case, *Mathematica* returns one solution, known as a *particular solution*. The full set of solutions is constructed by adding to the particular solution the set of *all* solutions of the corresponding homogeneous system, a.x = 0.

The set of all vectors, x, such that a.x = 0 is called the *null space* of a and is easily determined by the command NullSpace.

■ NullSpace[a] returns the basis vectors of the null space of a.

The nullity of a, the number of linearly independent vectors in the null space of a, can be found by computing Length[NullSpace[a]]. The rank of a, may be computed as n-Length[NullSpace[a]], where n represents the number of columns of a.

EXAMPLE 14

$2x + y + z = 7$, $x - 4y + 3z = 2$, $3x - 3y + 4z = 9$ has an infinite number of solutions.

$$a = \begin{pmatrix} 2 & 1 & 1 \\ 1 & -4 & 3 \\ 3 & -3 & 4 \end{pmatrix};$$

$$b = \begin{pmatrix} 7 \\ 2 \\ 9 \end{pmatrix};$$

`nullspacebasis = NullSpace[a]`
$\{\{-7, 5, 9\}\}$

> Because the null space contains a non-zero vector, there is no unique solution.

`particular = LinearSolve[a, b]`

$\left\{ \left\{ \dfrac{10}{3} \right\}, \left\{ \dfrac{1}{3} \right\}, \{0\} \right\}$ ← This is a particular solution.

The full set of solutions to the system is of the form

$$t * \text{nullspacebasis} + \text{particular}$$

where t is an arbitrary parameter.   However, to express as a single list, we must first flatten nullspacebasis and particular.

`generalsolution = t * Flatten[ nullspacebasis] + Flatten[particular]`

$$\left\{ \dfrac{10}{3} - 7t, \dfrac{1}{3} + 5t, 9t \right\}$$

As a check, we substitute our general solution back into the original system.

`a.x /. x → { 10/3 - 7t, 1/3 + 5t, 9t} // Expand`
$\{7, 2, 9\}$

The Gauss-Jordan method for solving the linear system $a.x = b$ is based upon the reduction of the augmented matrix `[a|b]` into *reduced row echelon form* by a series of *elementary row operations*. The three basic elementary row operations are:

 (*a*)   interchanging two rows.
 (*b*)   multiplying a row by a nonzero constant.
 (*c*)   replacing one row by itself plus a multiple of another row.

It is easily seen that elementary row operations have no affect upon the solution of the system.
   A matrix is said to be in *reduced row echelon form* if

 (*a*)   each leading nonzero entry is 1 (called a leading 1).
 (*b*)   each entry above or below a leading 1 is 0.
 (*c*)   If two rows have leading 1's, the lower row has its leading 1 further to the right.

To solve linear systems, we use elementary row operations to reduce the augmented matrix to reduced row echelon form.   The solution(s) of the system, or the fact that no solutions exist, may then be easily determined.
   Every student of linear algebra knows that row reduction is a time-consuming, tedious process which is highly prone to error.   However, the *Mathematica* command **RowReduce** automatically reduces any matrix to reduced row echelon form.

 ■   **RowReduce[*matrix*]** reduces *matrix* to reduced row echelon form.

EXAMPLE 15

Consider the 4×5 matrix whose general entry $a_{i,j} = |i - j|$.

```
a = Table[Abs[i - j], {i, 1, 4}, {j, 1, 5}];
a // MatrixForm
```

$$\begin{pmatrix} 0 & 1 & 2 & 3 & 4 \\ 1 & 0 & 1 & 2 & 3 \\ 2 & 1 & 0 & 1 & 2 \\ 3 & 2 & 1 & 0 & 1 \end{pmatrix}$$

```
RowReduce[a] // MatrixForm
```

$$\begin{pmatrix} 1 & 0 & 0 & 0 & \frac{1}{3} \\ 0 & 1 & 0 & 0 & 0 \\ 0 & 0 & 1 & 0 & 0 \\ 0 & 0 & 0 & 1 & \frac{4}{3} \end{pmatrix}$$

We now illustrate how row reduction can be used to solve a linear system. For comparison purposes we use the three examples previously considered in Examples 13 and 14.

EXAMPLE 16

(a) $2x + y + z = 7$, $x - 4y + 3z = 2$, $3x + 2y + 2z = 13$ (unique solution)

(b) $2x + y + z = 7$, $x - 4y + 3z = 2$, $3x - 3y + 4z = 10$ (no solution)

(c) $2x + y + z = 7$, $x - 4y + 3z = 2$, $3x - 3y + 4z = 9$ (infinite number of solutions)

We find the augmented matrix for each of the three systems.

$$a1 = \begin{pmatrix} 2 & 1 & 1 & 7 \\ 1 & -4 & 3 & 2 \\ 3 & 2 & 2 & 13 \end{pmatrix}; \qquad a2 = \begin{pmatrix} 2 & 1 & 1 & 7 \\ 1 & -4 & 3 & 2 \\ 3 & -3 & 4 & 10 \end{pmatrix}; \qquad a3 = \begin{pmatrix} 2 & 1 & 1 & 7 \\ 1 & -4 & 3 & 2 \\ 3 & -3 & 4 & 9 \end{pmatrix};$$

```
RowReduce[a1] // MatrixForm
```

$$\begin{pmatrix} 1 & 0 & 0 & 1 \\ 0 & 1 & 0 & 2 \\ 0 & 0 & 1 & 3 \end{pmatrix}$$

This reduced matrix, when interpreted as a system of equations, reads: $x = 1$, $y = 2$, $z = 3$.

```
RowReduce[a2] // MatrixForm
```

$$\begin{pmatrix} 1 & 0 & \frac{7}{9} & 0 \\ 0 & 1 & -\frac{5}{9} & 0 \\ 0 & 0 & 0 & 1 \end{pmatrix}$$

The bottom row reads $0x + 0y + 0z = 1$ which, of course, is impossible. This contradiction (a row of 0's and a final 1) reveals that no solution is possible.

```
RowReduce[a3] // MatrixForm
```

$$\begin{pmatrix} 1 & 0 & \frac{7}{9} & \frac{10}{3} \\ 0 & 1 & -\frac{5}{9} & \frac{1}{3} \\ 0 & 0 & 0 & 0 \end{pmatrix}$$

The bottom row of 0's is not a contradiction. However there cannot be a unique solution. If we let $z = t$, an independent parameter, the solution may be put into the form

$$x = \frac{10}{3} - \frac{7}{9}t, \qquad y = \frac{1}{3} + \frac{5}{9}t, \qquad z = t.$$

Note: Although the solution looks slightly different than the solution obtained in Example 14, it is equivalent in the sense that it describes precisely the same solution set.

Another popular method, *LU* decomposition, is useful, particularly if you have many systems, all having the same coefficient matrix. The idea behind the method is simple.

If A is a square matrix, it may be possible to factor $A = LU$ where L is lower triangular with 1's on the main diagonal and U is upper triangular. The system $A\mathbf{x} = \mathbf{b}$ then reads $(LU)\mathbf{x} = \mathbf{b}$ which can be written $L(U\mathbf{x}) = \mathbf{b}$. If we let $\mathbf{y} = U\mathbf{x}$, we can solve $L\mathbf{y} = \mathbf{b}$ for $\mathbf{y}$. Once we have determined $\mathbf{y}$, we solve $U\mathbf{x} = \mathbf{y}$ for $\mathbf{x}$.

Even though the solution of a system by LU decomposition involves solving two systems of equations, each involves a *triangular* matrix, so the computation is efficient.

Thus there are two steps to solving a system of equations by LU decomposition: factorization and back substitution. The corresponding *Mathematica* commands are **LUDecomposition** and **LUBackSubstitution**.

- **LUDecomposition[*matrix*]** finds the LU decomposition of *matrix*.
- **LUBackSubstitution[*data*, b]** uses the output of LUDecomposition[*matrix*] to solve the system matrix.x = b.

The output of LUDecomposition, *data*, consists of three parts: (i) the matrices L and U "packed" as a single matrix, (ii) a permutation vector, and (iii) the L^∞ condition number of the matrix. *data* is fed into LUBackSubstitution to solve the system.

The permutation vector rearranges the rows in order to ensure a maximum degree of numerical stability. The condition number will be of no concern to us in this book.

LUDecomposition and LUBackSubstitution cannot be used on systems which possess an infinite number of solutions.

EXAMPLE 17

To solve the system $2x + y + z = 7$, $x - 4y + 3z = 2$, $3x + 2y + 2z = 13$ using LU decomposition, we first obtain the matrix factorization of the coefficient matrix.

$$a = \begin{pmatrix} 2 & 1 & 1 \\ 1 & -4 & 3 \\ 3 & 2 & 2 \end{pmatrix} ; \quad b = \begin{pmatrix} 7 \\ 2 \\ 13 \end{pmatrix} ;$$

```
data = LUDecomposition[a]
```
$\{\{\{1, -4, 3\}, \{2, 9, -5\}, \{3, \frac{14}{9}, \frac{7}{9}\}\}, \{2, 1, 3\}, 1\}$

```
LUBackSubstitution[data, b]
```
$\{\{1\}, \{2\}, \{3\}\}$

Examples 18 and 19 illustrate the structure of *data*.

EXAMPLE 18

$$m = \begin{pmatrix} 2 & 3 & 4 \\ 4 & 11 & 14 \\ 6 & 29 & 43 \end{pmatrix} ;$$

```
{lu, p, cond} = LUDecomposition[m]
```
$\{\{\{2, 3, 4\}, \{2, 5, 6\}, \{3, 4, 7\}\}, \{1, 2, 3\}, 1\}$

In this example, no rearrangement of the rows was performed since the permutation vector, p, is $\{1, 2, 3\}$.

The first part of LUDecomposition[m] is given in a "packed" format. Since LU is known to be of the

$$\text{form} \begin{pmatrix} 1 & 0 & 0 \\ x & 1 & 0 \\ x & x & 1 \end{pmatrix} \begin{pmatrix} x & x & x \\ 0 & x & x \\ 0 & 0 & x \end{pmatrix} , \text{ only nine entries (represented by } x \text{) need be specified. The first part of}$$

LUDecomposition[m] specifies these nine numbers as a single matrix.

```
lu // MatrixForm
```

$$\begin{pmatrix} 2 & 3 & 4 \\ 2 & 5 & 6 \\ 3 & 4 & 7 \end{pmatrix}$$

The numbers, although combined into one matrix, are in their correct positions.   Hence

$$lu = \begin{pmatrix} 1 & 0 & 0 \\ 2 & 1 & 0 \\ 3 & 4 & 1 \end{pmatrix} \begin{pmatrix} 2 & 3 & 4 \\ 0 & 5 & 6 \\ 0 & 0 & 7 \end{pmatrix}$$

EXAMPLE 19

$$m = \begin{pmatrix} 2 & 1 & 1 \\ 1 & -4 & 3 \\ 3 & 2 & 2 \end{pmatrix};$$

```
{lu, p, cond} = LUDecomposition[m]
```

$$\left\{ \left\{ \{1, -4, 3\}, \{2, 9, -5\}, \left\{3, \frac{14}{9}, \frac{7}{9}\right\} \right\}, \{2, 1, 3\}, 1 \right\}$$

```
lu // MatrixForm
```

$$\begin{pmatrix} 1 & -4 & 3 \\ 2 & 9 & -5 \\ 3 & \frac{14}{9} & \frac{7}{9} \end{pmatrix}$$

If we proceed as in the previous example, we would be tempted to say that

$$l = \begin{pmatrix} 1 & 0 & 0 \\ 2 & 1 & 0 \\ 3 & \frac{14}{9} & 1 \end{pmatrix};$$

$$u = \begin{pmatrix} 1 & -4 & 3 \\ 0 & 9 & -5 \\ 0 & 0 & \frac{7}{9} \end{pmatrix};$$

Multiplying **l** by **u** does *not* give back the original matrix.

```
l.u //MatrixForm
```

$$\begin{pmatrix} 1 & -4 & 3 \\ 2 & 1 & 1 \\ 3 & 2 & 2 \end{pmatrix}$$

The permutation vector, p = {2, 1, 3}, indicates that rows 1 and 2 of the matrix have been interchanged. If we permute the rows of l, we should get back our original matrix upon multiplication by U.

$$l = \begin{pmatrix} 2 & 1 & 0 \\ 1 & 0 & 0 \\ 3 & \frac{14}{9} & 1 \end{pmatrix};$$

`l.u //MatrixForm`

$$\begin{pmatrix} 2 & 1 & 1 \\ 1 & -4 & 3 \\ 3 & 2 & 2 \end{pmatrix}$$

A more convenient way of reconstructing L and U is the command **LUMatrices**, found in the **LinearAlgebra`MatrixManipulation`** package.

- **LUMatrices[lu]** returns a list, in the form {l, u}, of the permuted lower and upper matrices in the LU decomposition of *matrix*, corresponding to lu, the first element of LUDecomposition[*matrix*].

EXAMPLE 20

`<<LinearAlgebra`MatrixManipulation` `

$$m = \begin{pmatrix} 2 & 1 & 1 \\ 1 & -4 & 3 \\ 3 & 2 & 2 \end{pmatrix};$$

`{lu, p, cond} = LUDecomposition[m]`

$$\left\{\left\{\{1, -4, 3\}, \{2, 9, -5\}, \left\{3, \frac{14}{7}, \frac{7}{9}\right\}\right\}, \{2, 1, 3\}, 1\right\}$$

`LUMatrices[lu]`

$$\left\{\left\{\{1, 0, 0\}, \{2, 1, 0\}, \left\{3, \frac{14}{9}, 1\right\}\right\}, \left\{\{1, -4, 3\}, \{0, 9, -5\}, \left\{0, 0, \frac{7}{9}\right\}\right\}\right\}$$

`l = LUMatrices[lu][[1]];`
`u = LUMatrices[lu][[2]];`
`l[[p]] //MatrixForm`   ←`l[[p]]` permutes the rows of matrix l according to vector p.

$$\begin{pmatrix} 2 & 1 & 0 \\ 1 & 0 & 0 \\ 3 & \frac{14}{9} & 1 \end{pmatrix}$$

`u // MatrixForm`

$$\begin{pmatrix} 1 & -4 & 3 \\ 0 & 9 & -5 \\ 0 & 0 & \frac{7}{9} \end{pmatrix}.$$

`l[[p]].u // MatrixForm`

$$\begin{pmatrix} 2 & 1 & 1 \\ 1 & -4 & 3 \\ 3 & 2 & 2 \end{pmatrix}$$

SOLVED PROBLEMS

12.23 Describe the set of vectors S spanned by $(1, 2, 1, 2, 1)$, $(1, 3, 2, 4, 2)$, and $(1, 4, 3, 6, 3)$.

SOLUTION

`a = {1, 2, 1, 2, 1};`
`b = {1, 3, 2, 4, 2};`
`c = {1, 4, 3, 6, 3};`
`m = {a, b, c};`

```
m // MatrixForm
```

$$\begin{pmatrix} 1 & 2 & 1 & 2 & 1 \\ 1 & 3 & 2 & 4 & 2 \\ 1 & 4 & 3 & 6 & 3 \end{pmatrix}$$

Form a matrix, **m**, using the given vectors as rows. The row space of **m** is the space spanned by **a**, **b**, and **c**. Then reduce the matrix to reduced row echelon form. The nonzero rows form a basis for the row space. Every vector in S is a linear combination of its basis vectors.

```
rref = RowReduce[m];
rref // MatrixForm
```

$$\begin{pmatrix} 1 & 0 & -1 & -2 & -1 \\ 0 & 1 & 1 & 2 & 1 \\ 0 & 0 & 0 & 0 & 0 \end{pmatrix}$$

```
s * rref[[1]] + t * rref[[2]]
{s, t, -s+t, -2s+2t, -s+t}
```
← Every vector in S is of this form.

12.24 A theorem of linear algebra says that every vector in the row space of A is orthogonal to every vector in the null space of A. Verify this result for

$$A = \begin{pmatrix} 1 & 2 & 3 & 4 & 5 \\ 6 & 7 & 8 & 9 & 10 \\ 11 & 12 & 13 & 14 & 15 \\ 16 & 17 & 18 & 19 & 20 \\ 21 & 22 & 23 & 24 & 25 \end{pmatrix}$$

SOLUTION

It suffices to show that each basis vector of the row space is orthogonal to every basis vector in the null space.

```
a = Partition[Range[25], 5];
a // MatrixForm
```

$$\begin{pmatrix} 1 & 2 & 3 & 4 & 5 \\ 6 & 7 & 8 & 9 & 10 \\ 11 & 12 & 13 & 14 & 15 \\ 16 & 17 & 18 & 19 & 20 \\ 21 & 22 & 23 & 24 & 25 \end{pmatrix}$$

```
rowspacebasis = RowReduce[a];
rowspacebasis // MatrixForm
```

$$\begin{pmatrix} 1 & 0 & -1 & -2 & -3 \\ 0 & 1 & 2 & 3 & 4 \\ 0 & 0 & 0 & 0 & 0 \\ 0 & 0 & 0 & 0 & 0 \\ 0 & 0 & 0 & 0 & 0 \end{pmatrix}$$

To show that every vector in rowspacebasis is orthogonal to every vector in nullspacebasis, we must show that every dot product is 0. The easiest way is to multiply rowspacebasis by the *transpose* of nullspacebasis. The zero rows at the bottom of the product may be safely ignored.

```
nullspacebasis = NullSpace[a];
nullspacebasis // MatrixForm
```

$$\begin{pmatrix} 3 & -4 & 0 & 0 & 1 \\ 2 & -3 & 0 & 1 & 0 \\ 1 & -2 & 1 & 0 & 0 \end{pmatrix}$$

```
rowspacebasis.Transpose[nullspacebasis] // MatrixForm
```

$$\begin{pmatrix} 0 & 0 & 0 \\ 0 & 0 & 0 \\ 0 & 0 & 0 \\ 0 & 0 & 0 \\ 0 & 0 & 0 \end{pmatrix}$$

12.25 Construct a 5×5 matrix of random digits **a** and a 5×1 matrix of random digits **b**, and solve the linear system **ax = b** using **LinearSolve**. Then verify that your solution is correct.

SOLUTION

```
a = Table[Random[Integer, {0, 9}], {i, 1, 5}, {j, 1, 5}];
a // MatrixForm
```

$$\begin{pmatrix} 0 & 9 & 5 & 0 & 2 \\ 7 & 7 & 5 & 9 & 3 \\ 0 & 6 & 6 & 2 & 6 \\ 5 & 8 & 4 & 3 & 9 \\ 3 & 0 & 8 & 1 & 5 \end{pmatrix}$$

```
Det[a] == 0
```
False   ← Since the determinant $\neq 0$, the system has a unique solution.

```
b = Table[Random[Integer, {0, 9}], {i, 1, 5}];
b // MatrixForm
```

$$\begin{pmatrix} 1 \\ 2 \\ 4 \\ 6 \\ 1 \end{pmatrix}$$

```
x = LinearSolve[a, b]
```

$$\left\{ \left\{ -\frac{58}{217} \right\}, \left\{ \frac{65}{651} \right\}, \left\{ -\frac{187}{651} \right\}, \left\{ \frac{111}{434} \right\}, \left\{ \frac{143}{186} \right\} \right\}$$

```
a.x == b
```
True

12.26 Find the general solution of the system $\begin{cases} w + 2x + 3y + 3z = 9 \\ 2w + x + 2y + 5z = 10 \\ 2w + 2x + y + 2z = 7 \\ 2w - x - 3y + z = -1 \end{cases}$.

SOLUTION

$$a = \begin{pmatrix} 1 & 2 & 3 & 3 \\ 2 & 1 & 2 & 5 \\ 2 & 2 & 1 & 2 \\ 2 & -1 & -3 & 1 \end{pmatrix}; b = \begin{pmatrix} 9 \\ 10 \\ 7 \\ -1 \end{pmatrix};$$

```
Det[a]
```
0   ← Since the determinant is 0, we anticipate either no solution or an infinite number of solutions.

```
nullspacebasis = NullSpace[a]      ← Since the null space contains a non-zero vector,
                                     there will be an infinite number of solutions.
{{-13, 11, -10, 7}}
particular = LinearSolve[a, b]
```

$$\left\{\left\{\frac{20}{7}\right\}, \left\{-\frac{4}{7}\right\}, \left\{\frac{17}{7}\right\}, \{0\}\right\}$$

```
generalsolution = t * Flatten[nullspacebasis] + Flatten[particular]
```

$$\left\{\frac{20}{7} - 13\ t,\ -\frac{4}{7} + 11\ t,\ \frac{17}{7} - 10\ t,\ 7\ t\right\}$$

12.27 Find the general solution of the system $\begin{cases} w + 2x + 3y + 3z = 9 \\ 3w + 4x + 4y + 5z = 16 \\ 2w + 2x + \ \ y + 2z = 7 \\ 4w + 6x + 7y + 8z = 25 \end{cases}$.

SOLUTION

$$a = \begin{pmatrix} 1 & 2 & 3 & 3 \\ 3 & 4 & 4 & 5 \\ 2 & 2 & 1 & 2 \\ 4 & 6 & 7 & 8 \end{pmatrix}; \ b = \begin{pmatrix} 9 \\ 16 \\ 7 \\ 25 \end{pmatrix};$$

```
Det[a] == 0
True
nullspacebasis = NullSpace[a]
{{1, -2, 0, 1}, {4, -5, 2, 0}}
particular = LinearSolve[a, b]
```

$$\left\{\{-2\}, \left\{\frac{11}{2}\right\}, \{0\}, \{0\}\right\}$$

```
generalsolution = s * nullspacebasis[[1]] +
                  t * nullspacebasis[[2]] + Flatten[particular]
```

$$\left\{-2 + s + 4\ t,\ \frac{11}{2} - 2\ s - 5\ t,\ 2\ t,\ s\right\}$$

12.28 Let A be a 7×7 tridiagonal matrix having 3's on the main diagonal and -1's on the diagonals adjacent to the main diagonal. Let e_i be the seven-dimensional vector having 1 in the ith position and 0's elsewhere. Solve $A\mathbf{x} = e_i$, $i = 1, \ldots, 7$.

SOLUTION

```
a = Table[If[Abs[i - j] == 1, -1, If[i == j, 3, 0]], {i, 1, 7}, {j, 1, 7}];
a // MatrixForm
```

$$\begin{pmatrix} 3 & -1 & 0 & 0 & 0 & 0 & 0 \\ -1 & 3 & -1 & 0 & 0 & 0 & 0 \\ 0 & -1 & 3 & -1 & 0 & 0 & 0 \\ 0 & 0 & -1 & 3 & -1 & 0 & 0 \\ 0 & 0 & 0 & -1 & 3 & -1 & 0 \\ 0 & 0 & 0 & 0 & -1 & 3 & -1 \\ 0 & 0 & 0 & 0 & 0 & -1 & 3 \end{pmatrix}$$

```
ludata = LUDecomposition[a];

b = Table[KroneckerDelta[i, j], {i, 1, 7}, {j, 1, 7}];

LUBackSubstitution[ludata, b] // TableForm
```

$\frac{377}{987}$	$\frac{48}{329}$	$\frac{55}{987}$	$\frac{1}{47}$	$\frac{8}{987}$	$\frac{1}{329}$	$\frac{1}{987}$
$\frac{48}{329}$	$\frac{144}{329}$	$\frac{55}{329}$	$\frac{3}{47}$	$\frac{8}{329}$	$\frac{3}{329}$	$\frac{1}{329}$
$\frac{55}{987}$	$\frac{55}{329}$	$\frac{440}{987}$	$\frac{8}{47}$	$\frac{64}{987}$	$\frac{8}{329}$	$\frac{8}{987}$
$\frac{1}{47}$	$\frac{3}{47}$	$\frac{8}{47}$	$\frac{21}{47}$	$\frac{8}{47}$	$\frac{3}{47}$	$\frac{1}{47}$
$\frac{8}{987}$	$\frac{8}{329}$	$\frac{64}{987}$	$\frac{8}{47}$	$\frac{440}{987}$	$\frac{55}{329}$	$\frac{55}{987}$
$\frac{1}{329}$	$\frac{3}{329}$	$\frac{8}{329}$	$\frac{3}{47}$	$\frac{55}{329}$	$\frac{144}{329}$	$\frac{48}{329}$
$\frac{1}{987}$	$\frac{1}{329}$	$\frac{8}{987}$	$\frac{1}{47}$	$\frac{55}{987}$	$\frac{48}{329}$	$\frac{377}{987}$

> KroneckerDelta [i, j] = 1 if $i = j$ and 0 otherwise.

The ith column of the table represents the solution of $A\mathbf{x} = \mathbf{e}_i$.

12.5 ORTHOGONALITY

Two vectors are *orthogonal* if their inner product is 0. Orthogonal vectors possess useful properties which make working with them convenient. For example, if $\mathbf{u}$ and $\mathbf{v}$ are orthogonal, they satisfy the (generalized) Theorem of Pythagoras: $\|\mathbf{u} + \mathbf{v}\|^2 = \|\mathbf{u}\|^2 + \|\mathbf{v}\|^2$.

Orthogonality also allows us to introduce the concept of projection. In $\mathbb{R}^2$ it is easy to visualize what projection means. If $\mathbf{a} = \vec{PQ}$ and $\mathbf{b} = \vec{PR}$ are two vectors with the same initial point P, then if S is the foot of the perpendicular from R to $\vec{PQ}$, the projection of $\mathbf{b}$ onto $\mathbf{a}$ is the vector $\vec{PS}$. This vector is often represented $\text{proj}_\mathbf{a}\mathbf{b}$.

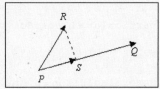

The projection vector can be computed using the *Mathematica* command **Projection**, which is contained within the **LinearAlgebra`Orthogonalization`** package.

- **Projection[*vector1*, *vector2*]** returns the orthogonal projection of *vector1* onto *vector2*.

EXAMPLE 21

The following sequence computes the projection of $(1, 2, 3)$ onto $(-2, 3, -1)$.

```
<< LinearAlgebra`Orthogonalization`

a = {1, 2, 3};

b = {-2, 3, -1};

Projection[a, b]
```

$$\left\{-\frac{1}{7}, \frac{3}{14}, -\frac{1}{14}\right\}$$

The concept of orthogonality depends upon the definition of inner product for the space under consideration. By default, *Mathematica* uses the Euclidean inner product (dot product) in linear algebra commands. However, this can be changed by including the option **InnerProduct**.

- **InnerProduct → *func***, where *func* is a pure function defines the inner product associated with the space being considered. (See Appendix for a discussion of pure functions.)

It can be shown that if c_1, c_2, and c_3 are positive real numbers, then $<\mathbf{a},\mathbf{b}> = c_1 a_1 b_1 + c_2 a_2 b_2 + c_3 a_3 b_3$ defines an inner product on $\mathbb{R}^3$. To compute the orthogonal projection of $(1,2,3)$ onto $(-2,3,-1)$ using this inner product, we must define an appropriate function describing the inner product. To do this, we compute the vector $\mathbf{a}*\mathbf{b}$ and then take its dot product with $\mathbf{c} = (c_1, c_2, c_3)$. The technique is illustrated in the next example.

EXAMPLE 22

Compute the projection of $(1,2,3)$ onto $(-2,3,-1)$ using the inner product $<\mathbf{a},\mathbf{b}> = 2a_1 b_1 + 3a_2 b_2 + 4a_3 b_3$

```
<< LinearAlgebra`Orthogonalization`
a = {1, 2, 3};
b = {-2, 3, -1};
Projection[a, b, InnerProduct → ({2, 3, 4}.(#1 * #2) &)]
```

$$\left\{-\frac{4}{39}, \frac{2}{13}, -\frac{2}{39}\right\}$$

EXAMPLE 23

A useful inner product often used in function spaces is $<f,g> = \int_{-1}^{1} f(x) g(x)\, dx$. Using this inner product, compute the projection of x^2 on $x^3 + 1$.

```
a = x²;
b = x³ + 1;
Projection[a, b, InnerProduct → (Integrate[#1 * #2, {x, -1, 1}] &)]
```

$$\frac{7}{24}(1 + x^3)$$

A finite dimensional vector space, by definition, has a finite basis. However, except for the trivial vector space which contains only the zero vector, an infinite number of different bases are possible.

The most convenient basis for any vector space is an orthonormal basis. The *Gram Schmidt orthogonalization process* provides a "recipe" for converting any basis into an orthonormal basis.

- `Normalize[`*vector*`]` converts *vector* into a unit vector.
- `GramSchmidt[`*vectorlist*`]` produces an orthonormal set of vectors whose span is *vectorlist*.

  - `Normalized → False`, included as an option in `GramSchmidt`, will produce an orthogonal, but not necessarily orthonormal, set of vectors.
  - `InnerProduct → `*func* may be used to define an inner product other than the default (dot product).

As with `Projection`, `Normalize` and `GramSchmidt` are found in the package `LinearAlgebra`Orthogonalization`.

EXAMPLE 24

To normalize $(3,4,12)$ with respect to the Euclidean inner product we type

```
<< LinearAlgebra`Orthogonalization`
Normalize[{3, 4, 12}]
```

$$\left\{\frac{3}{13}, \frac{4}{13}, \frac{12}{13}\right\}$$

To normalize with respect to the inner product $<\mathbf{a}, \mathbf{b}> = 2a_1 b_1 + 3a_2 b_2 + 4a_3 b_3$,

```
Normalize[{3, 4, 12}, InnerProduct → ({2, 3, 4} . (#1 * #2) &)]
```

$$\left\{ \sqrt{\frac{3}{214}}, \; 2\sqrt{\frac{2}{321}}, \; 2\sqrt{\frac{6}{107}} \right\}$$

EXAMPLE 25

Find an orthonormal basis for the space spanned by $(1, 1, 1, 0, 0)$, $(0, 1, 1, 1, 0)$, and $(0, 0, 1, 1, 1)$ and verify that the result is correct.

```
v = {{1, 1, 1, 0, 0}, {0, 1, 1, 1, 0}, {0, 0, 1, 1, 1}};
w = GramSchmidt[v]
```

$$\left\{ \left\{ \frac{1}{\sqrt{3}}, \frac{1}{\sqrt{3}}, \frac{1}{\sqrt{3}}, 0, 0 \right\}, \left\{ -\frac{2}{\sqrt{15}}, \frac{1}{\sqrt{15}}, \frac{1}{\sqrt{15}}, \sqrt{\frac{3}{5}}, 0 \right\}, \left\{ \frac{1}{2\sqrt{10}}, -\frac{3}{2\sqrt{10}}, \frac{1}{\sqrt{10}}, \frac{1}{2\sqrt{10}}, \frac{\sqrt{\frac{5}{2}}}{2} \right\} \right\}$$

To verify that the result is correct, we compute six dot products.

```
w[[1]].w[[1]]
```
1
```
w[[2]].w[[2]]
```
1
```
w[[3]].w[[3]]
```
1   The vectors are all unit length.
```
w[[1]].w[[2]]
```
0
```
w[[1]].w[[3]]
```
0
```
w[[2]].w[[3]]
```
0   Distinct vectors are orthogonal.

SOLVED PROBLEMS

12.29 Compute the norm of the vector $(1, 2, 3, 4, 5)$ with respect to (a) the Euclidean inner product and (b) $<\mathbf{u}, \mathbf{v}> = 2u_1 v_1 + 3u_2 v_2 + u_3 v_3 + 3u_4 v_4 + 2u_5 v_5$.

SOLUTION

(a)

```
u = {1, 2, 3, 4, 5};
norm = √u.u
```
$\sqrt{55}$

(b)
```
u = {1, 2, 3, 4, 5};
c = {2, 3, 1, 3, 2};
norm = √c.(u*u)
```
11

12.30 Find the projection of the vector (a, b, c) onto each of the coordinate axes.

SOLUTION

```
<< LinearAlgebra`Orthogonalization`
v = {a, b, c};
Projection[v, {1, 0, 0}]
```
{a, 0, 0}

```
Projection[v, {0, 1, 0}]
{0, b, 0}
Projection[v, {0, 0, 1}]
{0, 0, c}
```

12.31 Find a unit vector having the same direction as $(1, -2, 2, -3)$.

SOLUTION

```
<< LinearAlgebra`Orthogonalization`
Normalize[{1, -2, 2, -3}]
```

$$\left\{\frac{1}{3\sqrt{2}}, -\frac{\sqrt{2}}{3}, \frac{\sqrt{2}}{3}, -\frac{1}{\sqrt{2}}\right\}$$

12.32 If $\mathbf{a} = (1, 2, 3)$ and $\mathbf{b} = (1, -2, 5)$, compute the length of the vector $\mathbf{v}$ shown in the diagram below.

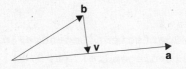

SOLUTION

Since $\mathbf{b} + \mathbf{v} = \text{proj}_a\mathbf{b}$, it follows that $\mathbf{v} = \text{proj}_a\mathbf{b} - \mathbf{b}$.

```
<< LinearAlgebra`Orthogonalization`
a = {1, 2, 3}; b = {1, -2, 5};
v = Projection[b, a] - b;
norm = Sqrt[v.v]
```

$$\sqrt{\frac{138}{7}}$$

12.33 Find an orthogonal basis for the space spanned by $(1, 2, 1, 3)$, $(2, 2, 2, 2)$, $(1, -1, 1, -1)$, and $(3, 4, 3, 5)$.

SOLUTION

Since we require only orthogonal vectors, the option **Normalized → False** is used in **GramSchmidt**.

```
<< LinearAlgebra`Orthogonalization`
v1 = {1, 2, 1, 3};
v2 = {2, 2, 2, 2};
v3 = {1, -1, 1, -1};
v4 = {3, 4, 3, 5};
v = {v1, v2, v3, v4};
w = GramSchmidt[v, Normalized → False]
```

$$\left\{\{1, 2, 1, 3\}, \left\{\frac{16}{15}, \frac{2}{15}, \frac{16}{15}, -\frac{4}{5}\right\},\right.$$

$$\left.\left\{\frac{2}{11}, -\frac{8}{11}, \frac{2}{11}, \frac{4}{11}\right\}, \{0, 0, 0, 0\}\right\}$$

> The set v was linearly dependent so
> GramSchmidt can only produce three
> basis vectors. {0, 0, 0, 0} can be
> disregarded.

12.34 Construct an orthonormal basis for P_5, the space of all polynomials of degree 5 or less with respect to the inner product $<\mathbf{p}, \mathbf{q}> = \int_{-1}^{1} p(x) q(x) \, dx$.

SOLUTION

One basis for P_5 is the set $v = \{1, x, x^2, x^3, x^4, x^5\}$. The vectors form a linearly independent set which spans P_5.

```
<< LinearAlgebra`Orthogonalization`
v = {1, x, x², x³, x⁴, x⁵};
GramSchmidt[v, InnerProduct → (Integrate[#1 #2, {x, -1, 1}] &),
         Normalized → False] // Expand
```

$$\left\{ 1, x, -\frac{1}{3} + x^2, -\frac{3x}{5} + x^3, \frac{3}{35} - \frac{6x^2}{7} + x^4, \frac{5x}{21} - \frac{10x^3}{9} + x^5 \right\}$$

Note: The functions obtained in this problem are the Legendre polynomials, normalized so that their leading coefficients are 1. These functions are represented in *Mathematica* by `LegendreP`.

```
normalizedlegendre :=
  LegendreP[n, x]/Coefficient[LegendreP[n, x], x, n]
  Table[{n, normalizedlegendre}, {n, 0, 5}] // Expand // TableForm
```

0	1
1	x
2	$-\frac{1}{3} + x^2$
3	$-\frac{3x}{5} + x^3$
4	$\frac{3}{35} - \frac{6x^2}{7} + x^4$
5	$\frac{5x}{21} - \frac{10x^3}{9} + x^5$

12.6 EIGENVALUES AND EIGENVECTORS

λ is said to be an eigenvalue of a square matrix A if there exists a nonzero vector $\mathbf{x}$ such that $A\mathbf{x} = \lambda\mathbf{x}$. As powerful as the eigenvalue concept is in linear algebra, the computation of eigenvalues and their corresponding eigenvectors can be extremely difficult if the matrix is large.

One way to determine the eigenvalues of a matrix is to solve the characteristic equation $\det(A - \lambda I) = 0$. Once the eigenvalues are determined, the eigenvectors can be found by solving a homogeneous linear system.

EXAMPLE 26

$$a = \begin{pmatrix} 4 & 1 & -1 \\ 2 & 5 & -2 \\ 1 & 1 & 2 \end{pmatrix};$$

```
len = Length[a];
Solve[Det[a - λ IdentityMatrix[len]] == 0, λ]
{{λ → 3}, {{λ → 3}, {λ → 5}}
```

The eigenvalues are 3, with multiplicity 2, and 5. To find the eigenvectors, we look at the null space of $A - \lambda I$:

```
NullSpace[a - 3 IdentityMatrix[len]]
{{1, 0, 1}, {-1, 1, 0}}
NullSpace[a - 5 IdentityMatrix[len]]
{{1, 2, 1}}
```

Of course, as one might expect, *Mathematica* contains commands which automatically compute eigenvalues, eigenvectors, and some other related items.

- **CharacteristicPolynomial[*matrix*, *var*]** returns the characteristic polynomial of *matrix* expressed in terms of variable *var*.
- **Eigenvalues[*matrix*]** returns a list of the eigenvalues of *matrix*.
- **Eigenvectors[*matrix*]** returns a list of the eigenvectors of *matrix*.
- **Eigensystem[*matrix*]** returns a list of the form {*eigenvalues*, *eigenvectors*}.

EXAMPLE 27

$$a = \begin{pmatrix} 4 & 1 & -1 \\ 2 & 5 & -2 \\ 1 & 1 & 2 \end{pmatrix};$$

```
CharacteristicPolynomial[a, x]
45 - 39 x + 11 x² - x³
Eigenvalues[a]
{3, 3, 5}
Eigenvectors[a]
{{1, 0, 1}, {-1, 1, 0}, {1, 2, 1}}
Eigensystem[a]
{{3, 3, 5}, {{1, 0, 1}, {-1, 1, 0}, {1, 2, 1}}}
```

If the entries of the matrix are expressed exactly, i.e., in nondecimal form, *Mathematica* tries to determine the eigenvalues and eigenvectors exactly. If any of the entries of the matrix are expressed in decimal form, *Mathematica* returns decimal approximations. Alternatively, one can use **N[*matrix*]** as the argument of CharacteristicPolynomial, Eigenvalue, Eigenvectors, and Eigensystem to force *Mathematica* to use decimal algorithms. If higher precision is desired, **N[*matrix*, k]** will return k significant digits.

Note: Because different algorithms are used for computing numerical eigenvalues, they sometimes emerge in a different order. Furthermore, since eigenvectors are not uniquely determined, the numerical eigenvectors may appear to be different than those obtained previously. However, they are valid eigenvectors.

EXAMPLE 28

$$a = \begin{pmatrix} 1 & 1 \\ 1 & 3 \end{pmatrix};$$

```
Eigenvalues[a]
```
$\left\{ 2 - \sqrt{2}, 2 + \sqrt{2} \right\}$
```
Eigenvectors[a]
```
$\left\{ \left\{ -1 - \sqrt{2}, 1 \right\}, \left\{ -1 + \sqrt{2}, 1 \right\} \right\}$

```
Eigenvalues[N[a]]
{3.41421, 0.585786}
Eigenvectors[N[a]]
{{-0.382683, -0.92388}, {-0.92388, 0.382683}}
Eigenvalues[N[a, 20]]
{3.4142135623730950488, 0.58578643762690495120}
Eigenvectors[N[a, 20]]
{{-0.38268343236508977173, -0.92387953251128675613},
   {-0.92387953251128675613, 0.38268343236508977173}}
```

SOLVED PROBLEMS

12.35 What is the characteristic polynomial of the matrix A, whose entries are the first 25 consecutive integers?

SOLUTION

```
a = Partition[Range[25], 5];
a // MatrixForm
```

$$\begin{pmatrix} 1 & 2 & 3 & 4 & 5 \\ 6 & 7 & 8 & 9 & 10 \\ 11 & 12 & 13 & 14 & 15 \\ 16 & 17 & 18 & 19 & 20 \\ 21 & 22 & 23 & 24 & 25 \end{pmatrix}$$

```
CharacteristicPolynomial[a, x]
```

$250 x^3 + 65 x^4 - x^5$

12.36 Consider the tridiagonal 5×5 matrix whose main diagonal entries are 4, with 1's on the adjacent diagonals. Show the eigenvalues and corresponding eigenvectors in a clear, unambiguous manner.

SOLUTION

```
m = Table[If[Abs[i - j] == 1, 1, If[i == j, 4, 0]], {i, 1, 5}, {j, 1, 5}];
m // MatrixForm
```

$$\begin{pmatrix} 4 & 1 & 0 & 0 & 0 \\ 1 & 4 & 1 & 0 & 0 \\ 0 & 1 & 4 & 1 & 0 \\ 0 & 0 & 1 & 4 & 1 \\ 0 & 0 & 0 & 1 & 4 \end{pmatrix}$$

```
data = Eigensystem[m]
```

$$\left\{\left\{3, 4, 5, 4 - \sqrt{3}, 4 + \sqrt{3}\right\}, \left\{\{-1, 1, 0, -1, 1\}, \{1, 0, -1, 0, 1\}, \{-1, -1, 0, 1, 1\}, \right.\right.$$
$$\left.\left.\left\{1, -\sqrt{3}, 2, -\sqrt{3}, 1\right\}, \left\{1, \sqrt{3}, 2, \sqrt{3}, 1\right\}\right\}\right\}$$

```
Do[Print["eigenvalue #", k, " is ", data[[1, k]],
  " with corresponding eigenvector: ", data[[2, k]]], {k, 1, 5}]
eigenvalue #1 is 3 with corresponding eigenvector : {-1, 1, 0, -1, 1}
eigenvalue #2 is 4 with corresponding eigenvector : {1, 0, -1, 0, 1}
```

eigenvalue #3 is 5 with corresponding eigenvector : {-1, -1, 0, 1, 1}

eigenvalue #4 is $4 - \sqrt{3}$ with corresponding eigenvector : $\left\{1, -\sqrt{3}, 2, -\sqrt{3}, 1\right\}$

eigenvalue #5 is $4 + \sqrt{3}$ with corresponding eigenvector : $\left\{1, \sqrt{3}, 2, \sqrt{3}, 1\right\}$

12.37 An important theorem in linear algebra, the *Cayley-Hamilton theorem*, says that every square matrix satisfies its characteristic equation. Verify the Cayley-Hamilton theorem for

$$A = \begin{pmatrix} 1 & 2 & 1 & 2 \\ 3 & -1 & 3 & -1 \\ 2 & 5 & 7 & 1 \\ 1 & 2 & 3 & 6 \end{pmatrix}$$

SOLUTION

```
a = 
```
$$\begin{pmatrix} 1 & 2 & 1 & 2 \\ 3 & -1 & 3 & -1 \\ 2 & 5 & 7 & 1 \\ 1 & 2 & 3 & 6 \end{pmatrix};$$

```
CharacteristicPolynomial[a, x]
```
$-196 + 161\,x + 15\,x^2 - 13\,x^3 + x^4$

```
-196 IdentityMatrix[4] + 161 a + 15 MatrixPower[a, 2]
  -13 MatrixPower[a, 3] + MatrixPower[a, 4] // MatrixForm
```

$$\begin{pmatrix} 0 & 0 & 0 & 0 \\ 0 & 0 & 0 & 0 \\ 0 & 0 & 0 & 0 \\ 0 & 0 & 0 & 0 \end{pmatrix}$$

12.38 Approximate the eigenvalues of the 10×10 Hilbert matrix: $h_{i,j} = \dfrac{1}{i + j - 1}$.

SOLUTION

```
n = 10;
hilbert = Table[1/(i + j - 1), {i, 1, n}, {j, 1, n}];
hilbert // MatrixForm
```

$$\begin{pmatrix}
1 & \frac{1}{2} & \frac{1}{3} & \frac{1}{4} & \frac{1}{5} & \frac{1}{6} & \frac{1}{7} & \frac{1}{8} & \frac{1}{9} & \frac{1}{10} \\
\frac{1}{2} & \frac{1}{3} & \frac{1}{4} & \frac{1}{5} & \frac{1}{6} & \frac{1}{7} & \frac{1}{8} & \frac{1}{9} & \frac{1}{10} & \frac{1}{11} \\
\frac{1}{3} & \frac{1}{4} & \frac{1}{5} & \frac{1}{6} & \frac{1}{7} & \frac{1}{8} & \frac{1}{9} & \frac{1}{10} & \frac{1}{11} & \frac{1}{12} \\
\frac{1}{4} & \frac{1}{5} & \frac{1}{6} & \frac{1}{7} & \frac{1}{8} & \frac{1}{9} & \frac{1}{10} & \frac{1}{11} & \frac{1}{12} & \frac{1}{13} \\
\frac{1}{5} & \frac{1}{6} & \frac{1}{7} & \frac{1}{8} & \frac{1}{9} & \frac{1}{10} & \frac{1}{11} & \frac{1}{12} & \frac{1}{13} & \frac{1}{14} \\
\frac{1}{6} & \frac{1}{7} & \frac{1}{8} & \frac{1}{9} & \frac{1}{10} & \frac{1}{11} & \frac{1}{12} & \frac{1}{13} & \frac{1}{14} & \frac{1}{15} \\
\frac{1}{7} & \frac{1}{8} & \frac{1}{9} & \frac{1}{10} & \frac{1}{11} & \frac{1}{12} & \frac{1}{13} & \frac{1}{14} & \frac{1}{15} & \frac{1}{16} \\
\frac{1}{8} & \frac{1}{9} & \frac{1}{10} & \frac{1}{11} & \frac{1}{12} & \frac{1}{13} & \frac{1}{14} & \frac{1}{15} & \frac{1}{16} & \frac{1}{17} \\
\frac{1}{9} & \frac{1}{10} & \frac{1}{11} & \frac{1}{12} & \frac{1}{13} & \frac{1}{14} & \frac{1}{15} & \frac{1}{16} & \frac{1}{17} & \frac{1}{18} \\
\frac{1}{10} & \frac{1}{11} & \frac{1}{12} & \frac{1}{13} & \frac{1}{14} & \frac{1}{15} & \frac{1}{16} & \frac{1}{17} & \frac{1}{18} & \frac{1}{19}
\end{pmatrix}$$

```
Eigenvalues[N[hilbert]]
```
{1.75192 , 0.34293 , 0.0357418 , 0.00253089 ,
 0.00012875 , 4.72969×10^{-6} , 1.22897×10^{-7} ,
 2.14744×10^{-9} , 2.26675×10^{-11} , 1.09325×10^{-13}}

12.39 Approximate the eigenvalues of the 10×10 matrix A such that

$$a_{i,j} = \begin{cases} i+j-1 & \text{if } i+j \le 11 \\ 21-i-j & \text{if } i+j > 11 \end{cases}$$

SOLUTION

```
f[i_, j_] := i + j - 1 /; i + j ≤ 11
f[i_, j_] := 21 - i - j /; i + j > 11
a = Array[f, {10, 10}];
a // MatrixForm
```

$$\begin{pmatrix} 1 & 2 & 3 & 4 & 5 & 6 & 7 & 8 & 9 & 10 \\ 2 & 3 & 4 & 5 & 6 & 7 & 8 & 9 & 10 & 9 \\ 3 & 4 & 5 & 6 & 7 & 8 & 9 & 10 & 9 & 8 \\ 4 & 5 & 6 & 7 & 8 & 9 & 10 & 9 & 8 & 7 \\ 5 & 6 & 7 & 8 & 9 & 10 & 9 & 8 & 7 & 6 \\ 6 & 7 & 8 & 9 & 10 & 9 & 8 & 7 & 6 & 5 \\ 7 & 8 & 9 & 10 & 9 & 8 & 7 & 6 & 5 & 4 \\ 8 & 9 & 10 & 9 & 8 & 7 & 6 & 5 & 4 & 3 \\ 9 & 10 & 9 & 8 & 7 & 6 & 5 & 4 & 3 & 2 \\ 10 & 9 & 8 & 7 & 6 & 5 & 4 & 3 & 2 & 1 \end{pmatrix}$$

```
Eigenvalues[N[a]]
```
{67.8404, -20.4317, 4.45599, -2.42592, 1.39587,
   -1., 0.756101, -0.629808, 0.55164, -0.512543}

12.7 DIAGONALIZATION AND JORDAN CANONICAL FORM

Given an $n \times n$ matrix A, if there exists a matrix P such that $A = PDP^{-1}$, where D is a diagonal matrix, we say that A is *diagonalizable*.

Not every matrix is diagonalizable. However, it can be shown that if A has a *complete* set of eigenvectors, i.e., a set of n linearly independent eigenvectors, then A is diagonalizable. P is the matrix whose columns are the eigenvectors of A, and D is the diagonal matrix whose main diagonal entries are their respective eigenvalues.

EXAMPLE 29

$$a = \begin{pmatrix} 18 & -51 & 27 & -15 \\ 8 & -24 & 14 & -8 \\ 15 & -48 & 28 & -15 \\ 15 & -47 & 25 & -12 \end{pmatrix};$$

```
Eigenvalues[a]
```
          ← Since the eigenvalues are distinct, the corresponding
{1, 2, 3, 4}   eigenvectors will be linearly independent.

```
Eigenvectors[a]
```
{{3, 2, 3, 2}, {0, 1, 3, 2}, {1, 0, 0, 1}, {3, 1, 2, 3}}

p = Transpose[%]   ← The transpose makes the eigenvectors columns.

{{3, 0, 1, 3}, {2, 1, 0, 1}, {3, 3, 0, 2}, {2, 2, 1, 3}}

d = DiagonalMatrix[Eigenvalues[a]]

{{1, 0, 0, 0}, {0, 2, 0, 0}, {0, 0, 3, 0}, {0, 0, 0, 4}}

p.d.Inverse[p] // MatrixForm

$$\begin{pmatrix} 18 & -51 & 27 & -15 \\ 8 & -24 & 14 & -8 \\ 15 & -48 & 28 & -15 \\ 15 & -47 & 25 & -12 \end{pmatrix}$$ ← p.d.Inverse[p] = a

To summarize,

MatrixForm[a] == MatrixForm[p].MatrixForm[d].MatrixForm[Inverse[p]]

$$\begin{pmatrix} 18 & -51 & 27 & -15 \\ 8 & -24 & 14 & -8 \\ 15 & -48 & 28 & -15 \\ 15 & -47 & 25 & -12 \end{pmatrix} == \begin{pmatrix} 3 & 0 & 1 & 3 \\ 2 & 1 & 0 & 1 \\ 3 & 3 & 0 & 2 \\ 2 & 2 & 1 & 3 \end{pmatrix} \cdot \begin{pmatrix} 1 & 0 & 0 & 0 \\ 0 & 2 & 0 & 0 \\ 0 & 0 & 3 & 0 \\ 0 & 0 & 0 & 4 \end{pmatrix} \cdot \begin{pmatrix} -1 & 4 & -2 & 1 \\ -1 & 2 & -1 & 1 \\ -5 & 15 & -9 & 6 \\ 3 & -9 & 5 & -3 \end{pmatrix}$$

Unfortunately, not every matrix can be diagonalized. However, there is a form, called the *Jordan canonical form*, that every matrix possesses.

A Jordan block is a square matrix whose elements are zero except for the main diagonal where all numbers are equal, and the superdiagonal, where all values are 1:

$$\begin{pmatrix} \lambda & 1 & 0 & 0 & .. & 0 \\ 0 & \lambda & 1 & 0 & .. & 0 \\ 0 & 0 & \lambda & 1 & .. & 0 \\ . & . & . & . & .. & . \\ 0 & 0 & 0 & 0 & .. & \lambda \end{pmatrix}$$

If A is any $n \times n$ matrix, there exists a matrix Q such that $A = QJQ^{-1}$ and

$$J = \begin{pmatrix} J_1 & 0 & 0 & .. & 0 \\ 0 & J_2 & 0 & .. & 0 \\ 0 & 0 & J_3 & .. & 0 \\ . & . & . & .. & . \\ 0 & 0 & 0 & .. & J_k \end{pmatrix}$$

The J_i's are Jordan blocks. The same eigenvalue may occur in different blocks. The number of distinct blocks corresponding to a given eigenvalue is equal to the number of independent eigenvectors corresponding to that eigenvalue.

EXAMPLE 30

$$a = \begin{pmatrix} 5 & 4 & 3 \\ -1 & 0 & -3 \\ 1 & -2 & 1 \end{pmatrix};$$

Eigensystem[a]

{{-2, 4, 4}, {{-1, 1, 1}, {1, -1, 1}, {0, 0, 0}}}

The eigenvalues are -2 and 4 with eigenvectors, respectively, $(-1, 1, 1)$ and $(1, -1, 1)$. The vector {0, 0, 0} is not an eigenvector; its presence simply indicates a third linearly independent eigenvector cannot be found. To construct Q, the standard procedure is to find a vector **x** such that $(A - 4\,\mathrm{I})\,\mathbf{x} = (1, -1, 1)$. **x** is called a *generalized* eigenvector.

```
LinearSolve[a - 4IdentityMatrix[3], {1, -1, 1}]
```
{1, 0, 0}

The matrices Q and J may now be constructed:

$$q = \begin{pmatrix} -1 & 1 & 1 \\ 1 & -1 & 0 \\ 1 & 1 & 0 \end{pmatrix};$$

> The columns of q are the eigenvectors and generalized eigenvectors of a.

$$j = \begin{pmatrix} -2 & 0 & 0 \\ 0 & 4 & 1 \\ 0 & 0 & 4 \end{pmatrix};$$

```
q.j.Inverse[q] // MatrixForm
```

$$\begin{pmatrix} 5 & 4 & 3 \\ -1 & 0 & -3 \\ 1 & -2 & 1 \end{pmatrix}$$

- **JordanDecomposition[*matrix*]** computes the Jordan canonical form of *matrix*. The output is a list {q, j} where q and j correspond to Q and J as described above.

EXAMPLE 31

$$a = \begin{pmatrix} 5 & 4 & 3 \\ -1 & 0 & -3 \\ 1 & -2 & 1 \end{pmatrix};$$

```
{q, j} = JordanDecomposition[a];
q // MatrixForm
```

$$\begin{pmatrix} -1 & 1 & 1 \\ 1 & -1 & 0 \\ 1 & 1 & 0 \end{pmatrix}$$

```
j // MatrixForm
```

$$\begin{pmatrix} -2 & 0 & 0 \\ 0 & 4 & 1 \\ 0 & 0 & 4 \end{pmatrix}$$ Of course this agrees with the results obtained in Example 30.

EXAMPLE 32

$$a = \begin{pmatrix} 18 & -51 & 27 & -15 \\ 8 & -24 & 14 & -8 \\ 15 & -48 & 28 & -15 \\ 15 & -47 & 25 & -12 \end{pmatrix};$$

```
{q, j} = JordanDecomposition[a];
q // MatrixForm
```

$$\begin{pmatrix} 3 & 0 & 1 & 3 \\ 2 & 1 & 0 & 1 \\ 3 & 3 & 0 & 2 \\ 2 & 2 & 1 & 3 \end{pmatrix}$$

`j // MatrixForm`

$$\begin{pmatrix} 1 & 0 & 0 & 0 \\ 0 & 2 & 0 & 0 \\ 0 & 0 & 3 & 0 \\ 0 & 0 & 0 & 4 \end{pmatrix}$$

> In this example the Jordan matrix reduces to a diagonal matrix since a has a complete set of eigenvectors.

Other decompositions such as **QR** decomposition and Schur decomposition are available in *Mathematica*, but shall not be discussed in this book. Their respective command names are `QRDecomposition` and `SchurDecomposition`.

SOLVED PROBLEMS

12.40 Construct a 5×5 matrix whose eigenvalues are $-2, -1, 0, 1,$ and 2 with respective eigenvectors $(1, 1, 0, 0, 0)$, $(0, 1, 1, 0, 0)$, $(0, 0, 1, 1, 0)$, $(0, 0, 0, 1, 1)$, and $(1, 0, 0, 0, 1)$.

SOLUTION

```
d = DiagonalMatrix[{ - 2, - 1, 0, 1, 2}];
p = Transpose[{{1, 1, 0, 0, 0}, {0, 1, 1, 0, 0}, {0, 0, 1, 1, 0},
          {0, 0, 0, 1, 1}, {1, 0, 0, 0, 1}}];
a = p.d.Inverse[p];
a // MatrixForm
```

$$\begin{pmatrix} 0 & -2 & 2 & -2 & 2 \\ -\frac{1}{2} & -\frac{3}{2} & \frac{1}{2} & -\frac{1}{2} & \frac{1}{2} \\ \frac{1}{2} & -\frac{1}{2} & -\frac{1}{2} & \frac{1}{2} & -\frac{1}{2} \\ -\frac{1}{2} & \frac{1}{2} & -\frac{1}{2} & \frac{1}{2} & \frac{1}{2} \\ \frac{1}{2} & -\frac{1}{2} & \frac{1}{2} & -\frac{1}{2} & \frac{3}{2} \end{pmatrix}$$

12.41 Show that the matrix $\begin{pmatrix} a & b \\ c & d \end{pmatrix}$ has real eigenvalues if and only if $a^2 + 4bc - 2ad + d^2 \geq 0$.

SOLUTION

```
m = {{a, b}, {c, d}};
Eigenvalues[{{a, b}, {c, d}}]
```

$$\left\{ \frac{1}{2} \left(a+d - \sqrt{a^2 + 4bc - 2ad + d^2} \right), \frac{1}{2} \left(a+d + \sqrt{a^2 + 4bc - 2ad + d^2} \right) \right\}$$

The eigenvalues will be real if and only if the expression inside the radical symbol is nonnegative.

12.42 Construct a 7×7 matrix of random digits and show that the sum of its eigenvalues is equal to its trace and the product of its eigenvalues is equal to its determinant.

SOLUTION

```
a = Table[Random[Integer, {0, 9}], {i, 1, 7}, {j, 1, 7}];
a // MatrixForm
```

$$\begin{pmatrix} 6 & 8 & 7 & 0 & 9 & 5 & 3 \\ 5 & 6 & 3 & 6 & 9 & 0 & 4 \\ 5 & 3 & 3 & 0 & 0 & 1 & 9 \\ 1 & 3 & 9 & 9 & 9 & 9 & 5 \\ 5 & 0 & 2 & 8 & 4 & 1 & 3 \\ 9 & 5 & 7 & 4 & 9 & 9 & 3 \\ 1 & 3 & 1 & 5 & 5 & 1 & 3 \end{pmatrix}$$

```
eigen = Eigenvalues[N[a]]
```
{31.4349 , 6.08322 + 4.84653 i, 6.08322 − 4.84653 i, −6.43529 ,
 1.52463 + 3.69404 i, 1.52463 − 3.69404 i, −0.215269 }

$$\sum_{i=1}^{7} \texttt{eigen[[i]]} \quad \texttt{or} \quad \texttt{Sum[eigen[[i]], \{i, 1, 7\}]}$$

40. + 0. i
```
Tr[a]
```
40

$$\prod_{i=1}^{7} \texttt{eigen[[i]]} \quad \texttt{or} \quad \texttt{Product[eigen[[i]], \{i, 1, 7\}]}$$

42072. + 0. i
```
Det[a]
```
42072

12.43 A matrix P is said to be orthogonal if $P^{\mathrm{T}}P = \mathrm{I}$. (The inverse of an orthogonal matrix is its transpose.) If it is possible to find an orthogonal matrix P which diagonalizes A, then A is said to be *orthogonally diagonalizable*. Only symmetric matrices are orthogonally diagonalizable. (A matrix is symmetric if $A^{\mathrm{T}} = A$. If A is symmetric, it can be shown that the eigenvectors corresponding to distinct eigenvalues are orthogonal.) Find a matrix P which orthogonally diagonalizes

$$A = \begin{pmatrix} 3 & 1 & 0 & 0 & 0 \\ 1 & 3 & 0 & 0 & 0 \\ 0 & 0 & 2 & 1 & 1 \\ 0 & 0 & 1 & 2 & 1 \\ 0 & 0 & 1 & 1 & 2 \end{pmatrix}$$

SOLUTION

```
<< LinearAlgebra`Orthogonalization`
```

$$a = \begin{pmatrix} 3 & 1 & 0 & 0 & 0 \\ 1 & 3 & 0 & 0 & 0 \\ 0 & 0 & 2 & 1 & 1 \\ 0 & 0 & 1 & 2 & 1 \\ 0 & 0 & 1 & 1 & 2 \end{pmatrix};$$

```
{values, vectors} = Eigensystem[a]
```
```
{{1, 1, 2, 4, 4}, {{0, 0, -1, 0, 1}, {0, 0, -1, 1, 0},
   {-1, 1, 0, 0, 0}, {0, 0, 1, 1, 1}, {1, 1, 0, 0, 0}}}
```
```
eigenspace1 = {vectors[[1]], vectors[[2]]};
```
```
eigenspace2 = {vectors[[3]]};
```
```
eigenspace3 = {vectors[[4]], vectors[[5]]};
```
```
v1 = GramSchmidt[eigenspace1];
```
```
v2 = GramSchmidt[eigenspace2];
```
```
v3 = GramSchmidt[eigenspace3];
```
```
<<LinearAlgebra`MatrixManipulation`
```
```
orthogonalmatrix = Transpose[AppendColumns[v1, v2, v3]];
```
```
orthogonalmatrix // MatrixForm
```

There are five eigenvectors, two of which have multiplicity 2. We group them into three eigenspaces and apply the Gram-Schmidt process to each. The orthogonal matrix is the matrix whose columns are the orthogonal vectors from Gram-Schmidt.

$$
\begin{pmatrix}
0 & 0 & -\dfrac{1}{\sqrt{2}} & 0 & \dfrac{1}{\sqrt{2}} \\[2mm]
0 & 0 & \dfrac{1}{\sqrt{2}} & 0 & \dfrac{1}{\sqrt{2}} \\[2mm]
-\dfrac{1}{\sqrt{2}} & -\dfrac{1}{\sqrt{6}} & 0 & \dfrac{1}{\sqrt{3}} & 0 \\[2mm]
0 & \sqrt{\dfrac{2}{3}} & 0 & \dfrac{1}{\sqrt{3}} & 0 \\[2mm]
\dfrac{1}{\sqrt{2}} & -\dfrac{1}{\sqrt{6}} & 0 & \dfrac{1}{\sqrt{3}} & 0
\end{pmatrix}
$$

```
p = orthogonalmatrix;
Transpose[p].p // MatrixForm
```

$$
\begin{pmatrix}
1 & 0 & 0 & 0 & 0 \\
0 & 1 & 0 & 0 & 0 \\
0 & 0 & 1 & 0 & 0 \\
0 & 0 & 0 & 1 & 0 \\
0 & 0 & 0 & 0 & 1
\end{pmatrix}
$$

← Just a check to see if the matrix is orthogonal.

```
d = DiagonalMatrix[values];
p.d.Transpose[p] // MatrixForm
```

$$
\begin{pmatrix}
3 & 1 & 0 & 0 & 0 \\
1 & 3 & 0 & 0 & 0 \\
0 & 0 & 2 & 1 & 1 \\
0 & 0 & 1 & 2 & 1 \\
0 & 0 & 1 & 1 & 2
\end{pmatrix}
$$

← This gives us back our original matrix.

```
MatrixForm[a] == MatrixForm[p].MatrixForm[d].MatrixForm[Transpose[p]]
```

$$\begin{pmatrix} 3 & 1 & 0 & 0 & 0 \\ 1 & 3 & 0 & 0 & 0 \\ 0 & 0 & 2 & 1 & 1 \\ 0 & 0 & 1 & 2 & 1 \\ 0 & 0 & 1 & 1 & 2 \end{pmatrix} =$$

$$\begin{pmatrix} 0 & 0 & -\dfrac{1}{\sqrt{2}} & 0 & \dfrac{1}{\sqrt{2}} \\[2mm] 0 & 0 & \dfrac{1}{\sqrt{2}} & 0 & \dfrac{1}{\sqrt{2}} \\[2mm] -\dfrac{1}{\sqrt{2}} & -\dfrac{1}{\sqrt{6}} & 0 & \dfrac{1}{\sqrt{3}} & 0 \\[2mm] 0 & \sqrt{\dfrac{2}{3}} & 0 & \dfrac{1}{\sqrt{3}} & 0 \\[2mm] \dfrac{1}{\sqrt{2}} & -\dfrac{1}{\sqrt{6}} & 0 & \dfrac{1}{\sqrt{3}} & 0 \end{pmatrix} \cdot \begin{pmatrix} 1 & 0 & 0 & 0 & 0 \\ 0 & 1 & 0 & 0 & 0 \\ 0 & 0 & 2 & 0 & 0 \\ 0 & 0 & 0 & 4 & 0 \\ 0 & 0 & 0 & 0 & 4 \end{pmatrix} \cdot \begin{pmatrix} 0 & 0 & -\dfrac{1}{\sqrt{2}} & 0 & \dfrac{1}{\sqrt{2}} \\[2mm] 0 & 0 & -\dfrac{1}{\sqrt{6}} & \sqrt{\dfrac{2}{3}} & -\dfrac{1}{\sqrt{6}} \\[2mm] -\dfrac{1}{\sqrt{2}} & \dfrac{1}{\sqrt{2}} & 0 & 0 & 0 \\[2mm] 0 & 0 & \dfrac{1}{\sqrt{3}} & \dfrac{1}{\sqrt{3}} & \dfrac{1}{\sqrt{3}} \\[2mm] \dfrac{1}{\sqrt{2}} & \dfrac{1}{\sqrt{2}} & 0 & 0 & 0 \end{pmatrix}$$

APPENDIX

A.1 PURE FUNCTIONS

A function is a correspondence between two sets of numbers, A and B, such that for each number in A there corresponds a unique number in B. For example, the "squaring" function: for each real number there corresponds a unique nonnegative real number called its square.

Although it is customary to write $f(x) = x^2$, one must understand that there is no special significance to the letter x; it is the process of *squaring* which defines the function.

Although *Mathematica* allows a function to be defined in terms of a variable, such as in $f[x\_] = x^2$, the variable x acts as a "dummy" variable and is insignificant. The function would be the same had we used y, z, or any other symbol.

A "pure" function is defined without a reference to any specific variable. Its arguments are labeled successively #1, #2, #3, and so forth. To distinguish a pure function from any other *Mathematica* construct, an ampersand, &, is used at the end of its definition. Once defined, we can deal with a pure function as we would any other function.

EXAMPLE 1

```
f =#1²&;
f[3]
9
f[x]
x²
f[a+b]
(a+b)²
```

EXAMPLE 2

```
g = #1#2² + 3 &;
g[3, 4]
51
g[u, v]
3 + u v²
```

Although the concept of a pure function is a natural one, it is possible to use *Mathematica* and never be concerned with it. Occasionally, however, *Mathematica* will express an answer as a pure function and it is therefore worthy of a brief mention at this point. The interested reader can find more information in *The Mathematica Book* by Stephen Wolfram or the Help menu.

SOLVED PROBLEMS

A.1 Express as a pure function the process of adding the square of a number to its square root and compute its value at 9.

SOLUTION

```
f = #1^2 + Sqrt[#1] &;
f[9]
84
```

A.2   A number is formed from two other numbers by adding the square of their sum to the sum of their squares.   Express this operation as a pure function and compute its value for the given numbers 3 and 4.

SOLUTION

```
g = (#1 + #2)^2 + #1^2 + #2^2 &;
g[3, 4]
74
```

A.3   Express the derivative of the function Sin as a pure function and compute its value at $\pi/6$.

SOLUTION

```
f = Sin'
Cos[#1] &
f[π/6]
```
$$\frac{\sqrt{3}}{2}$$

A.4   Define $f(x) = (1 + x + x^2)^5$ and express its second derivative as a pure function.

SOLUTION

```
f[x_] = (1 + x + x²)⁵;
f''
20 (1 + 2 #1)² (1 + #1 + #1²)³ + 10 (1 + #1 + #1²)⁴ &
```

A.5   Express the solution of the differential equation

$$\frac{d^2y}{dx^2} + y = 0,\ y'(0) = y(0) = 1$$

as a pure function and evaluate it for $x = \pi/2$.

SOLUTION

```
DSolve[{y''[x] + y[x] == 0, y[0] == 1, y'[0] == 1}, y, x]
{{y → (Cos[#1] + Sin[#1] &)}}
solution = %[[1, 1, 2]]
Cos[#1] + Sin[#1] &
solution[π/4]
```
$$\sqrt{2}$$

A.2   CONTEXTS

It is common practice to define symbols using names that are reminiscent of the symbol's purpose.   Sometimes, however, the names get unwieldy and cumbersome to work with.   *Contexts* are used as a tool to help organize the symbols used in a *Mathematica* session.

The complete name of a symbol is divided into two parts, a context and a shorter name, separated by a backquote character (`). Used for this purpose the backquote is called a *context mark*.

EXAMPLE 3

atomicnumber`au and atomicweight`au are two distinct symbols with a common short name, au (the chemical symbol for gold).

```
atomicnumber`au = 79;
atomicweight`au = 196.967;
atomicnumber`au
79
atomicweight`au
196.967
```

When you begin a *Mathematica* session, the default context is Global`. Thus, for example, the symbol obj is equivalent to Global`obj. The default can be changed by redefining the symbol $Context.

- **$Context** is the current default context.
- **Context [*symbol*]** returns the context of *symbol*.

EXAMPLE 4

```
au = "gold";
atomicnumber`au = 79;
atomicweight`au = 196.967;
au
gold
$Context = "atomicnumber`";    ←Context names are strings; quotes are important.
au
79
$Context = "atomicweight`";
au
196.967
$Context = "Global`";
au
gold
```

"Built-in" *Mathematica* symbols have context System`.

EXAMPLE 5

```
Context[Pi]
System`
```

It is common for symbols in different contexts to have the same short name. If only the short name is referenced, *Mathematica* decides which is called by its position in a list called $ContextPath. The default $ContextPath is {Global`, System`}.

- **$ContextPath** is the current context search path.
- **Context[ ]** returns a list of all contexts in the current *Mathematica* session.

EXAMPLE 6

```
au = "gold";
atomicnumber`au = 79;
atomicweight`au = 196.967;
$ContextPath
```
{Global`, System`}   ←This is the default context path.
```
$ContextPath = Join[$ContextPath, {"atomicweight`"}, {"atomicnumber`"}]
```
{Global`, System`, atomicweight`, atomicnumber`}
```
au
```
gold   ←Global` is the first element of $ContextPath.
```
Remove[Global`au]
au
```
196.967   ←AtomicWeight` is now the first element of
                                        $ContextPath in which au appears.
```
Remove[atomicweight`au]
au
```
79   ←AtomicNumber` is now the first element of
                                        $ContextPath in which au appears.

A common error when using packages is to invoke the name of a command or function before the package is loaded. If the package is subsequently loaded, the command or function appears to be inaccessible.

EXAMPLE 7

The function DaysBetween, located in the package Miscellaneous`Calendar`, gives the number of days between two specified dates.

```
DaysBetween[{1999, 1, 1}, {2000, 1, 1}]
```                                     ←The function is called before the package is loaded.
```
DaysBetween[{1999, 1, 1}, {2000, 1, 1}]

<< Miscellaneous`Calendar`
```
DaysBetween::shdw : Symbol DaysBetween appears in multiple
    contexts {Miscellaneous`Calendar`, Global`}; definitions in context
    Miscellaneous`Calendar` may shadow or be shadowed by other definitions.
```
DaysBetween[{1999, 1, 1}, {2000, 1, 1}]
```
DaysBetween[{1999, 1, 1}, {2000, 1, 1}]

> *Mathematica* defaults to (undefined)
> Global`DaysBetween.

If the entire context path is specified, the correct calculation is performed:

```
Miscellaneous`Calendar`DaysBetween[{1999, 1, 1}, {2000, 1, 1}]
```
365

If Global`DaysBetween is removed, the function in Miscellaneous`Calendar` can be accessed:

```
Remove[DaysBetween]
DaysBetween[{1999, 1, 1}, {2000, 1, 1}]
```
365

A.3 PATTERNS

You have certainly noticed the use of the underscore (_) character when defining functions in *Mathematica*. The use of the underscore is an important concept in *Mathematica* called *pattern matching*.

A pattern is an expression such as x_ which contains an underscore character. The pattern can stand for any expression. Thus, f[x_] specifies how the function f should be applied to *any* argument. When you define a function such as f[x_] = x², you are telling *Mathematica* to automatically apply the transformation rule f[x_] → x² whenever possible.

In contrast, a transformation rule for f[x] without an underscore specifies only how the literal expression f[x] should be transformed, and does not say anything about the transformation of f[y], f[z], etc.

EXAMPLE 8

```
Clear[f]
f[x_] = x²;
f[x]
x²
f[y]
y²
f[a + b]
(a + b)²
```

| x_ is matched by any expression. |

```
Clear[f]
f[x] = x²;
f[x]
x²
f[y]
f[y]
f[a + b]
f(a + b)²
```

| x is matched only by x. |

EXAMPLE 9

```
1 + x^p + x^q /. x^q_ → Log[q]
1 + Log[p] + Log[q]
```
← All exponentials are transformed to Log.

```
1 + x^p + x^q /. x^q → Log[q]
1 + x^p + Log[q]
```
← Only x^q is transformed.

Patterns can specify the *type* of an expression as well as its format. For example, _Integer stands for an integer pattern. Similarly, _Rational, _Real, and _Complex are acceptable patterns representing other types of numbers.

EXAMPLE 10

The *Mathematica* function **Factorial[n]** computes n! if n is an integer and Γ[1 + n] if n is a positive real number. For certain applications it might be useful to leave the factorial of a noninteger undefined.

```
fact[n_Integer] = n!;
fact[n_Real] = "undefined";
fact[5]
120
fact[5.5]
undefined
```

```
Factorial[5]
120
Factorial[5.5]
287.555
```

EXAMPLE 11

This example defines a function

$$f(x, y) = \begin{cases} xy & \text{if both } x \text{ and } y \text{ are integers} \\ x + y & \text{if both } x \text{ and } y \text{ are real} \\ x - y & \text{if } x \text{ or } y \text{ is an integer and the other is real} \end{cases}$$

```
f[a_Integer, b_Integer] = a b;
f[a_Real, b_Real] = a + b;
f[a_Real, b_Integer] = f[a_Integer, b_Real] = a - b;
f[2, 3]
```

6

```
f[2.,3.]
```
5.
```
f[2.,3]
```
-1.
```
f[2,3.]
```
-1.
```
f[I,1]
```
f[i,1]   ← f is undefined for complex arguments.